Fast Pyrolysis of Biomass

Advances in Science and Technology

Green Chemistry Series

Editor-in-chief:
James H. Clark, *Department of Chemistry, University of York, UK*

Series editors:
George A. Kraus, *Iowa State University, USA*
Andrzej Stankiewicz, *Delft University of Technology, The Netherlands*
Peter Siedl, *Federal University of Rio de Janeiro, Brazil*

Titles in the series:
1: The Future of Glycerol: New Uses of a Versatile Raw Material
2: Alternative Solvents for Green Chemistry
3: Eco-Friendly Synthesis of Fine Chemicals
4: Sustainable Solutions for Modern Economies
5: Chemical Reactions and Processes under Flow Conditions
6: Radical Reactions in Aqueous Media
7: Aqueous Microwave Chemistry
8: The Future of Glycerol: 2nd Edition
9: Transportation Biofuels: Novel Pathways for the Production of Ethanol, Biogas and Biodiesel
10: Alternatives to Conventional Food Processing
11: Green Trends in Insect Control
12: A Handbook of Applied Biopolymer Technology: Synthesis, Degradation and Applications
13: Challenges in Green Analytical Chemistry
14: Advanced Oil Crop Biorefineries
15: Enantioselective Homogeneous Supported Catalysis
16: Natural Polymers Volume 1: Composites
17: Natural Polymers Volume 2: Nanocomposites
18: Integrated Forest Biorefineries
19: Sustainable Preparation of Metal Nanoparticles: Methods and Applications
20: Alternative Solvents for Green Chemistry: 2nd Edition
21: Natural Product Extraction: Principles and Applications
22: Element Recovery and Sustainability
23: Green Materials for Sustainable Water Remediation and Treatment
24: The Economic Utilisation of Food Co-Products
25: Biomass for Sustainable Applications: Pollution Remediation and Energy
26: From C–H to C–C Bonds: Cross-Dehydrogenative-Coupling
27: Renewable Resources for Biorefineries
28: Transition Metal Catalysis in Aerobic Alcohol Oxidation
29: Green Materials from Plant Oils
30: Polyhydroxyalkanoates (PHAs) Based Blends, Composites and Nanocomposites

31: Ball Milling Towards Green Synthesis: Applications, Projects, Challenges
32: Porous Carbon Materials from Sustainable Precursors
33: Heterogeneous Catalysis for Today's Challenges: Synthesis, Characterization and Applications
34: Chemical Biotechnology and Bioengineering
35: Microwave-Assisted Polymerization
36: Ionic Liquids in the Biorefinery Concept: Challenges and Perspectives
37: Starch-based Blends, Composites and Nanocomposites
38: Sustainable Catalysis: With Non-endangered Metals, Part 1
39: Sustainable Catalysis: With Non-endangered Metals, Part 2
40: Sustainable Catalysis: Without Metals or Other Endangered Elements, Part 1
41: Sustainable Catalysis: Without Metals or Other Endangered Elements, Part 2
42: Green Photo-active Nanomaterials
43: Commercializing Biobased Products: Opportunities, Challenges, Benefits, and Risks
44: Biomass Sugars for Non-Fuel Applications
45: White Biotechnology for Sustainable Chemistry
46: Green and Sustainable Medicinal Chemistry: Methods, Tools and Strategies for the 21st Century Pharmaceutical Industry
47: Alternative Energy Sources for Green Chemistry
48: High Pressure Technologies in Biomass Conversion
49: Sustainable Solvents: Perspectives from Research, Business and International Policy
50: Fast Pyrolysis of Biomass: Advances in Science and Technology

How to obtain future titles on publication:
A standing order plan is available for this series. A standing order will bring delivery of each new volume immediately on publication.

For further information please contact:
Book Sales Department, Royal Society of Chemistry, Thomas Graham House, Science Park, Milton Road, Cambridge, CB4 0WF, UK
Telephone: +44 (0)1223 420066, Fax: +44 (0)1223 420247
Email: booksales@rsc.org
Visit our website at www.rsc.org/books

Fast Pyrolysis of Biomass
Advances in Science and Technology

Edited by

Robert C. Brown
Iowa State University, USA
Email: rcbrown3@iastate.edu

Kaige Wang
RTI International, USA
Email: kaigew@gmail.com

ROYAL SOCIETY OF CHEMISTRY

THE QUEEN'S AWARDS FOR ENTERPRISE: INTERNATIONAL TRADE 2013

Green Chemistry Series No. 50

Print ISBN: 978-1-78262-618-3
PDF eISBN: 978-1-78801-024-5
EPUB eISBN: 978-1-78801-186-0
ISSN: 1757-7039

A catalogue record for this book is available from the British Library

The Royal Society of Chemistry is a charity, registered in England and Wales, Number 207890, and a company incorporated in England by Royal Charter (Registered No. RC000524), registered office: Burlington House, Piccadilly, London W1J 0BA, UK, Telephone: +44 (0) 207 4378 6556.

For further information see our web site at www.rsc.org

Printed in the United Kingdom by CPI Group (UK) Ltd, Croydon, CR0 4YY, UK

Acknowledgements

George Kraus, series editor for the Green Chemistry Series at the Royal Society of Chemistry (RSC), suggested the idea for this book to RCB early in 2015. After convincing KW, who had recently accepted a new job at RTI International, to serve as co-editor, we started in earnest to line up a distinguished group of pyrolysis researchers to contribute chapters on topics that would synthesize recent progress in understanding the fundamental mechanisms of fast pyrolysis. We were fortunate in convincing Jacques Lédé, Joon Weon Choi, Xianglan Bai, David Dayton, Shurong Wang, Kwang Ho Kim, Anja Oasmaa, Dietrich Meier, Wolter Prins, Stuart Daw and Tristan Brown to accept our invitations to assume leadership roles in preparing chapters along with their multiple co-authors.

We appreciate the patient assistance of Lindsay McGregor at the RSC in preparing the book proposal and getting it approved within the RSC, and Catriona Clarke's help in assembling the chapters and checking on copyright permissions, resolution of figures, and the other myriad details that escape the notice of amateur book editors like ourselves.

RCB would like to acknowledge the staff at Iowa State University's Bioeconomy Institute, especially its deputy directors, Ryan Smith and Jill Euken, who took on additional responsibilities as RCB juggled editing deadlines. He also appreciates the love and patience of his wife Carolyn. KW would like to thank David Dayton and Raghubir Gupta at RTI International for allowing him to take on the role of co-editor in addition to his regular duties at RTI. KW would like to especially thank his wife Guanqun and their soon-to-be-born son Xiaomai for allowing KW to spend many weekends and evenings editing chapters. We hope the knowledge and ideas contributed by the several authors of this book contribute toward a cleaner and brighter future for the next generation.

Robert C. Brown, Iowa State University
Kaige Wang, RTI International

Green Chemistry Series No. 50
Fast Pyrolysis of Biomass: Advances in Science and Technology
Edited by Robert C. Brown and Kaige Wang

Published by the Royal Society of Chemistry, www.rsc.org

Contents

Green Chemistry Series No. 50
Fast Pyrolysis of Biomass: Advances in Science and Technology
Edited by Robert C. Brown and Kaige Wang

Published by the Royal Society of Chemistry, www.rsc.org

CHAPTER 1

Prospects for Fast Pyrolysis of Biomass

KAIGE WANG*[a] AND ROBERT C. BROWN[b]

[a]Energy Technology Division, RTI International, Research Triangle Park, NC, 27709, USA; [b]Bioeconomy Institute, Iowa State University, Ames, IA 50011, USA
*E-mail: kaigewang@rti.org

1.1 Introduction

The past decade has seen increasing interest in production of fuels and chemicals from biomass. Based on the types of feedstock used, biofuels are classified as either first or second-generation.[1] First-generation biofuels include ethanol produced from sugars and starch crops such as maize and sugarcane and biodiesel from seed oils. In contrast, second-generation biofuels are produced from cellulosic and lipid-rich plant materials that are not food crops. These include agricultural and forestry residues, dedicated energy crops like hybrid poplar and switchgrass, algae and municipal solid waste. Although the commercial production of first-generation biofuels has grown tremendously in the last decade, they have been challenged for their limited greenhouse gas reductions compared to petroleum-based fuels and concerns that their production diverts these crops from food production,

Green Chemistry Series No. 50
Fast Pyrolysis of Biomass: Advances in Science and Technology
Edited by Robert C. Brown and Kaige Wang

Published by the Royal Society of Chemistry, www.rsc.org

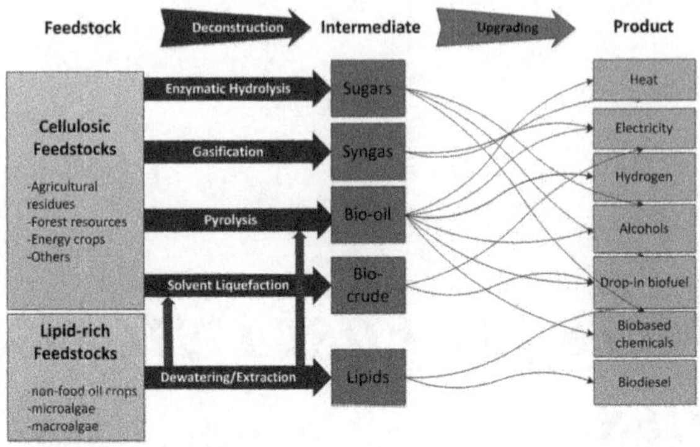

Figure 1.1 Second-generation pathways for converting cellulosic and lipid-rich biomass into power, fuels and chemicals.

the so-called food-*vs.*-fuel debate.[1] Second-generation biofuels offer the prospect of overcoming both of these challenges compared to first-generation biofuels.

Second-generation biofuels can be produced by thermochemical or biochemical processes.[2] Thermochemical processing utilizes heat and catalysts while biochemical processing employs enzymes and microorganisms to convert biomass into fuels and chemicals. Hybrid processing, which combines aspects of thermochemical and biochemical processing, is of growing interest.[3,4] Figure 1.1 summarizes the conversion of cellulosic and lipid-rich feedstocks by biochemical and thermochemical processes into diverse products.

Thermochemical processes, operating at significantly higher temperatures than biochemical processing, are usually very fast, measured in seconds or minutes compared to hours or days for biochemical processes. On the other hand, thermochemical processes can be less selective than biochemical processes, which can unfavorably affect yields of desired products. However, this lack of selectivity often means that more kinds of feedstock molecules are converted, resulting in higher overall yields of drop-in fuels from lignocellulosic feedstocks, for example.

Thermochemical processes can be classified into gasification, pyrolysis, and solvent liquefaction.[5] Gasification converts solid feedstocks into flammable gases known as producer gas or syngas. Pyrolysis converts solid feedstocks into mostly liquid products. Solvent liquefaction resembles pyrolysis in some respects, producing mostly liquid products, but occurs in the presence of a solvent. Of these three thermochemical technologies, pyrolysis has received the most attention in the last few years for its potential to convert lignocellulosic biomass into a liquid intermediate that can be upgraded to drop-in (hydrocarbon) fuels using technologies familiar to the petroleum industry. It also has prospects for distributed processing of biomass, which can simplify the logistics of providing feedstock to a processing plant.[6,7]

1.2 Biomass Fast Pyrolysis Technology

1.2.1 Basic Concepts

Pyrolysis is the thermal decomposition of organic substances in the absence of oxygen to form liquids, solids, and non-condensable gases. The rate of pyrolysis profoundly affects product distributions. Slow pyrolysis, developed centuries ago to produce charcoal for heating purposes, occurs over periods measured in hours or even days. In contrast, fast pyrolysis both rapidly heats the feedstock and quenches the products, usually in the order of seconds, with the goal of producing an energy-rich liquid, known as bio-oil, from the vapors as the primary product. Although originally produced for use as heating oil or electric power generation, bio-oil has been increasingly regarded as an intermediate for the production of drop-in biofuels, biobased chemicals, and hydrogen fuel. To maximize bio-oil production (up to 75 wt% of biomass), several conditions must be met during pyrolysis:[8,9]

- the biomass must be rapidly heated, in the order of a few seconds;
- the products of pyrolysis must be rapidly removed from the reaction zone and cooled, in the order of a few seconds;
- optimum reaction temperature is thought to be between 400–500 °C.

The solid product of fast pyrolysis, known as biochar, consists mostly of carbon but also contains ash originating from biomass. Biochar, which represents 12–15 wt% of the products of fast pyrolysis, can be used as boiler fuel but more intriguing applications include soil amendment, carbon sequestration agent, and activated carbon.[10] Non-condensable gases from fast pyrolysis, yielding 13–25 wt%, are a flammable mixture of carbon monoxide, hydrogen, carbon dioxide, and light hydrocarbons suitable for generating process heat.[9]

1.2.2 Fast Pyrolysis Feedstock

Many kinds of lignocellulosic biomass, ranging from agricultural residues, forestry waste, and energy crops, have been tested for suitability as fast pyrolysis feedstock. The three major components of lignocellulosic biomass, illustrated in Figure 1.2, are cellulose, hemicellulose, and lignin. Cellulose, the most abundant polymer on the planet, constitutes 30–50% of lignocellulosic biomass. It is a structural polysaccharide consisting of pyranose rings linked by glycosidic bonds. Hemicellulose is a heteropolysaccharide of random, amorphous structure cross-linked to cellulose and lignin. Lignin is a highly branched phenol-based polymer bound to cellulose and lignin to form a lignocellulosic matrix.[2] Each of these components produce distinctive products under fast pyrolysis.

Fast pyrolysis has also been used to thermally deconstruct other kinds of biomass feedstocks such as algae and a variety of mixed wastes including

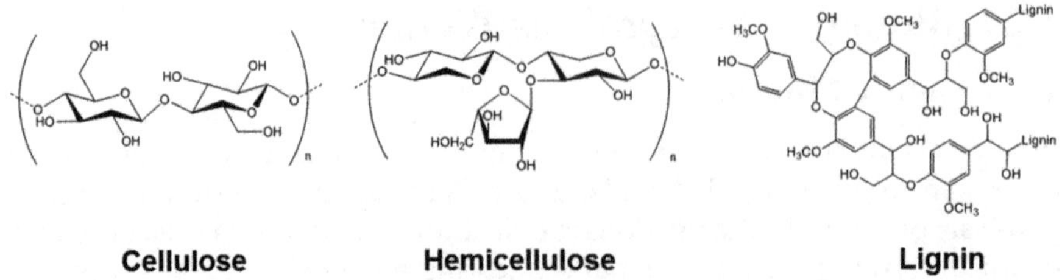

Figure 1.2 Major components of lignocellulosic biomass.

municipal solid waste, sewage sludge, manure, food processing waste, and organic by-products from manufacturing.[9,11,12] These feedstocks often contain relatively large fractions of starch, lipids and proteins compared to lignocellulosic biomass, as well as significant ash. The compositional complexities of alternative feedstocks adds to the difficulties of pyrolyzing them.[13,14]

Drying biomass feedstocks to less than 10 wt% moisture and comminution to less than 3 mm particle diameter are important steps to successful fast pyrolysis. Despite this preprocessing, the low bulk density, irregular particle shapes and cohesive/adhesive behavior of many kinds of biomass can lead to bridging, blockage, and other feeding difficulties.[15]

1.2.3 Types of Fast Pyrolysis Reactors

Although fast pyrolysis was first investigated as early as 1875,[16] significant progress in developing it for bio-oil production only dates from the 1980s. A variety of reactors were investigated with the goal of heating biomass to temperatures exceeding 400 °C in a few seconds. Suitable reactors include bubbling fluidized beds, circulating fluidized beds, rotating cone reactors, auger reactors, entrained flow reactors and ablative reactors.[8,17] Key features of these different classes of fast pyrolysis reactors are summarized in Table 1.1.

Among the various kinds of reactors, fluidized beds have received the most attention for fast pyrolysis due to excellent heat and mass transfer characteristics, simplicity of operation, and relative ease of scale-up. Bio-oil yields of 60–75% from fast pyrolysis of lignocellulosic biomass have been reported.[8] Canadian-based Dynamotive commercialized the bubbling fluidized bed fast pyrolysis technology with a capacity of 200 tons of biomass per day (ton day^{-1}).[18] Another Canadian company, Ensyn Technologies Inc. has developed and commercialized its Rapid Thermal Processing (RTP) process since the early 1980s, which is based on circulating fluidized bed technology. Several RTP™ biomass pyrolysis plants are in commercial operation, with capacity ranging from 1 to 200 ton day^{-1}.[18]

Rotating cone reactors were originally developed at Twente University, with the aim of eliminating carrier gas requirements while maintaining a high throughput and rapid heat transfer similarly to fluidized bed reactors. Instead of using inert gas, biomass is mechanically mixed with a solid heat transfer medium in a rotating cone. The rotating cone reactor concept has been commercialized by Biomass Technology Group (BTG) in the Netherlands, building

Table 1.1 An overview of different fast pyrolysis reactor technologies (reproduced from ref. 17 with permission from Dr Anthony V. Bridgwater).

Reactor type	Development Status	Max. yield wt%	Complexity	Feed size specification	Inert gas requirements	Specific reactor size	Scale-up
Bubbling fluidized bed	Commercial	75	Medium	High	High	Medium	Easy
Circulating fluidized bed	Commercial	70	High	High	High	Medium	Easy
Rotating cone	Commercial	70	High	High	Low	Low	Easy
Auger	Pilot	60	Medium	Medium	Low	Low	Medium
Entrained flow	Laboratory	60	Medium	High	High	Medium	Easy
Ablative	Laboratory	75	High	Low	Low	Low	Difficult

several reactors with capacities of up to 120 ton day^{-1}.[8,18] The auger reactor, which also features mechanical mixing of biomass with a solid heat transfer medium, has been developed by ABRI-Tech of Canada and the Karlsruhe Institute of Technology (KIT). A pilot scale twin-screw auger reactor with 12 ton day^{-1} capacity is now operational at KIT.[19] Other types of reactor, including entrained flow, vacuum moving bed, ablative reactor, and microwave pyrolysis reactors are still in relatively early stage of technology development.

1.2.4 Bio-Oil from Fast Pyrolysis

Quenching of the condensable vapors and aerosols from fast pyrolysis produces bio-oil. The rapid removal and cooling of pyrolysis products to minimize secondary reactions that can crack vapors are critical to achieving high liquid yields. Traditionally this is accomplished by spraying cold liquid hydrocarbon or recycling bio-oil into the pyrolysis vapor stream. The bio-oil collected in this fashion is an emulsion of lignin-derived phenolic compounds in an aqueous phase containing mostly carbohydrate-derived compounds.[18] Unfortunately, the high reactivity of the bio-oil makes it difficult to distill into separate components, limiting opportunities for upgrading it to diverse value-added products. More recently, efforts have been made to develop fractionating bio-oil collection systems based on boiling points of bio-oil constituents. One such system consists of pairs of temperature-controlled condensers and electrostatic precipitators operated in series to remove vapors and the aerosols as heavy ends (anhydrosugars and phenolic oligomers), a middle fraction (furans and phenolic monomers), and light ends (water and light oxygenates).[20]

Although the dark-brown liquid collected from fast pyrolysis is called "bio-oil" or "bio-crude", it has few similarities with petroleum. Unlike the hydrocarbons found in petroleum, it contains very high levels of oxygen, comparable to that of the biomass from which it is derived. On a mass basis, its heating value is comparable to biomass. In fact, it is sometimes characterized as "liquid biomass". Bio-oil is composed of a complex mixture of water, volatile oxygenates, anhydrosugars, and non-volatile oligomers. Many of these compounds are extremely reactive, leading to polymerization of bio-oil even at ambient temperatures. These physiochemical properties make bio-oil unsuitable as transportation fuel without upgrading. A number of upgrading processes have been explored, including catalytic cracking, hydroprocessing, aqueous phase reforming, and fermentation, to produce transportation fuels from fast pyrolysis process.[21]

1.3 Recent Advances in Fast Pyrolysis Research and Development

The pace of research and development in fast pyrolysis has accelerated considerably in the past decade. As shown in Figure 1.3, scientific publications increased from a dozen or so per year in the early 1980s to over 700 per year in 2016. Most of this growth occurred in the last decade in a period when

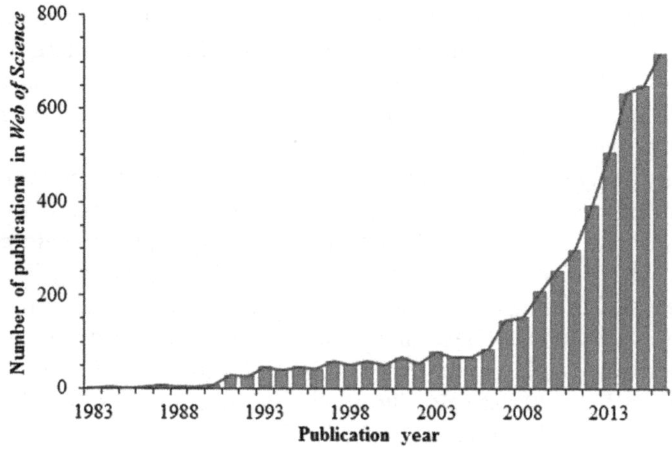

Figure 1.3 Number of scientific journals published between 1983–2016 with the topic of "fast pyrolysis" indexed in the ISI Web of Science™ (www.webofknowledge.com).

development of advanced biofuels became a priority in the United States and many other countries around the world.

This rapid growth in research and development has not yet translated into notable successes in the commercial deployment of fast pyrolysis technologies.[22] A number of factors have contributed to the slow pace of commercialization. The meteoric rise in petroleum prices in the first decade of the twenty first century was followed in the next decade by a dramatic price slump, hurting the economics of advanced biofuels. Failure of many markets to assign a cost to carbon emissions from fossil fuels has also diminished the incentive for the commercialization of low carbon fuels. However, in some respects, commercialization efforts launched by the U.S. Energy Security Act of 2005 got ahead of the state-of-the art technology in advanced biofuels: crucial scientific questions and engineering challenges related to pyrolysis as well as other advanced biofuels technologies had not been resolved. Much has changed in the past decade. This book covers recent advances in fast pyrolysis science and technology, with chapters on reaction chemistry of thermal deconstruction, computational modeling, product utilization, catalytic pyrolysis and economic evaluation of fast pyrolysis technology.

1.3.1 Reaction Chemistry of Fast Pyrolysis

As the most abundant polymer on the planet and the major component of lignocellulosic biomass, the pyrolysis of cellulose has been studied since the early 20th century.[23] Although it is generally accepted that levoglucosan is the primary product of cellulose fast pyrolysis,[13] the mechanism by which it forms and its role in producing secondary products continues to be debated. It has been reported recently that cellulose pyrolysis proceeds through the formation of a liquid intermediate, from which volatile products are formed.[24]

Multiphase reactions are involved in cellulose pyrolysis, including solid–liquid, liquid–liquid, liquid–gas, and gas–gas reactions. Moreover, reaction intermediates may exist for only fractions of a second,[24] further complicating experiment investigations of cellulose pyrolysis. Chapter 2 is a historical review of the many studies of cellulose pyrolysis covering the several theories and unresolved questions around the primary reactions of cellulose depolymerization.

As described in Chapter 3, lignin is the most complex and difficult to pyrolyze component of lignocellulosic biomass. Lignin melts before it depolymerizes. The liquid products of lignin pyrolysis include volatile phenolic monomers and non-volatile phenolic oligomers with molecular weights ranging up to 2500 Da. These phenolic compounds are extremely reactive, often condensing and dehydrating to char and light gas before they can volatilize. These characteristics of lignin can cause agglomeration in reactors and clogging of condensers.

Due to the inherent reactivity of pyrolysis products, various secondary reactions could occur during biomass pyrolysis, leading to a lower yield of bio-oil with undesirable properties. Chapter 4 reviews recent studies related to secondary reactions of pyrolysis of cellulose, lignin, and whole biomass. Heat and mass transfer phenomena during fast pyrolysis are also addressed in this chapter.

The role of free radicals in pyrolysis has been the subject of speculation for many years. The short half-life of radicals is among the challenges in studying them. Recent advances in analytical techniques such as *in situ* Electron Paramagnetic Resonance have been employed to probe free radical chemistry during pyrolysis. Recent studies[25,26] have revealed that free radicals participate in a variety of reactions, especially the depolymerization of lignin. Chapter 7 reviews studies that have detected free radicals in bio-oil and biochar and investigated their role in fast pyrolysis chemistry. The role of radicals in bio-oil stability is also discussed.

1.3.2 Computational Modeling of Fast Pyrolysis

Improved understanding of the physics and chemistry of fast pyrolysis arising from experimental studies has encouraged multi-scale modeling of pyrolysis processes. Intra-particle chemistry and transport processes are receiving particular attention from the scientific community as their role in determining overall yields and selectivity of pyrolysis products becomes increasingly evident. Chapter 11 describes recent advances in modeling pyrolysis in individual particles and provides readers the state-of-the-art in predicting pyrolysis kinetics. Limitation of current modeling efforts and future opportunities are also discussed.

1.3.3 Utilization of Bio-Oil

Fast pyrolysis proceeds through a complex set of primary and secondary reactions leading to a large number of organic compounds with a wide range of properties, making characterization of bio-oil challenging.

Detailed characterization of bio-oil is valuable for both interpreting pyrolysis reaction chemistry and advancing its upgrading to transportation fuels or refining to biobased chemicals. Chapter 6 covers the physiochemical properties of bio-oil and the corresponding characterization methods.

One of the most distinguishing characteristics of bio-oil is its high reactivity compared to petroleum, which presents one of the chief challenges in using it as feedstock in petroleum refineries. Unlike the hydrocarbons in petroleum, bio-oil is highly oxygenated and includes alcohols, aldehydes, ketones, carboxylic acids, furans, sugars, and phenolic compounds. These compounds can participate in a wide variety of reactions contributing to a phenomena referred to as "instability" or "aging", characterized by an increase in viscosity, molecular weight, and water content of the bio-oil and ultimately phase separation.[27] In ambient conditions, instability can manifest itself over time periods as short as a few weeks. At the elevated temperatures employed for upgrading bio-oil, this instability can occur within minutes. Stability problems need to be solved to produce bio-oil with uniform characteristics in commercial scale. Chapter 8 reviews the complex phenomenon of bio-oil instability and summarizes stabilization methods including dilution, neutralization, esterification, mild hydrotreating, and stage-fractionation.

The production of chemicals has received significant attention as a way to valorize bio-oil. However, its complex chemical composition and reactivity makes the separation of bio-oil molecules by functional groups challenging.[9] Chapter 9 provides an overview of relevant separation and enrichment techniques for bio-oil, including distillation, liquid–liquid extraction, adsorption/desorption, and fractional condensation. A specific example of chemical recovery from bio-oil at an industrial scale is also briefly reported.

1.3.4 Catalytic Fast Pyrolysis

Although fast pyrolysis bio-oil has prospects for being upgraded into hydrocarbon fuels, raw bio-oil is presently incompatible with processing at traditional petroleum refineries. One of the challenges is that the elemental composition of bio-oil more closely resembles biomass than petroleum. Some of the chemical and physical properties of crude bio-oil that prevent its processing in petroleum refineries include its high oxygen content and acidity and poor thermal stability. In an effort to improve the quality of bio-oil for upgrading, recent research has focused on catalytically upgrading pyrolysis vapors before they are condensed to bio-oil. Known as catalytic fast pyrolysis, the goal is to substantially, if not completely, deoxygenate the products, producing molecules that are suitable as blendstock for refining to hydrocarbon transportation fuels.[28] Among the catalysts available, inexpensive zeolites are particularly attractive due to their ability to deoxygenate bio-oil molecules without the addition of hydrogen.[29] Chapter 10 explores the ability of zeolite catalysts to produce a stable, upgradable product from bio-oil.

Low carbon efficiency and excessive coke formation on catalysts are two of the hurdles in the further development of catalytic fast pyrolysis. Due to the hydrogen-deficient nature of biomass, the liquid product from catalytic pyrolysis can become even more hydrogen deficient as dehydration occurs. In order to improve the carbon efficiency and bio-oil quality, the addition of reactive gases, such as hydrogen and carbon monoxide, to the pyrolysis process, with or without deoxygenation catalysts, has been proposed. Very recently, several fast pyrolysis technologies involving reactive gases have been developed.[30–33] One of these novel processes is tail gas recycling pyrolysis (TGRP), which recycles the pyrolysis gases for fluidization.[32] Deoxygenated bio-oil with improved quality was observed for TGRP in a reducing environment. However, the detailed mechanism of bio-oil deoxygenation *via* tail gas recycling is still unclear. Another example is the IH^2 process developed by the Gas Technology Institute,[30,31] which integrates hydropyrolysis of biomass with close-coupled hydroprocessing of the hydropyrolysis vapors to produce gasoline and diesel. Chapter 5 provides an overview of the recent studies in the role of reactive gases in improving the products of fast pyrolysis.

1.3.5 Economics of Fast Pyrolysis

There are no successful commercial projects from which to estimate the cost of fuels and chemicals from fast pyrolysis. Techno-economic analyses based on limited yield data and well-established process engineering principles have been conducted to estimate the economic feasibility of fast pyrolysis of biomass in comparison to other fuel and chemical pathways including biochemical processing of biomass and conventional petroleum refining.[34,35] Chapter 12 reviews recent economic assessments of the major pyrolysis pathways for advanced biofuel production. Suggestions are also made to improve the accuracy of existing deterministic techno-economic models.

References

1. R. C. Brown and T. R. Brown, *Why Are We Producing Biofuels?* Brownia LLC, Ames, IA, 2012.
2. R. C. Brown and T. R. Brown, *Biorenewable Resources: Engineering New Products from Agriculture*, John Wiley & Sons, 2014.
3. D. Choi, A. A. Dispirito, D. C. Chipman and R. C. Brown, in *Thermochemical Processing of Biomass: Conversion into Fuels, Chemicals and Power*, ed. R. C. Brown, John Wiley & Sons, 2011, p. 280.
4. Y. Shen, L. Jarboe, R. Brown and Z. Wen, *Biotechnol. Adv.*, 2015, **33**, 1799.
5. R. C. Brown, in *Thermochemical Processing of Biomass: Conversion into Fuels, Chemicals and Power*, ed. R. C. Brown, John Wiley & Sons, 2011, p. 1.
6. J. Rogers and J. G. Brammer, *Biomass Bioenergy*, 2009, **33**, 1367.
7. M. M. Wright, R. C. Brown and A. A. Boateng, *Biofuels, Bioprod. Biorefin.*, 2008, **2**, 229.
8. A. V. Bridgwater, *Biomass Bioenergy*, 2012, **38**, 68.

9. D. Mohan, C. U. Pittman and P. H. Steele, *Energy Fuels*, 2006, **20**, 848.
10. D. A. Laird, R. C. Brown, J. E. Amonette and J. Lehmann, *Biofuels, Bioprod. Biorefin.*, 2009, **3**, 547.
11. A. V. Bridgwater, *Chem. Eng. J.*, 2003, **91**, 87.
12. P. Manara and A. Zabaniotou, *Renewable Sustainable Energy Rev.*, 2012, **16**, 2566.
13. P. R. Patwardhan, J. A. Satrio, R. C. Brown and B. H. Shanks, *Bioresour. Technol.*, 2010, **101**, 4646.
14. J. Zhang, Y. S. Choi, C. G. Yoo, T. H. Kim, R. C. Brown and B. H. Shanks, *ACS Sustainable Chem. Eng.*, 2015, **3**, 293.
15. J. Dai, H. Cui and J. R. Grace, *Prog. Energy Combust. Sci.*, 2012, **38**, 716.
16. J. Lédé, *J. Anal. Appl. Pyrolysis*, 2012, **94**, 17.
17. http://www.pyne.co.uk/?_id=69, accessed on Jan 2017.
18. R. Venderbosch and W. Prins, *Biofuels, Bioprod. Biorefin.*, 2010, **4**, 178.
19. N. Dahmen, E. Dinjus, T. Kolb, U. Arnold, H. Leibold and R. Stahl, *Environ. Prog. Sustainable Energy*, 2012, **31**, 176.
20. R. C. Brown, S. T. Jones and A. Pollard, Google Patents, 2013.
21. A. V. Bridgwater, in *Thermochemical Processing of Biomass: Conversion into Fuels, Chemicals and Power*, ed. R. C. Brown, John Wiley & Sons, 2011, p. 157.
22. T. R. Brown and R. C. Brown, *Biofuels, Bioprod. Biorefin.*, 2013, **7**, 235.
23. M. S. Mettler, D. G. Vlachos and P. J. Dauenhauer, *Energy Environ. Sci.*, 2012, **5**, 7797.
24. P. J. Dauenhauer, J. L. Colby, C. M. Balonek, W. J. Suszynski and L. D. Schmidt, *Green Chem.*, 2009, **11**, 1555.
25. C. Bährle, V. Custodis, G. Jeschke, J. A. van Bokhoven and F. Vogel, *ChemSusChem*, 2016, **9**, 2397.
26. K. H. Kim, X. Bai, S. Cady, P. Gable and R. C. Brown, *ChemSusChem*, 2015, **8**, 894.
27. A. Oasmaa, T. Sundqvist, E. Kuoppala, M. Garcia-Perez, Y. Solantausta, C. Lindfors and V. Paasikallio, *Energy Fuels*, 2015, **29**, 4373.
28. C. Liu, H. Wang, A. M. Karim, J. Sun and Y. Wang, *Chem. Soc. Rev.*, 2014, **43**, 7594.
29. T. R. Carlson, Y.-T. Cheng, J. Jae and G. W. Huber, *Energy Environ. Sci.*, 2011, **4**, 145.
30. T. L. Marker, L. G. Felix, M. B. Linck and M. J. Roberts, *Environ. Prog. Sustainable Energy*, 2012, **31**, 191.
31. T. L. Marker, L. G. Felix, M. B. Linck, M. J. Roberts, P. Ortiz-Toral and J. Wangerow, *Environ. Prog. Sustainable Energy*, 2014, **33**, 762.
32. C. A. Mullen, A. A. Boateng and N. M. Goldberg, *Energy Fuels*, 2013, **27**, 3867.
33. K. Murugappan, C. Mukarakate, S. Budhi, M. Shetty, M. R. Nimlos and Y. Román-Leshkov, *Green Chem.*, 2016, **18**, 5548.
34. M. M. Wright, D. E. Daugaard, J. A. Satrio and R. C. Brown, *Fuel*, 2010, **89**, S2.
35. T. R. Brown, *Bioresour. Technol.*, 2015, **178**, 166.

CHAPTER 2

Primary Reactions of Cellulose Pyrolysis

J. LÉDÉ[a]

[a]Laboratoire Réactions et Génie des Procédés, CNRS-Université de Lorraine, 1 rue Grandville – BP 20451 – 54001 NANCY Cedex, France
*E-mail: Jacques.Lede@univ-lorraine.fr

2.1 Introduction

Biomass has long been a major source of energy, chemicals and materials. However, biomass must undergo major biochemical or thermochemical processes to be transformed into useful products. One of the advantages of the thermal routes of transformation is that all the components of biomass, not just the carbohydrate, can be upgraded. The various processes are usually classified as pyrolysis, gasification or combustion.[1] In each case, the chemical phenomena begin with primary steps of thermal decomposition followed by various further transformations. Despite the extraordinarily large number of papers and reports published (representing many hundreds of communications), uncertainties and disagreements remain in the literature due to the complex chemical and physical properties of biomass and its components, and the great number of reactions that can occur at different operating conditions. For these reasons, this chapter focuses on cellulose, a major component of biomass (mass fractions of 40–50% on a dry basis). Cellulose[2] is a high molecular weight linear polymer of β-(1-4)-D-glucopyranose units

Green Chemistry Series No. 50
Fast Pyrolysis of Biomass: Advances in Science and Technology
Edited by Robert C. Brown and Kaige Wang

Published by the Royal Society of Chemistry, www.rsc.org

linked together by (1-4)glycosidic bonds.[3] These repeating units are cellobiose units (two glucose units). Cellulose has a crystalline structure including amorphous regions. Its variable degrees of polymerization (DP) can reach several thousands. Groups of cellulose chains make microfibril sheets which are the basis of complex fibers.[3]

The purpose of this chapter is to review the main published works related to the primary mechanisms of cellulose pyrolysis with some consideration of secondary reactions of the primary products. We shall also consider problems related to the chemical and physical properties of intermediate primary species formed under these conditions, a problem which continues to give rise to numerous discussions in the literature. This review follows a chronological description to show progress in the field and to remind readers that many important results were discovered several decades ago.

2.2 Thermal Conditions of Cellulose Pyrolysis[4]

Temperature T and heating rate T' are often used to distinguish between two types of experimental conditions: slow pyrolysis (SP) and fast pyrolysis (FP). In SP, reaction is assumed to occur under relatively mild conditions (lower values of T and T') than for FP. Published studies on pyrolysis were conducted under a wide range of experimental conditions, making comparisons among them difficult, especially when T and T' are inaccurately measured or defined.

2.2.1 Experimental and Conceptual Difficulties in the Accurate Knowledges of T and T'

Values of T and T' are difficult to measure during FP in different cases:

- Small, fast moving particles are undergoing fast reactions
- Reactions are occurring in ablation conditions, *i.e.* at the blind interface between the reacting particle and a hot wall
- Particles are heated by more or less concentrated radiation
- Reaction front is moving inside a thick particle

Other difficulties and sources of errors can arise from:

- Thermocouple contact resistances and response time
- Inability of pyrometry to measure temperature inside particles
- Influence of radiation on particle temperature measurements
- Optical properties of particles changing during reaction (for example cellulose producing liquids, char, *etc.*)

Depending on the type of reactor (transient, steady-state, *etc.*) the solid may be heated by various types of processes: radiation; contact with heated walls; exchange with a hot carrier gas; interaction with inert or catalytic particles.

In the literature, many of the reported values of T and T' are actually those of the reactor rather than the pyrolyzing particles. These measurements, thus, have little value in interpreting the progress of pyrolysis reactions.

2.2.2 Unclear Frontiers Between SP and FP

An overview of the literature recently revealed[4] the absence of consensus and coherence in the quantitative definitions of necessary conditions for FP on the basis of T and T'. Often repeated from reference to reference, they are not clearly justified. Also, simple qualitative notions of moderate and high T as well as those of process T and pyrolysis T are of little value for kinetic studies. Also, the differences between "fast" and "flash" are not clear, as well as the undefined notion of intermediate pyrolysis. So, significant inconsistences exist in the definitions of frontiers between SP and FP.

2.2.3 Modeling of Solid Particle Pyrolysis

In a recent paper, Lédé and Authier[4] published the results of a simple mathematical model of a solid particle (cellulose) undergoing a simple first order pyrolysis reaction. The authors considered two cases for reactor temperature T_R:

- Fixed T_R as in a continuous reactor fed by a constant flow of cellulose particles or in a batch reactor with a fixed mass of cellulose introduced into the reactor
- Increasing T_R as in a thermogravimetric analysis (TGA) or electrically heated surface device

The operating conditions, such as T_R, heat transfer coefficient at particle surface, and particle size, have been chosen to cover all practical situations. The results show:

- For T_R higher than about 600–700 K, cellulose reacting temperature T_C is always lower than T_R. The thermal lags ($T_R - T_C$) may exceed several 100 K mainly for high values of T_R and T'_R. $T_C = T_R$ only in mild TGA conditions as shown by Narayan and Antal.[5] Accordingly, chemical interpretations relying on reactor T_R should be hence avoided
- As soon as the particle begins to react, T_C rapidly stabilizes below about 780 K regardless of the conditions, even for very fine samples (down to 2 μm). Thus, it is ineffective to increase T_R in the hope of reaching high T_C values. It can also be inadvisable to define different chemical regimes relying on small T_R variations (low sensitivity of T_C with T_R). Due to possible important thermal lags between T_R and T_C, the use of T_R in place of T_C leads to lower calculated values of activation energies, giving a possible explanation of the compensation effect[5]
- The reaction is never strictly isothermal even for fixed T_R values. T_C can increase several tens of degrees Kelvin for particle conversion increasing from 5% to 95%

- The actual particle heating rate T'_C is not constant during heating and reaction. It varies with time and conversion, with passage through a low minimum during the reaction (because of T_C stabilization) even in FP
- The values of T'_C are always much lower than those of T'_R except in mild TGA conditions

2.2.4 Summary and Foreword

The difficulties in measuring T and T', and the inconsistencies revealed in the literature and modeling results, show that differentiation between SP and FP cannot be accurately nor quantitatively defined on the basis of the notions of temperature and heating rate. Other more reliable parameters have been discussed,[4] such as heat transfer coefficient at particle level and efficiency of (liquid and/or solid) primary product removal from the particle. In the following text, the reported values of "reaction temperature" and "heating rate" given by authors should be considered mainly as qualitative indications, except in a few experimental cases (low heating rate TGA). Prudence is also required in comparisons of works performed in different types of reactors, and in the definition of distinct chemical regimes supposed to occur inside narrow domains of temperatures (sometimes of only few tens of K).

Both results obtained in SP and FP will be reported, because of the existence of close complementarities and the conceptual difficulties in distinguishing SP from FP. The literature survey shows that the same word "pyrolysis" is used for describing: the elementary primary reactions of cellulose; the whole chain of reactions involving consecutive and competitive reactions underwent by species issued from primary reactions; and the whole process including these reactions, the whole reactor and products treatments. In this chapter, the word "pyrolysis" will be used for designing the first steps of the thermal deconstruction of cellulose.

2.3 Primary Cellulose Pyrolysis

2.3.1 Research Prior to the Mid-1960s

As described in the review by Antal and Gronli,[6] biomass thermal degradation has been used by mankind for several centuries in various domains such as energy, chemical preparation and art; for example, cave drawings dating to 30 000 years BP, and char for ore reductions and various industries of Greek and Roman civilizations.[7] As early as the beginning of the 19th century, Violette[8,9] and Mollerat[10] published first quantitative results with special emphasis on wood distillation. Already at that time, Gruner[11] observed that "slow increase of low heat intensity" produced much charcoal while "intense heat rapidly applied" increased oils, tars and gases. Clearly the notions of SP and FP were already discovered. Such observations were also reported later on by Klar.[12]

Cited in 1923 in the book of Hawley,[13] Klason[14] showed the extraordinary increase of tar fractions (by a factor higher than 4) for experiments performed under vacuum. These tars could have different colors. The existence of primary and secondary reactions, exothermic aspects of some reactions and existence of several types of charcoals were already evidenced by these authors more than 100 years ago!

In a report in 1918, Pictet and Sarasin[15] already showed the role of catalysts in the pyrolysis of cellulose. They also confirmed an increase (up to 76%) of distillate yields under fast heating (at 523–623 K). The "yellow fraction" was attributed to levoglucosan (LVG) already discovered in 1894. LVG was assumed to be an intermediate during cellulose pyrolysis. Clement and Rivière[16] summarized these results in 1920.

2.3.2 First Qualitative Evidence of the Formation of Liquid Products During Cellulose Pyrolysis Between the Mid-1960s and Mid-1970s

In 1963, Goring[17] observed thermal softening of fine cellulose powder compressed in a capillary and subjected to a prescribed heating rate. The softening point, defined as the temperature at which the powder collapsed into a solid plug, ranged between 504 and 526 K. The phenomenon was shown to be due to the crystalline nature of cellulose rather than a glass transition or melting. Weight loss was observed only above this temperature range (between 528 and 538 K). In 1967, Back *et al.*[18] measured the modulus of elasticity of different types of papers. The authors showed a glass transition temperature of around 483 K for the amorphous part of cellulose. Thermal hardening was then observed in the 573–593 K range. In 1974, Nordin *et al.*[19] submitted cellulose to a rapid (0.1 millisecond) heating through the use of a laser, followed by rapid quenching (liquid N_2). Photographs clearly showed the presence of small bubbles in place of the initial fibrillar structure of cellulose. The decrease in crystallinity was interpreted as resulting from cellulose melting but not due to its degradation. Using an empirical correlation of Boyer,[20] the authors calculated a theoretical melting point of cellulose of between 690 and 751 K. These values were confirmed several decades later by other authors. These pioneering results, obtained about 50 years ago, continue to be topical subjects of papers in the present literature!

2.3.3 First Kinetic and Chemical Interpretations Published Before the Mid-1980s

Several references and results can be found in a 1956 paper of Stamm[21] who studied the thermal degradation of cellulose for reaction times ranging from minutes to several days between 366 and 573 K. Activation energy of about 110 kJ mol^{-1} was proposed for a first order type reaction. At the same time, Madorsky *et al.*[22] suggested a two-step decomposition pathway including depolymerization.

Martin[23] and Lincoln[24,25] carried out pioneering experiments in conditions of fast thermal decomposition. Martin used a concentrated radiation source with exposure times of between 0.4 and 8 s. He showed that ignition of cellulose occurred when the surface temperature reached 873–923 K (independent of the flux density), LVG being the main product in the tars. Lincoln used carbon arcs, xenon flash lamps and lasers to irradiate cellulose samples for times as short as 0.5 milliseconds. The results confirmed a reaction temperature of around 873 K, independent of heat flux. Both authors showed that char fractions decreased as available flux increased.

Other extensively published results during this time used TGA, differential thermal analysis (DTA) and mass spectrometric thermal analysis (MTA). Kilzer and Broido,[26] Arsenau,[27] Broido and Nelson[28] proposed a general kinetic scheme for cellulose pyrolysis relying on the existence of two competitive reactions occurring directly from cellulose. The first one corresponded to a dehydration process followed by formation of char and gas species. The second one transformed cellulose into an intermediate and unstable compound that ultimately yielded LVG. Later on, Antal and Varhegyi[29] found that vapor–solid interactions could have favored char formation. In 1968, Shafizadeh[30] proposed a scheme involving three concurrent reactions with a fourth reaction added to take into account high temperatures processes.

Pacault and Sauret[31] used viscosimetric methods to show that between 443 and 503 K the average DP decreased from 1360 to about 300 after several hours of heating. They showed that the reaction order was zero at the boundaries between amorphous and crystalline parts of cellulose. Between 543 and 583 K, Chatterjee and Conrad[32] (TGA) measured an activation energy of 227 kJ mol^{-1}. Halpern and Patai[33] studied the reaction under vacuum between 523 and 573 K. From viscosity measurements, they reported that the initial DP value of 1800 rapidly decreased to 200 (with only a small mass loss) from the cleavage of non-crystalline sites. Golova[34] confirmed these findings (rapid decrease of DP in amorphous regions). In 1973, Broido *et al.*[35] using TGA similarly found decomposition into smaller polymers without mass loss at crystalline-amorphous boundaries. Similar conclusions were also reported by Broido and Weinstein.[36]

From TGA experiments under vacuum, Broido *et al.*[37] proposed a model including an initial high activation energy step during which cellulose gave rise to active cellulose (AC) that subsequently decomposed according to two concurrent processes (formation of volatile tars and char). From experiments performed from 532 to 614 K under reduced pressure, Bradbury *et al.*[38] observed that in the low temperature domain, a change in the mass loss regime was due to a high activation energy (242 kJ mol^{-1}) process resulting from the passage of cellulose to an AC form without mass loss. AC subsequently decomposed *via* two reactions (one giving volatiles, the other char plus gases).[2] This kinetic scheme, usually described as the "Broido Shafizadeh" model (BS), is well known in the literature and continues to be often used and cited. This paper[38] mentioned the works of Shafizadeh and Bradbury[39] reporting that initial depolymerization led to DP molecules of around 200. It should be noted that Allan *et al.*[40] showed that rapid heating led to different selectivities.

At the end of 1980, a key event was the Specialists' Workshop on Fast Pyrolysis of Biomass held at Copper Mountain, Colorado, USA during which many pioneering works were presented. Many of them remain of interest today. Summarizing the various presentations, a general kinetic model was proposed by Diebold.[41] It included the primary decomposition of biomass which was supposed to occur *via* two possible processes giving rise to small fractions of char and to main fractions of AC which, in turn, subsequently decomposed into complex secondary processes. Among the 24 presentations of this workshop, the "Hot Wire" (HW) experiments of Diebold[42] were of particular interest. An electrically heated wire was able to "saw" a piece of biomass with formation of visible smoke without charcoal formation. Sawing produced a lubricated feeling. Microscopic observations revealed the formation of a thin brownish tar layer (similar to a varnish), which was attributed to a melted phase. This product was a liquid under reaction temperature but a solid at room temperature (unlike usual bio-oils).

2.3.4 Developments from the Mid-1980s to Mid-1990s

Following these HW experiments, Lédé *et al.*[43] published results of "spinning disk" (SD) experiments where a rod of wood was applied (under 0.1–3.5 MPa pressure) on a hot spinning disk (773–1173 K). Rate of consumption of the rod and thickness of the pyrolysis zone (10–100 μm) were measured as functions of several experimental parameters. The lubricating effect, similar to that observed in HW experiments, confirmed the primary formation of a liquid. Similar observations have been made later by Peacocke.[44] In 1987, Lédé *et al.*[45] published other SD results obtained with true melting materials (such as ice) showing similar qualitative behaviors. A simple theoretical approach led to the conclusion that wood behaved as during a fusion with an estimated equivalent melting temperature of 739 K. These phenomena have been modeled by Martin *et al.*[46] who proposed a universal relationship allowing the calculation of an ablation rate as a function of all experimental parameters for any melting material (including biomass). All these experiments as well as HW observations were at the origin of the notion of fast ablative pyrolysis.

During this period Antal and co-workers[47] published many results of biomass (including cellulose) FP performed under concentrated radiant energy provided by solar concentrators and also laboratory-scale image furnaces. The authors underlined that these conditions enhanced the formation of either liquid syrups or gases, LVG being the major product. The results were modeled on the basis of various chemical kinetics schemes relying either on one step or on concurrent reactions mechanisms. The authors also designed an original fast TGA apparatus for FP studies. Mok and Antal[48] measured the heat demand in cellulose pyrolysis, explaining that high pressure as well as low flowrate increased char formation. Their model supposed that primarily formed AC competitively gave rise to anhydrocellulose and LVG.

Using molecular-beam mass spectroscopy (MBMS), Evans and Milne[49] studied the products formed during FP of wood components. In the case of cellulose, the authors discussed the formation of AC corresponding to a decrease of DP (1000–2000) to 200 between 493 and 523 K. Weight loss began to be observed at 553 K. At around 573 K, the transglycosylation process led to the formation of high LVG yields, with a strong influence of inorganic matter. From experiments made under vacuum and at 623 K, Richards[50] evidenced that LVG was not a major product and that hydroxyacetaldehyde (HAA) was concurrently formed with the important influence of NaCl pretreatment.

Radlein *et al.*[51] and Piskorz *et al.*[52] studied the liquid products formed in a fluidized bed reactor (Waterloo fast pyrolysis process) under various temperatures (762–793 K) and at 1 atm. The authors first evidenced LVG as a main product and also the formation of a dimer (cellobiosan). They confirmed also that LVG and HAA were concurrently formed, the latter being favored after pretreatment. From their results, they proposed a kinetic model (Waterloo model): cellulose was first decomposed into "cellulose low DP" which, in turn, gave rise to two reactions forming HAA (with other low molecular weight oxygenated compounds) and LVG.

From experiments under vacuum performed between 623 and 773 K, Essig *et al.*[53] confirmed two concurrent pathways leading to LVG and HAA. LVG decreased at high temperature and in the presence of inorganics. At the same time, Powels *et al.*[54] showed the formation of anhydro-oligosaccharides resulting from primary cellulose depolymerization. LVG would be the result of subsequent reactions and, hence, not a primary product.

2.3.5 New Results, Controversies and Models Published from the 1990s, Until the Beginning of the 2000s

In 1993, Vladars-Usas[55] studied the fast thermal decomposition of Avicel cellulose between 1123 and 1473 K inside a downflow transport flow reactor and for residence times of a few tens of milliseconds. Depolymerization of cellulose was clearly evidenced with the formation of a water-soluble molten state containing LVG, cellobiosan and anhydrosaccharides. Reactions occurred in the range 773–823 K. Confirming these results, Piskorz *et al.*[56] showed that these primary products contained significant yields of LVG and oligosaccharides with DP between 2 and 7. Photographs clearly revealed the liquid nature of these products.

Boutin *et al.*[57] submitted cellulose powder to high radiant flux provided by an image furnace (5 kW xenon lamp). For exposure times between 0.3 and 1 s, cellulose passed through a yellowish liquid phase which was solid after rapid cooling. This water-soluble solid did not correspond to molten cellulose but to partially degraded species giving rise to vapors and gases for longer exposure times. No char was observed. The BS model[38] was shown to apply. Boutin and Lédé[58] calculated the life-times of these primary liquids to be a few tens of milliseconds under heat flux densities of 2×10^7 W m^{-2}. These results were confirmed in 1995 by Piskorz *et al.*[59] who suggested that the reaction would be close to a simple true melting phenomenon. Lédé *et al.*[45] calculated a "fusion like reaction temperature" of 739 K, a value very close to the simple previous calculation of Back *et al.*[18] reported above.

Pyrolysis of cellulose pellets under controlled flash times (<1 s) of a concentrated radiation (available flux densities of 10^5 to 7.4×10^6 W m^{-2}) were published in 2002 by Boutin *et al.*[60] All the products were recovered allowing the establishment of mass balances. After each flash, a yellow layer made of agglomerated melted particles was observed. This material (solid at room temperature and soluble in water) was called "Intermediate Liquid Compound" (ILC). A layer of char was observed upon the ILC layer only for flux densities under 9×10^5 W m^{-2}. The chemical nature of all the products was reported by Lédé *et al.*[61] The gases did not contain CO_2 but did contain CO, H_2 and significant mass fractions of light hydrocarbons. The compositions of ILC and condensed vapors were analyzed by HPLC/MS analysis. For ILC, high DP species (until septoasan) were in the great majority while cellobiosan (11%) and LVG (only 5%) were of minor importance. Much higher fractions of cellobiosan and LGV were observed in condensed vapors. These results suggested, unlike many other studies in the literature, that LVG was not the main primary product. Higher DP species would be produced prior to LVG. HAA has never been observed under these conditions.[61] In 2001, Banyasz *et al.*[62] studied the pyrolysis under conditions that increased temperature by 653 to 1023 K within about 1 min. From FTIR analysis of gaseous products, it was shown that formaldehyde, HAA and CO increased with heating rate while CO_2 decreased. The authors proposed a kinetic scheme where cellulose would decompose according to two competing reactions. The first one produced anhydrocellulose leading to char (1), the second one corresponded to a depolymerization followed by two other competing reactions: formation of CO_2, tar (containing mainly LVG) and char (2), and formation of intermediates decomposing into HAA or (fast) formaldehyde. This model as well as the simple BS model led to same results in terms of weight losses.

In 1993, Varhegyi *et al.*[63] scrutinized the results of Mok *et al.*[64] with crystalline cellulose under high pressure which gave rise to high char yields. They proposed a model that could mimic the results. It did not take into account an initiation step but relied on only two cellulose primary reactions, the first giving rise to char, H_2O, gases and volatiles, and the second to intermediates, that ultimately yield similar products.

In a paper entitled "Is the BS model for cellulose pyrolysis true?", Varhegyi *et al.*[65] questioned the validity of the BS model, mainly on the basis of TGA experiments performed between 523 and 643 K. In the absence of prolonged thermal pretreatment, a simple irreversible reaction (activation energy of 238 kJ mol^{-1}) appeared sufficient to mimic weight loss experiments for heating rates above 2 K min^{-1}. The authors claimed that the reaction giving AC in the BS model was superfluous in this temperature range. It is however interesting to notice that the activation energy for the BS model[38] (242 kJ mol^{-1}) is quite close to that calculated by the authors for mass loss experiments.[65] They also reported that the only source of char resulted from heterogeneous vapor–solid interactions. Minute amounts of inorganics considerably modified the pyrolytic behavior.

In 1995, Milosavljevic and Suuberg[66] showed that in their TGA experiments, the kinetics were highly sensitive to the heating rate. When cellulose was rapidly heated above 600 K, activation energies were found between 140 and 155 kJ mol^{-1}. On the other hand, in conditions of slow heating below 600 K, activation energy increased to 218 kJ mol^{-1}. They proposed a new model relying on three parallel reactions occurring directly from cellulose without passage through AC intermediates. Soon after, Milosavljevic *et al.*[67] showed that according to the importance of mass transfer limitations, the main pathway could be endothermic or exothermic. In their 1998 review, Antal *et al.*[68] explained the disagreement between their results and those of Milosavljevic and co-workers on the basis of inaccuracies in the temperature measurements resulting from the use of large particle sizes.

In 1999, the results of a Round Robin study of cellulose pyrolysis (heating rate between 5 and 40 K min^{-1}) were published by Gronli and Antal.[69] At high heating rates, larger scatters were observed because of uncertainties in temperatures. AC was not mentioned and all data could be represented by a single step process with activation energy of 244 kg mol^{-1} (close to that of the activation process producing AC in the BS model[38]).

In order to interpret the experimental observations, mathematical models are needed. The aim is to take into account numerous physical properties and processes, such as hydrodynamics and transfer limitations, which could occur in competition with pure chemical phenomena. Depending upon levels of complexities, the models include different numbers of parameters. Chemical kinetics may or may not include a fast initial step of cellulose depolymerization (AC formation). We have shown that in very severe experimental conditions, the direct measurement of reaction temperature is very difficult[4] in which case reaction temperature can only be obtained from models. Unfortunately, the accuracy of these calculated temperatures is not excellent because the results depend on the chosen values of several poorly-known physical parameters that can change during pyrolysis.

Diebold[70,71] reported the results of a model that calculated the fractions of products as a function of numerous experimental parameters and covering very large conditions of heating rates from SP to FP (0.01 to 60 000 K min^{-1}). The model confirmed that primary reaction would occur in the range 723–773 K and that AC fractions could sometimes exceed 90% during FP. Operation under vacuum minimized char fractions.

In 1996, Miller and Bellan[72] reported a model for the pyrolysis of porous cellulose on the basis of the BS model. They suggested the existence of three distinct regimes: initial heating, primary reaction (600–650 K regardless of conditions) and final heating.

Boutin *et al.*[60] proposed a model to explain their steady-state experiments of cellulose pyrolysis subjected to concentrated radiation. Formerly established under SP conditions, the BS model appeared to be also valid for FP. Under steady-state conditions, ILC thickness was estimated to be between 100 and 200 μm, with an ILC lifetime of around 20 milliseconds. The cellulose/ILC interface was calculated to be at 716 K and the external ILC layer at 947 K (an estimate of ILC vaporization temperature). Contact and radiant ablative pyrolysis were compared by Lédé[73] with evidence of different behaviors arising from different mass transfer efficiencies and residence times of ILC.

The ablation experiments of Lédé *et al.*[43] were modeled[74] on the basis of the BS model. The evolution of ILC layer and the formation of very low char fractions could be well represented. In the same year, Di Blasi[75] investigated the coupled effects of particle size and heating conditions on cellulose particle pyrolysis. On the basis of the BS model, her model showed that the primary step (AC formation) could not be ignored in conditions of high temperature and heating rate. It was shown also that the sample could not reach the final reactor temperature, with reaction occurring between 650 and 750 K. In the following year, the extended BS model was also used for studying the influence of several physical parameters.[76] At the same time, Lanzetta *et al.*[77] published results of a thin layer of cellulose submitted to radiant heating. The authors interpreted their results on the basis of the BS model but where AC formation was dropped according to the conclusions of Varhegyi *et al.*[65] However, it was pointed out that a change of properties (melting) could be important in the specific case of ablative pyrolysis.

In 1998, Di Blasi[78] used the BS model to study the degradation of anisotropic solids. Both one dimensional and two dimensional models were considered. In the same year, several kinetic schemes were considered for conditions of chemical reaction control.[79] They included the BS model with and without AC. The simulations were made for high and slow heating rates. The results were similar in terms of tar, gas and char fractions, each one appearing as a good approximation.

In her modeling of cellulose FP performed in a fluid bed reactor, Di Blasi[80] took into account all relevant processes (chemistry, heat and mass transfers and hydrodynamics). The BS model was extended to secondary cracking reactions. One of the results showed that actual reaction temperature varied inside the narrow range of 600–725 K. As shown above, the formation of primary intermediate species accompanied by phase change phenomena has not been always taken into account in spite of experimental evidence of its importance. Lédé *et al.*[81,82] published detailed discussions related to this problem on the basis of several experimental and simple theoretical considerations. The main conclusions were as follows:

Pioneering (mid-1960s to mid-1970s) experiments have clearly shown the presence of a liquid. The existence of an initiation step without mass loss has been evidenced by Broido *et al.*[37] and Bradbury *et al.*[38] in SP conditions. In FP conditions of HW[42] and SD[43] experiments the friction against hot metallic surfaces gave the feeling of smooth lubrication, resulting from the presence of a fluid (liquid or gas). No lubrication effect was evidenced for SP.

The physical nature of the fluid was determined from the comparison of two kinetic schemes: the BS model of Bradbury *et al.*[38] (with AC) and the one step model proposed by Antal and Varhegyi[29] (without AC). Simple mass balance calculations evidenced the impossibility for a vapor to be directly formed for two reasons. The theoretical escaping velocity of a vapor would be much higher than the velocity of sound. The estimated fluid viscosity would be about 1000 times higher than for a vapor.

In SD experiments, char observed on the disk was obviously issued from a liquid (in favor of BS model). Assuming the validity of the kinetic constants of Bradbury *et al.*[38] (obtained at low temperature and under vacuum), the ratio of kinetic constants giving rise to vapors and char respectively greatly exceeds 10 above 723 K (confirming that vapors would be favored in FP) while the ratio of the constants giving respectively rise to AC and vapors is much greater than one. Hence the step giving vapors would be controlling in the conditions of FP, allowing AC to be produced in high yields. Conversely, in SP, the step producing AC would be difficult to observe, as AC formation is rate controlling and hence would not accumulate. Consequently, a simple one step mechanism would be sufficient in SP. In these low temperature conditions, AC would be a slightly-deformable plastic material rather than a true liquid, explaining how char apparently keeps the initial structure of cellulose and that no lubrication effect is observed.

It was difficult to conclude that, in FP, AC would be a true melted form of cellulose, as AC is water-soluble, unlike cellulose. Another question was whether the vapors were formed *via* a simple process of AC boiling. According to the values of boiling temperatures of LVG, cellobiosan and triosan, it was concluded that in TGA, LVG would vaporize unlike other oligomers. In FP, it was concluded that as the vapors contained mainly LVG, AC would be made of oligomers undergoing cracking inside the liquid phase, as suggested by Powels *et al.*[54] The presence of higher oligomers in condensed vapors could result from further recondensation of monomers or through the trapping of aerosols droplets directly issued from AC.

2.3.6 New Results Published After the Beginning of the 2000s

New complex kinetics models have been proposed and new analytical methods have given evidence of several new products, providing new insight into the mechanisms of primary cellulose deconstruction. From their FP experiments and vapor GC-MS analysis, Luo *et al.*[83] proposed a chemical pathway which postulated AC competitively gave rise to LVG, HAA and 1-hydroxy-2propanone. A model relying on kinetic constants given by the literature[38,62,84,85] led to a reaction temperature of around 673 K for a heat source temperature of 883 K.

Following the works of Boutin *et al.*[60] and Authier *et al.*,[86] Liu *et al.*[87] studied the formation and evolution of AC in an image furnace apparatus. The formation of a yellowish layer on the surface of the sample was confirmed. Their HPLC analysis evidenced the formation of polysaccharides (average DP of 14), methylglyoxal, cellobiose, glucose and LVG with fractions depending on flash times and imposed heat fluxes. An improved new mechanism was proposed including the direct and fast formation of low DP (14) cellulose followed by two concurrent reactions. The first one gave methylglyoxal, the second led to AC (cellobiose, cellotriose, glucose, fructose) giving rise, in turn, to LVG, char and gases.

In 2005, Schroeter and Felix[88] studied the "Melting of Cellulose" under rapid heating. Cellulose samples were plastically deformed by the use of mechanical shear, uniaxial pressure and laser radiation. The resulting disks were examined by IR spectroscopy and microscopy. It was shown that cellulose had not been chemically modified, leading the authors to claim that they managed to melt cellulose, even if they recommended further studies of the DP and structure of the produced material. Several practical applications were suggested.

Zickler *et al.*[89] studied *in situ* X-ray diffraction using synchrotron radiation during thermal decomposition. Between 573 and 633 K, the authors evidenced a reduction of the microfibril diameters with random breaking of the fibers due to anisotropy of the material. In the same period, Wooten *et al.*[90] reported results of cellulose pyrolysis performed at low temperatures (573–598 K) and reaction times of 5–210 min. Analyzed by ^{13}C cross-polarization magic-angle spinning (CPMAS) NMR spectroscopy, it was shown that the residue contained unreacted crystalline cellulose, intermediate cellulose (IC), a low DP amorphous cellulose and a disordered final carbohydrate which was the final product after additional heating. IC appeared as a reaction intermediate composed of oligosaccharides which disappeared above 573 K. According to the authors, IC would correspond to the ILC observed in FP.[60,61] The authors concluded that these intermediate species, which were evidenced in FP, would be also found under low temperature conditions.

On the basis of theoretical considerations, Mamleev *et al.*[91] offered new interpretations of the experimental observations. They demonstrated the involvement of acid-base catalysis. The authors assumed the formation of encapsulated and growing cavities containing tars (LVG) resulting from transglycosylation and playing the role of electrolyte. Dissolved acids in the tars accelerated various reactions. Chain ends located at the cellulose–tar interface migrated into the tars. The authors proposed a model valid under TGA conditions and including two competing reactions directly issued from cellulose: β elimination and transglycosylation, each one being the source of two different kinds of chars. The model did not include the formation of ILC nor any dehydration reactions. However, it was underlined that the problems connected with ILC (and depolymerization into LVG) remained unresolved. The authors recommended the further developments of theories to describe the cellulose-ILC interface.

In the framework of a theoretical study on biomass pyrolysis kinetics, Ranzi *et al.*[92] took into account the large number of physicochemical parameters relevant to the reactor environment. The three main biomass components were considered. For cellulose, a multistep lumped mechanism was chosen, including the intermediate formation of AC even if about 5% of char plus H_2O could be formed in parallel directly from cellulose. AC decomposed according to two competitive reactions giving respectively LVG (at low temperature) and various products (such as HAA and glyoxal). Products fractions and kinetic constants were taken from previous TGA experiments. The kinetic constants for the process producing AC (8×10^{13} s^{-1} and 192 kJ mol^{-1}) were significantly lower than those of Bradbury *et al.*[38] which are often adopted in the literature. Six years later, Corbetta *et al.*[93] presented an updated model for the pyrolysis of centimeter-scale wood particles. The model took into account the coupled effect of kinetic, transport resistances and secondary gas phase reactions. In the case of cellulose, the kinetic model was similar as that of Ranzi *et al.*[92] but the kinetic constants were upgraded. For the reaction giving AC (4×10^{13} s^{-1} and 188.37 kJ mol^{-1}) the values were always significantly lower than those of Bradbury *et al.*[38]

In situ rheology measurements and ^{1}H NMR analysis were used for the first time in pyrolysis studies by Dufour *et al.*[94,95] The experiments evidenced softening phenomena during the pyrolysis of biomass and its components, in real time. A viscous and mobile material was formed for cellulose at 608 K even under the slow heating rate of 5 K min^{-1}. The authors showed that cellulose tended to impose its visco-elastic behavior to the polymer network during native biomass pyrolysis. The effect of minerals was also discussed. Very recently and in the framework of a study of fixed carbon yields, Van Wesenbeeck *et al.*[96] reported results of Avicel cellulose pyrolysis under pressure in sealed vessels. The SEM image of char products clearly evidenced that above 3.55 MPa and at 533 K, the coke-like product undergoes a molten phase, while at lower temperatures, char resembles the Avicel powder feedstock. These two references showed that phase change can be observed in mild SP conditions (low temperature and heating rate) and not just for FP.

Based on a literature review, Garcia-Perez[97] proposed in 2008 a detailed model including several types of reactions. Low DP species (AC) were initially formed giving rise to several types of processes: dehydration and depolymerization (in SP); depolymerization at high heating rate (FP); open ring-catalyzed (inorganic salts) reactions; acid-catalyzed dehydration reactions; polymerization of anhydrosugars; cross linking reactions and evolution of polyaromatics from char. Five years later, Bai *et al.*[98–100] published several investigations on the behavior of LVG under pyrolysis conditions. Under TGA/DSC (differential scanning calorimetry) conditions LVG was shown to undergo two competing reactions of evaporation and polymerization (formation of oligosaccharides that could subsequently decompose into light gases and vapors).[98] In other experiments performed in a micropyrolyser, the authors[99] showed the large influence of experimental conditions on reactions of LVG. Finally, and on the basis of specific experiments,[100] the authors proposed a physicochemical model for LVG pyrolysis allowing the prediction of the conditions promoting the LVG evaporation process.

Mattonai *et al.*[101] reported a study on the mechanisms of LVG formation. In a literature review, the authors noted that two types of mechanisms were usually proposed: ionic mechanisms (for example works of Patwardhan *et al.*[102]) and radical mechanisms (for example works of Shen and Gu[103]). The authors, then, described an original strategy based upon the co-pyrolysis of polysaccharides with radical scavenging species in order to see if, under these conditions, a loss of LVG was observed and if pyrolysis products composition had changed.

Several papers have recently been published on the effect of cellulose structure on pyrolysis reactions, mainly during SP. Yu *et al.*[104] and Liu *et al.*[105] studied the water-soluble intermediates obtained under SP conditions. These water-soluble intermediates were recovered by washing of the residues. In the first paper,[104] cellulose was subjected to temperatures ranging from 373 to 623 K at 10 K min^{-1} and a holding time of 30 min. Even at 373 K, sugar oligomers were detected, suggesting that they were produced from weak bonds (H bonds) in amorphous regions of cellulose. At temperatures reaching 623 K, all sugar-ring-containing oligomers disappeared while cellulose weight loss occurred. The results suggested that anhydro-sugar oligomers would result from various cleavages of glucosic bonds of cellulose rather than dehydration of the intermediate. In the second paper[105] and under similar experimental conditions, the authors showed markedly different behaviors between amorphous and crystalline celluloses. The strong H bonds in crystalline cellulose explained the much higher water-soluble intermediates produced from amorphous cellulose. It also explained the various sugar oligomers (DP ranging from 1 to 14) present after pyrolysis at 373 K. At higher temperatures, the intermediates produced by amorphous cellulose were mainly non-sugar products (from the destruction of pyranose rings).

During the same period, Wang *et al.*[106] studied the effect of crystallinity on the formation of liquid intermediate by TGA and Py-GC-MS techniques with Avicel crystalline cellulose and ball-milled Avicel cellulose. In the TGA studies, low crystallinity cellulose degraded at low temperature with a low activation energy of 162 kJ mol^{-1} instead of 222 kJ mol^{-1} for cellulose having higher crystallinity, while the fibrous structure disappeared (liquid). Py-GC-MS experiments showed that the reactions products were different with each type of cellulose (for example, more LVG with higher crystallinity samples). These differences were interpreted as the result of the acceleration of dehydration reactions in the presence of a liquid phase. Wang *et al.*[107] also reported results related to solid/liquid reactions in the formation of carbonaceous residues, with the same types of celluloses (high crystallinity and amorphous). The liquid intermediate caused agglomerations under 573 K. The amorphous cellulose completely melted while the crystalline cellulose kept its fibrous structure.

A new reaction mechanism[107] was proposed for cellulose under SP conditions. It included two parallel pathways for each type of cellulose. The primary species were: low DP cellulose for crystalline cellulose and liquid oligosaccharides for amorphous cellulose. Each one gave rise to the formation of numerous secondary products including LVG and cellobiosan.

The same year, Matsuoka *et al.*[108] discussed results related to AC, with a specific approach based on the reactivity of cellulose reducing ends under TGA conditions. In the case of low DP cellulose, the effect of elimination of the reducing ends with $NaBH_4$ was to prolong the induction period and to lower the fraction of water-soluble oligosaccharides. Conversely, and in the case of high initial DP cellulose, these treatments had no effect on weight loss behavior. An explanation was that the decomposition of reducing end groups activated the pyrolysis reactions. H bond rearrangement was proposed as the molecular mechanism for activation of the surface molecules of cellulose crystallites.

Kawamoto *et al.*[109] presented the results of a study of the influence of added aromatic substances on char formation during cellulose pyrolysis. No effect was observed under SP, while strong inhibition was observed under FP conditions, regardless of the type of added aromatics. The interpretation was that for FP, liquid intermediate species could more easily interact with the aromatic substances and lead to complete devolatilization (mainly LVG) without char formation. A model relying on cellulose crystallite surface and decomposition temperature was proposed.

Dauenhauer and co-workers have published numerous papers related to the primary species formed during FP. The impingement of microcrystalline cellulose particles (around 300 μm) onto a hot catalytic surface (973–1073 K) was observed by high speed photography.[110] The passage through a liquid intermediate, boiling into vapors (without char formation) was clearly evidenced. Similar observations were made with rods of cellulose pressed onto a hot surface (same experiments as Lédé *et al.*[43]). In the continuity of these works, Teixeira *et al.*[111] reported experiments and CFD modeling results showing that ejected aerosols resulted from the fragmentation of a liquid jet. The primary liquid was designed as "molten cellulose", which included LVG, cellobiosan, cellotriosan and longer anhydro-oligomers, according to literature. The model relied on two possible pathways of aerosols formations: evaporation of molten cellulose (followed by nucleation/condensation) and also direct ejection of liquids capable of transporting non-volatile materials. Surface tension and viscosities were estimated as well as aerosol velocity between 0.1 and 0.5 m s^{-1}.

In a study of wood FP, Paulsen *et al.*[112] used the new experimental technique of spatiotemporally resolved diffuse reflectance for measuring the sample composition both temporally and spatially with 10 μm and 1 microsecond resolutions, above 673 K. A 2D reaction-transport model was developed on the basis of three kinetic schemes for each of the three biomass main components. Each one relied on the primary formation of intermediate species followed by two possible concurrent processes.

Very recently, Krumm *et al.*[113] introduced a new thermal-pulsing reactor offering the possibility of heating rates much higher than in conventional TGA (10^6 C min^{-1}). The aim was to measure the millisecond-resolved evolution of crystalline cellulose chain scission, and its volatile products between 673 and 823 K, after rapid quenching. Samples sizes were in the order of a few tens of μm. By comparison of cellulose conversion and furan product formation kinetics, the authors showed that both models of chain-end cleavage (low temperature) and intra-chain scission (high temperature) could occur. A distinct kinetic transition between the first and second model was clearly observed at 740 K, a value quite consistent with reactive melting point of cellulose ("fusion like behavior"[2,45]).

Three other recent papers brought new evidence of liquid intermediate species through the measurement of some specific properties. Teixeira *et al.*[114] discovered that crystalline cellulose particles could lift off a heated surface (higher than 673 K) through high speed photography. Such a phenomenon was similar to the well-known Leidenfrost effect observed with liquids droplets over a hot surface. It was shown that the introduction of macroporosity to the hot surface could inhibit these affects providing a way to control particle-heat transfer rates. As a consequence, Krumm *et al.*[115] could show that cellulose particles on a micro-ratcheted hot surface could spontaneously move orthogonal to ratchet walls above 1023 K. It was noticed that such a phenomenon could provide a means of biomass conveyor without any moving parts in high temperature biofuels reactors. Finally, Teixeira *et al.*[116] evaluated cellulose pyrolysis with and without doped calcium for the generation of aerosols formed by boiling ejection of AC. They showed that approximately 3% of the initial feed (organic and inorganic) was transported to the gas phase as aerosols, bringing a possible explanation for the presence of non-volatile compounds often found in bio-oils. Such a transport mechanism for inorganics conformed with the results of Jendoubi *et al.*[117] They also confirmed the assumptions of Lédé *et al.*[81,82] for explaining the presence of heavy and non-volatile organic molecules in the condensed biomass pyrolysis products.

In 2015, Burnham *et al.*[118] published a critical literature review on cellulose pyrolysis. The objective was to identify correct global chemical kinetics models for cellulose thermal decomposition. The authors proposed an apparent activation energy close to 196 kJ mol^{-1}, if one allows catalytic acceleration of the initiation reaction (prior to volatile generation). It was underlined that fitting data at a single heating rate to a first order reaction model gave incorrect results. The problems connected to thermal transients and lags were also discussed.

Finally, in 2015, Le Brech[119] summarized his review of literature in terms of a general mechanism of cellulose pyrolysis. It included a first primary step giving rise to AC (through probable H bond cleavage and partial cellulose depolymerization). Then AC could give rise to three types of concurrent reactions: dehydration (mainly in SP); depolymerization through transglycosylation reactions (mainly in FP, *via* ionic or radical mechanisms); ring openings giving rise to low molar mass molecules. Typical products of cellulose pyrolysis (LVG, HAA, furans) would not result from a single reaction but from the combination of these various mechanisms. The author pointed out that exact mechanism of AC formations remained unknown.

2.4 Current Understanding of Cellulose Pyrolysis

Several hundred papers have been published in the last two centuries on biomass pyrolysis with special focus on cellulose pyrolysis. The experimental conditions are usually classified as either SP or FP. Due to poorly defined temperature and heating rates or measured parameters for distinguishing these two kinds of regimes, new more specific and quantitative criteria are desired.[4]

In spite of the many studies on the kinetics of SP and FP, no consensus has been reached on a universal mechanism of the primary and subsequent secondary processes (consecutive and/or concurrent, homogeneous and/or heterogeneous reactions). Simple first order models devised decades ago mostly from TGA experiments continue to be used today. After the first publication of the BS model in the late 1970s, most subsequent models include an initiation step with softening and/or phase change phenomena and formation of primary intermediate species. Several names have been given for these short-lived species (AC, low DP, ILC, IC and MC). For the sake of simplicity, the simple and single acronym IAC (Intermediate Active Cellulose) will be used to described these short-lived intermediates.[2,96] Subsequent reactions of IAC include dehydration, depolymerization (ionic and/or radical processes) and fragmentation (ring openings). The study of primary reactions to form IAC continue to give rise to numerous and sometimes contradictory papers. Difficulties in the study of these processes result from the rapidity of FP, inaccuracies in temperature measurements, and the short-lived nature of IAC. LVG was discovered more than 120 years ago as a major product of cellulose pyrolysis. However, its origin remains a matter of controversy, as it is potentially a product of either primary or secondary reactions.

The different roles of amorphous and crystalline cellulose in the formation of LVG, sugar oligomers and non-sugar products, as well as in the softening phenomena (during SP) and liquid formation (during FP) have been clearly evidenced. Inorganics have major effects either in IAC formation and/or IAC subsequent reactions. The various mechanisms of interactions between solid catalyst, solid sample and liquid IAC should be more clearly elucidated. In FP, liquid IAC becomes solid at room temperature. Microscopic observations suggest that char arises from this liquid state. In SP, there is evidence for softening of the IAC even though the char may retain the initial structure of cellulose. IAC is sometimes described as molten cellulose, which is inappropriate because the product has a smaller DP than cellulose and is soluble in water. Compositions of IAC are different in SP and FP. The mechanisms of DP decrease during the pyrolysis initiation step are subject to numerous interpretations. The chemical compositions of pyrolysis products are often determined after they have cooled down to room temperature during which time secondary reactions may have occurred. Quenching methods and efficiencies should be better specified. It is also difficult to prevent undesired reactions during reheating of IAC in order to study its behavior at high temperatures. The implementation of *in situ* measurements under pyrolysis conditions and using short response time apparatus and analysis methods is recommended. The measurement of actual cellulose reaction temperature also should be improved. IAC can be the source of various products such as chars of various porosities according to the type of pyrolysis (SP, FP). The migration of IAC through the pores of the particle gives rise to poorly understood secondary reactions.

The presence of a thin, lubricating liquid layer of IAC on the surface of reacting particles is the basis of ablative pyrolysis reactors. However, in any other type of reactor, IAC is a potential source of aerosols that can transport inorganics and non-volatile organics from the reaction zone and appear in the bio-oil. More generally, the formation of vapors depends on the IAC lifetime which is a function of complex competitions between its rates of formation, vaporization, repolymerization, dehydration and char formation. The presence of a liquid layer on the reacting particle modifies transfer and hydrodynamic phenomena. Sliding against the walls of a reactor is improved but also causes particles to stick together. Leidenfrost phenomena can also occur. Char formation from IAC may also cause agglomeration and fouling in the reactor and prevent steady-state operations. Clearly, the control of a continuous pyrolysis reactor depends on reactor properties but also on IAC elementary chemical reactions and of its poorly-known physical properties.

Few research studies on pyrolysis take into account the close interactions between chemistry and hydrodynamic and transfer processes whether for individual particles or the reactor as a whole. Erroneous chemical interpretations can be made if heat and mass transfer processes are rate controlling. For example, important thermal lags between heat source and particle explain important errors made in Arrhenius parameters determinations. Also, because of the differences in rate controlling processes in SP and FP, it may be inappropriate to use kinetic parameters obtained in TGA for FP conditions. Because of the complex composition of IAC, activation energies distributions might best be employed. Enthalpies of reaction essential for accurate modeling of pyrolysis are sparse in the literature. Physical properties of many pyrolysis species (including IAC) at reaction conditions are poorly known. They include thermal conductivity, density, heat capacity, radiative properties, surface tension and viscosity. In contrast, the existence of plastic or phase change phenomena during coal and synthetic polymers pyrolysis is relatively well known and could provide insight for understanding biomass pyrolysis.

2.5 Conclusions

Despite the numerous scientific studies of cellulose pyrolysis, no consensus has been reached on the mechanism by which cellulose decomposes to levoglucosan and other products. Simple first order models to describe the process have given way to much more complex ones. The primary reactions that depolymerize cellulose include the formation of intermediate species with softening and/or liquefaction phenomena although even the description of these processes is the subject of continuing debate in the literature. These primary steps are of importance because they control secondary very complex reactions as well as pyrolysis reactor performances.

The study of the fast initiation steps as well as the formation and properties of intermediate species is challenged by numerous experimental difficulties, including accurate measurement of reacting particle temperature. New experimental procedures, designs and analytical methods will be important for progress in this field.

Heat and mass transfer processes may be rate controlling during experiments designed to measure the chemical kinetics of pyrolysis, leading to frequent disagreements in the literature. Mathematical modeling of the close interactions between the elementary phenomena (chemistry transfers, hydrodynamic) occuring at the levels of the individual pyrolysing particles and of the reactor can provide guidance in devising experiments and analyzing kinetic data.

Glossary

AC	Active Cellulose
BS	Broido Shafizadeh model
DP	Degree of Polymerization
DTA	Differential Thermal Analysis
FP	Fast Pyrolysis
FTIR	Fourier Transform Infrared Spectroscopy
GC	Gas Chromatography
HAA	Hydroxyacetaldehyde
HPLC	High Performance Liquid Chromatography
HW	Hot Wire experiments
IAC	Intermediate Active Cellulose
IC	Intermediate Cellulose
ILC	Intermediate Liquid Compound (name representing any type of intermediate species and including AC, low DP, ILC, IC and MC[2,96])
LVG	Levoglucosan
MC	Molten Cellulose
MS	Mass Spectrometry
MTA	Mass Spectrometric Thermal Analysis
SD	Spinning Disk experiments
SP	Slow Pyrolysis
T	Temperature
T'	Heating rate
T_C	Reacting cellulose temperature
T'_C	Reacting cellulose heating rate
T_R	Reactor temperature
T'_R	Reactor heating rate
TGA	Thermogravimetric Analysis

References

1. *Thermal Biomass Conversion*, ed. A. V. Bridgwater, H. Hofbauer and S. Van Loo, CPL Press, Newbury, UK, 2009, vol. 1, 11, p. 37.
2. J. Lédé, *J. Anal. Appl. Pyrolysis*, 2012, **94**, 17.
3. D. Mohan, C. V. Pittmann and P. H. Steele, *Energy Fuels*, 2016, **30**, 480.
4. J. Lédé and O. Authier, *J. Anal. Appl. Pyrolysis*, 2015, **113**, 1.
5. R. Narayan and M. J. Antal, *Ind. Eng. Chem. Res.*, 1996, **35**, 1711.
6. M. J. Antal and M. Gronli, *Ind. Eng. Chem. Res.*, 2003, **42**, 1619.

7. R. Dumon and M. Gelus, *La valorisation chimique du bois*, Masson, Paris, 1982.
8. M. Violette, *Ann. Chim. Phys.*, 1853, **32**, 304.
9. M. Violette, *Ann. Chim. Phys.*, 1855, **39**, 291.
10. J. B. Mollerat, *Sur la distillation du bois et l'emploi de ses produits*, Institut de France, Paris, 11 Jan 1808, p. 1.
11. M. L. Gruner, in *Traité de Métallurgie*, ed. Dunod, Paris, 1875.
12. M. Klar, *The Technology of Wood Distillation*, Chapman and Hall, London, 1925.
13. L. F. Hawley, *Wood Distillation*, Am. Chem. Soc. Monograph Series, New York, 1923.
14. P. Klason, *J. für Prakt. Chem.*, 1914, **90**(2), 413.
15. A. Pictet and J. Sarasin, *Helv. Chim. Acta*, 1918, **87**, 1.
16. L. Clement and C. Rivière, in *La Cellulose et les éthers cellulosiques*, ed. Ch. Beranger, Librairie Polytechnique, Paris et Liège, 1920.
17. D. A. I. Goring, *Pulp Pap. Mag. Can.*, 1963, **12**, 517.
18. E. Back, M. Thoung, M. Jackson and F. Johanson, *Tappi*, 1967, 542.
19. S. B. Nordin, J. O. Nyren and E. L. Back, *Text. Res. J.*, 1974, 152.
20. R. F. Boyer, *Rubber Chem. Technol.*, 1963, **36**, 1303.
21. A. J. Stamm, *Ind. Eng. Chem.*, 1956, **48**, 413.
22. S. L. Madorsky, V. E. Hart and S. Strana, *J. Res. Natl. Bur. Stand.*, 1956, **56**, 343.
23. S. B. Martin, *10th International Symposium on Combustion*, The Combustion Institute, 1965, p. 877.
24. K. A. Lincoln, *Anal. Chem.*, 1965, **37**(4), 541.
25. K. A. Lincoln, *Proccedings of Specialist's Workshop on Fast Pyrolysis of Biomass*, (SERI/CP-622-1096), Copper Mountain (CO), October 19-22, 1980, p. 153.
26. K. J. Kilzer and A. Broido, *Pyrodynamics*, 1965, **2**, 151.
27. D. F. Arsenau, *Can. J. Chem.*, 1971, **49**, 632.
28. A. Broido and M. A. Nelson, *Combust. Flame*, 1975, **24**, 263.
29. M. J. Antal and G. Varhegyi, *Ind. Eng. Chem. Res.*, 1995, **34**, 703.
30. F. Shafizadeh, *Adv. Carbohydr. Chem.*, 1968, **23**, 419.
31. A. Pacault and G. Sauret, *C. R. Acad. Sci.*, 1958, **246**, 608.
32. P. K. Chatterjee and C. M. Conrad, *Text. Res. J.*, 1966, **36**(6), 487.
33. Y. Halpern and S. Patai, *Isr. J. Chem.*, 1969, **7**, 673.
34. O. P. Golova, *Russ. Chem. Rev.*, 1975, **44**(8), 687.
35. A. Broido, A. C. Javier-Son, A. C. Ouano and E. M. Barell, *J. Appl. Polym. Sci.*, 1973, **17**, 3627.
36. A. Broido and M. Weinstein, in *Proceedings 3rd International Conference Thermal Analysis*, ed. H. G. Wiedeman, Birkhauser Verlag, Basel, 1972, p. 285.
37. A. Broido, in *Thermal Uses and Properties of Carbohydrates and Lignins*, ed. F. Shafizadeh, K. V. Sarkauernard and D. A. Tallmann, Academic Press, New York, 1976, p. 19.

38. A. G. W. Bradbury, Y. Sakai and F. Shafizadeh, *J. Appl. Polym. Sci.*, 1979, **23**, 3271.
39. F. Shafizadeh and A. G. W. Bradbury, *J. Appl. Polym. Sci.*, 1979, **23**, 1431.
40. G. G. Allan, B. B. Kireger and D. W. Work, *J. Appl. Polym. Sci.*, 1980, **25**, 1839.
41. J. P. Diebold, *Proceedings of Specialist's Workshop on Fast Pyrolysis of Biomass*, (SERI/CP-622-1096), Copper Mountain, Co, October 19–22, 1980, p. 3.
42. J. P. Diebold, *Proceedings of Specialist's Workshop on Fast Pyrolysis of Biomass*, (SERI/CP-622-1096), Copper Mountain, Co, October 19-22, 1980, p. 237.
43. J. Lédé, J. Panagopoulos, H. Z. Li and J. Villermaux, *Fuel*, 1985, **64**, 1514.
44. G. V. C. Peacocke, PhD Thesis, University of Aston, Birmingham, 1994.
45. J. Lédé, H. Z. Li, J. Villermaux and H. Martin, *J. Anal. Appl. Pyrolysis*, 1987, **10**, 291.
46. H. Martin, J. Lédé, H. Z. Li, J. Villermaux, C. Moyne and A. Degiovanni, *Int. J. Heat Mass Transfer*, 1986, **29**(9), 1407.
47. J. Lédé, Special Issue in honor of Michael Antal, *Energy Fuels*, 2016, **30**(10), 7896.
48. W. S. L. Mok and M. J. Antal, *Thermochim. Acta*, 1983, **68**, 165.
49. R. J. Evans and T. A. Milne, *Energy Fuels*, 1987, **1**(2), 123.
50. G. N. Richards, *J. Anal. Appl. Pyrolysis*, 1987, **10**, 251.
51. D. A. G. Radlein, A. Grinshpun, J. Piskorz and D. S. Scott, *J. Anal. Appl. Pyrolysis*, 1987, **12**, 39.
52. J. Piskorz, D. A. G. Radlein, D. S. Scott and S. Czernik, in *Research in Thermochemical Biomass Conversion*, ed. A. V. Bridgwater and J. L. Kuester, Elsevier Applied Science, London, New-York, 1988, p. 557.
53. M. Essig, G. N. Richards and E. Schenck, in *Chemistry and Wood Chemistry Technology*, ed. C. Scherch, J. Wiley and Sons, New York, 1989, p. 841.
54. A. D. Powels, G. B. Hejkel, P. W. Arisz and J. J. Boon, *J. Anal. Appl. Pyrolysis*, 1989, **15**, 71.
55. A. Vladars-Usas, Ph.D. Thesis, University of Waterloo, Ontario, Canada, 1993.
56. J. Piskorz, D. Majerski, D. Radlein, A. Vladars-Usas and D. S. Scott, *J. Anal. Appl. Pyrolysis*, 2000, **56**, 145.
57. O. Boutin, M. Ferrer and J. Lédé, *J. Anal. Appl. Pyrolysis*, 1998, **47**, 13.
58. O. Boutin and J. Lédé, in *Fast Pyrolysis of Biomass : A Handbook*, ed. A. V. Bridgwater, CPL Press, Newbury, UK, 2002, vol. 2, p. 363.
59. J. Piskorz, D. S. Scott and D. Radlein, *Meeting Frontiers of Pyrolysis : Biomass Conversion and Polymer Recycling*, Breckenridge, CO, USA, June 25–30, 1995.
60. O. Boutin, M. Ferrer and J. Lédé, *Chem. Eng. Sci.*, 2002, **57**, 15.
61. J. Lédé, F. Blanchard and O. Boutin, *Fuel*, 2002, **81**, 1269.
62. J. L. Banyasz, S. Li, J. Lyons-Hart and K. H. Shafer, *Fuel*, 2001, **80**, 1757.

63. A. Varhegyi, P. Szalo, W. S. L. Mok and M. J. Antal, *J. Anal. Appl. Pyrolysis*, 1993, **26**, 159.
64. W. S. L. Mok, M. J. Antal, P. Szabo, G. Varhegyi and B. Zelei, *Ind. Eng. Chem. Res.*, 1992, **31**(4), 1162.
65. G. Vahregyi, E. Jakab and M. J. Antal, *Energy Fuels*, 1994, **8**(6), 1345.
66. J. Milosavljevic and E. M. Suuberg, *Ind. Eng. Chem. Res.*, 1995, **34**, 1081.
67. J. Milosavljevic, V. Oja and E. M. Suuberg, *Ind. Eng. Chem. Res.*, 1996, **35**, 653.
68. M. J. Antal, G. Varhegyi and E. Jakab, *Ind. Eng. Chem. Res.*, 1998, **37**, 1267.
69. M. Gronli and M. J. Antal, *Ind. Eng. Chem. Res.*, 1999, **38**, 2238.
70. J. P. Diebold, *Biomass Bioenergy*, 1994, 7(1–6), 75.
71. J. P. Diebold, Ph.D. Thesis, T-3007, Colorado School of Mines, 1985.
72. R. S. Miller and J. Bellan, *Combust. Sci. Technol.*, 1996, **119**, 331.
73. J. Lédé, *J. Anal. Appl. Pyrolysis*, 2003, **70**, 601.
74. C. Di Blasi, *Chem. Eng. Sci.*, 1996, **51**, 2211.
75. C. Di Blasi, *Ind. Eng. Chem. Res.*, 1986, **35**, 37.
76. C. Di Blasi, *Fuel*, 1997, **76**(10), 957.
77. M. Lanzetta, C. Di Blasi and F. Buonauno, *Ind. Eng. Chem. Res.*, 1997, **36**, 542.
78. C. Di Blasi, *Int. J. Heat Mass Transfer*, 1998, **41**, 4139.
79. C. Di Blasi, *J. Anal. Appl. Pyrolysis*, 1998, **47**, 43.
80. C. Di Blasi, *Chem. Eng. Sci.*, 2000, **55**, 5999.
81. J. Lédé, J. P. Diebold, G. V. C. Peacocke and J. Piskorz, in *Fast Pyrolysis of Biomass : A Handbook*, ed. A. V. Bridgwater, *et al.*, CPL Press, Newbury, UK, 1999, p. 51.
82. J. Lédé, J. P. Diebold, G. V. C. Peacocke and J. Pizkorz, in *Developments in Thermochemical Biomass Conversion*, ed. A. V. Bridgwater and D. G. B. Boocock, Blackie Academic and Professional, London, 1997, p. 27.
83. Z. Luo, S. Wang, Y. Liao and K. Cen, *Ind. Eng. Chem.Res.*, 2004, **43**, 4605.
84. G. Varhegyi, E. Jakab and M. J. Antal, *Energy Fuels*, 1994, **8**, 1345.
85. E. Shin, M. R. Nimlos and R. J. Evans, *Fuel*, 2001, **80**, 1697.
86. O. Authier, M. Ferrer, G. Mauviel, A. E. Khalfi and J. Lédé, *Ind. Eng. Chem. Res.*, 2009, **48**, 4796.
87. Q. Liu, S. Wang, K. Wang, X. Guo, Z. Luo and K. Cen, *Acta Phys.-Chim. Sin.*, 2008, **11**, 1957.
88. J. Schroeter and F. Felix, *Cellulose*, 2005, **12**, 159.
89. G. A. Zickler, W. Wagermaier, S. S. Funari, M. Burghammer and O. Paris, *J. Anal. Appl. Pyrolysis*, 2007, **80**, 143.
90. J. B. Wooten, J. J. Seemann and M. R. Hajaligol, *Energy Fuels*, 2004, **18**(1), 1.
91. V. Mamleev, S. Bourbigot, M. Le Bras and J. Yvan, *J. Anal. Appl. Pyrolysis*, 2009, **84**, 1.
92. E. Ranzi, A. Cuoci, T. Faravelli, A. Frassoldati, G. Migliavacca, S. Pierucci and S. Sommariva, *Energy Fuels*, 2008, **22**(6), 4292.

93. M. Corbetta, A. Frassoldati, H. Benmadji, K. Smith, M. J. Serapiglia, G. Gauthier, T. Telkion, E. Ranzi and E. M. Fisher, *Energy Fuels*, 2014, **28**, 3884.
94. A. Dufour, M. Castro-Diaz, P. Marchal, N. Brosse, R. Olcese, M. Bouroukba and C. Snape, *Energy Fuels*, 2012, **26**, 6432.
95. A. Dufour, M. Castro-Diaz, N. Brosse, M. Bouroukba and C. Snape, *ChemSusChem.*, 2012, **5**(7), 1258.
96. S. V. Wesenbeeck, C. Higashi, M. Legarra, L. Weng and M. J. Antal, *Energy Fuels*, 2016, **30**, 480.
97. M. Garcia-Perez, *The Formation of Polyaromatic Hydrocarbons and Dioxins During Pyrolysis : A Review of the Literature with Descriptions of Biomass Composition, Fast Pyrolysis Technologies and Thermochemical Reactions*, Washington State University, June, 2008.
98. X. Bai, P. Johnston, S. Sadula and R. Brown, *J. Anal. Appl. Pyrolysis*, 2013, **99**, 58.
99. X. Bai, P. Johnston and R. Brown, *J. Anal. Appl. Pyrolysis*, 2013, **99**, 130.
100. X. Bai and R. Brown, *J. Anal. Appl. Pyrolysis*, 2014, **105**, 363.
101. M. Mattonai, E. Ribechini and M. P. Colombini, *21th International Symposium on Analytical and Applied Pyrolysis*, Pyro16, Nancy, France, May 9-12, 2016.
102. P. R. Patwardhan, J. A. Satrio, R. C. Brown and B. H. Shanks, *J. Anal. Appl. Pyrolysis*, 2009, **86**, 323.
103. D. K. Shen and G. Gu, *Bioresour. Technol.*, 2009, **100**, 6496.
104. Y. Yu, D. Lin and H. Wu, *Energy Fuels*, 2012, **26**, 7331.
105. D. Liu, Y. Yu and H. Wu, *Energy Fuels*, 2013, **27**, 1371.
106. Z. Wang, A. G. McDonald, R. J. M. Westerhof, S. R. A. Kirsten, C. M. Cuba-Torres, S. Ha, B. Pecha and M. Garcia-Perez, *J. Anal. Appl. Pyrolysis*, 2013, **100**, 56.
107. Z. Wang, B. Pecha, R. J. M. Westerhof, S. R. A. Kirsten, C. Z. Li, A. G. McDonald and M. Garcia-Perez, *Ind. Eng. Chem. Res.*, 2014, **53**, 2940.
108. S. Matsuoka, H. Kawamoto and S. Saka, *J. Anal. Appl. Pyrolysis*, 2014, **106**, 138.
109. H. Kawamoto, T. Shoji and S. Saka, *21th International Symposium on Analytical and Applied Pyroysis*, Pyro 16, Nancy, France, May 9–12, 2016.
110. P. J. Dauenhauer, J. L. Colby, C. Balonek, W. J. Snazynski and L. D. Schmidt, *Green Chem.*, 2009, **11**, 1555.
111. A. R. Teixeira, K. G. Mooney, J. S. Kruger, C. L. Williams, W. J. Suszynski, L. D. Schmidt, D. P. Schmidt and P. J. Dauenhauer, *Energy Environ. Sci.*, 2011, **4**(10), 4306.
112. A. D. Paulsen, B. R. Hough, C. L. Williams, A. R. Teixeira, D. T. Schwartz, J. Pfaendtner and P. J. Dauenhauer, *ChemSusChem.*, 2014, **7**, 765.
113. C. Krumm, J. Pfaendtner and P. J. Dauenhauer, *Chem. Mater.*, 2016, **28**, 3108.
114. A. R. Teixeira, C. Krumm, K. P. Vinter, A. D. Paulsen, C. Zhu, S. Maduskar, K. E. Joseph, K. Greco, M. Stellata, E. Davis, B. Vincent, R. Hermann,

W. Suszynski, L. O. Schmidt, W. Fan, J. P. Rothstein and P. J. Dauenhauer, *Sci. Rep.*, 2015, **5**, 11238.
115. C. Krumm, S. Maduskar, A. D. Paulsen, A. D. Anderson, N. L. Barberio, J. N. Damen, C. A. Breach, S. Kunar and P. J. Dauenhauer, *Energy Environ. Sci.*, 2016, **9**, 1645.
116. A. R. Teixeira, R. Gantt, K. E. Joseph, S. Maduskar, A. D. Paulsen, G. Krumm, C. Zhu and P. J. Dauenhauer, *ChemSusChem.*, 2016, **9**, 1.
117. N. Jendoubi, F. Broust, J. M. Commandre, G. Mauviel, M. Sardin and J. Lédé, *J. Anal. Appl. Pyrolysis*, 2011, **92**, 59.
118. A. K. Burnham, X. Zhou and L. J. Broadbelt, *Energy Fuels*, 2015, **29**, 2906.
119. Y. Le Brech, PhD Thesis, Université de Lorraine, France, 2015.

CHAPTER 3

Lignin Depolymerization/ Deconstruction Reactions During Fast Pyrolysis

J. Y. KIM[a] AND J. W. CHOI*[b]

[a]Seoul National University, Department of Forest Sciences, 599 Gwanak-ro, Gwanak-gu, Seoul, 151-921, Korea; [b]Seoul National University, Graduate School of International Agricultural Technology, 1447 Pyeongchang-ro, Pyeongchang, 232-916, Korea
*E-mail: cjw@snu.ac.kr

3.1 Introduction

Lignins are the second most abundant polymer material in nature after cellulose, and account for 10–30 wt% of lignocellulosic biomass. Structurally, lignins are three-dimensional amorphous polymers composed of three types of monolignols (Figure 3.1) with various inter-unit linkages, involving ether linkages (*e.g.* α-O-4, 5-O-4, and β-O-4) and condensed carbon–carbon linkages (*e.g.* 5-5, β-β, β-5, and β-1 linkages).[1] Lignins are generally irregular structures that form highly condensed cross-linked polymer networks, providing biomass with mechanical strength as well as the rigidity to resist external forces.[2] Lignins can bind physically/chemically to cellulose/hemicellulose through the formation of benzyl-ether, benzyl-ester, and phenyl-glycoside covalent bonds, resulting in formation of lignin-carbohydrate complexes in

Green Chemistry Series No. 50
Fast Pyrolysis of Biomass: Advances in Science and Technology
Edited by Robert C. Brown and Kaige Wang

Published by the Royal Society of Chemistry, www.rsc.org

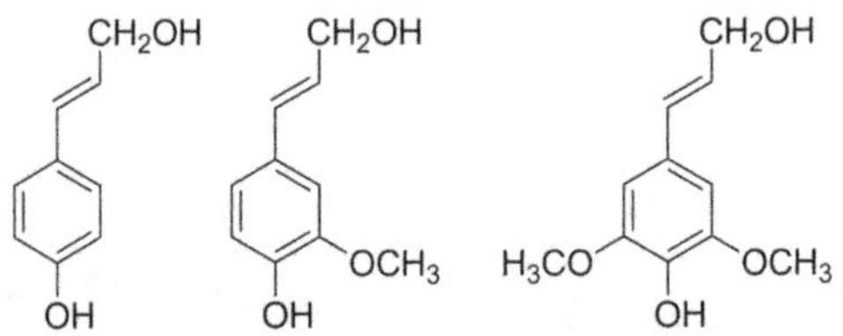

Figure 3.1 Three types of monolignols (*p*-coumaryl alcohol, coniferyl alcohol and sinapyl alcohol; from left to right).

plant cell walls.[3] Due to the complexity of this class of materials, it has been difficult to isolate lignins in their native form without physical or chemical modifications, and their structure has also been difficult to determine. Traditionally, milled wood lignin (ML) has been used as a representative source of native lignin, in spite of a small structural alteration that occurs during the ball-milling process. A structural fragment of lignin, proposed by Brunow *et al.*, is displayed in Figure 3.2.[4]

There are several different processes by which lignins can be isolated from biomass. Organosolv pulping or pretreatment, typically using solvents such as methanol, ethanol, acetone and ethylene glycol, is an efficient pretreatment method because organic solvents can be recycled.[5] Ionic liquid pretreatment is another promising method because of its eco-friendly characteristics and non-flammable and recyclable capacities, which can also enhance the accessibility of cellulolytic enzymes to the biomass by reducing both the crystallinity and lignin content.[6,7] The Arkenol process is a typical hydrolysis technology that uses highly concentrated sulfuric acid to hydrolyze polysaccharides in biomass.[8] In these processes, the diverse chemical and structural properties of lignin, such as molecular weight (M_w), the polydispersity index (PDI), the frequency of β-O-4 linkages, thermal stability, the content of functional groups (methoxyl: OMe and phenolic hydroxyl: Phe-OH) and the degree of condensation, vary considerably with the extraction conditions.[9] Elemental composition, functional group content and other chemical features of four lignins extracted from different processes are presented in Table 3.1.[10]

Pyrolysis is a thermochemical conversion technology that provides high yields of bio-oil (of up to 75 wt%) and can be used directly in a variety of applications.[11] Lignin pyrolysis for the efficient utilization of lignin byproducts as liquid fuel or value-added chemical resources has been proposed and developed. However, due to the structural complexity of lignins, various kinds of phenolic compounds are produced during lignin pyrolysis, resulting in bio-oils with a wide compositional distribution. In addition, some problems associated with the process, such as feeder blocking, bed agglomeration, and subsequent defluidization, have not yet been solved.[12] Recent developments in the field of lignin pyrolysis have renewed interest in the application of catalysts to avoid these drawbacks.[13,14] Among various catalysts, HZSM-5, a representative zeolite, is an excellent deoxygenation catalyst for the production

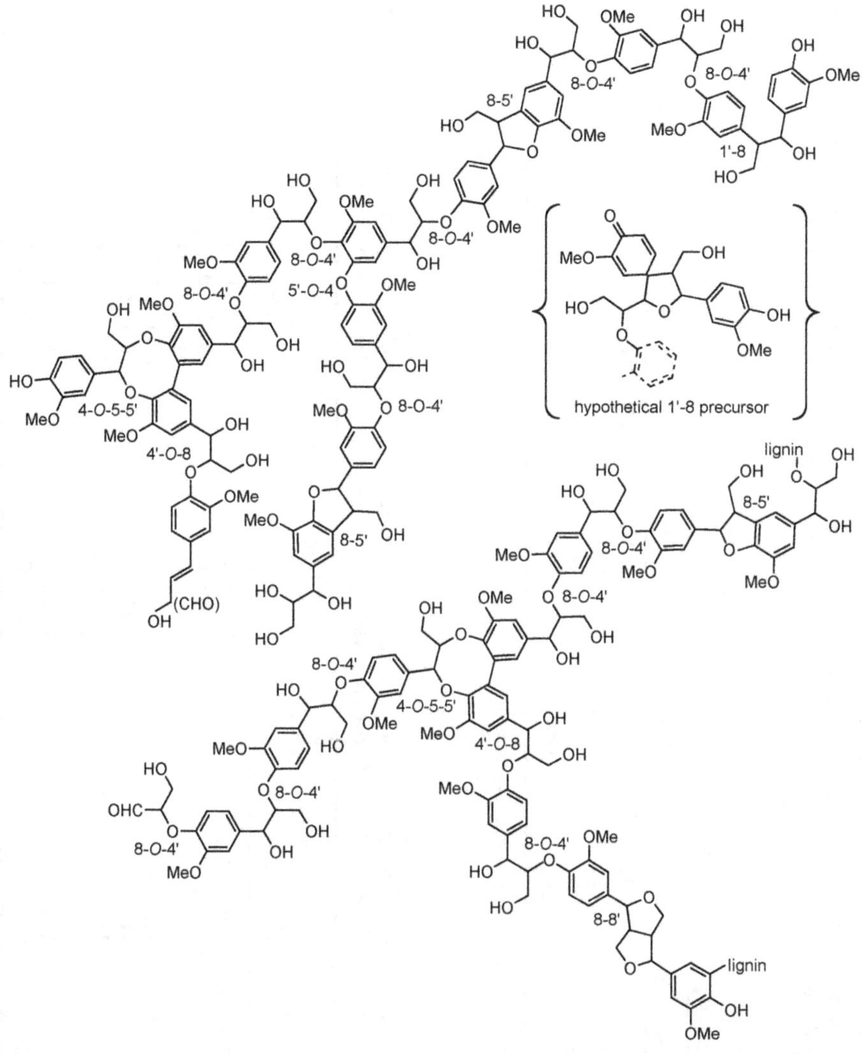

Figure 3.2 A structural fragment of lignin proposed by Brunow *et al.* Reprinted with permission from ref. 4. Copyright (1998) American Chemical Society.

of low oxygen-containing bio-oil and value-added aromatic hydrocarbons, such as benzene, toluene, and *p*-xylene.[15,16]

Over the past few years, research into the production of chemicals, liquid fuels, and platform compounds from lignins has grown rapidly because of the importance of lignins in biorefineries. However, lignins have not yet been effectively utilized in high value-added products because of their structural complexity. In fact, it has been estimated that less than 2% of lignin byproducts from the paper industry are used for high value products.[17] Several previous studies have reported applications for lignins, either directly or indirectly,

Table 3.1 Structural features of various lignins extracted from poplar wood. Reprinted from Polymer Degradation and Stability, **98**(9), J.-Y. Kim, S. Oh, H. Hwang, U.-J. Kim, J. W. Choi, Structural features and thermal degradation properties of various lignin macromolecules obtained from polar wood (*Populus albaglandulosa*), 1671–1678, Copyright (2013) with permission from Elsevier.[10]

	Elemental composition (wt%)					Functional groups (wt%)			
Lignin[a]	C	H	N	O[b]	S	OMe	Phe-OH	C_9 formula	M_w (Da)
ML	58.6	5.8	0.1	35.5	-	14.4 ± 0.1	6.3 ± 0.2	$C_9H_{8.12}O_{2.72}(OH)_{0.76}(OCH_3)_{0.94}$	10002
OL	65.4	5.9	0.3	23.4	-	14.7 ± 0.4	3.6 ± 0.3	$C_9H_{7.74}O_{1.46}(OH)_{0.36}(OCH_3)_{0.81}$	4169
IL	60.1	5.9	1.2	32.8	-	15.5 ± 1.5	6.7 ± 0.2	$C_9H_{7.90}O_{2.32}N_{0.15}(OH)_{0.79}(OCH_3)_{1.00}$	6347
KL	60.8	5.9	0.2	32.9	0.2	13.6 ± 0.6	0.7 ± 0.0	$C_9H_{8.76}O_{3.07}(OH)_{0.08}(OCH_3)_{0.85}$	-

[a]ML: milled wood lignin, OL: organosolv lignin, IL: ionic liquid lignin, KL: klason lignin.
[b]Determined by difference.

Table 3.2 Kinetic parameters derived for the various lignins. Reproduced from Thermochimica Acta, **498**(1–2), G. Jiang, D. J. Nowakowski, A. V. Bridgwater, A systematic study of the kinetics of lignin pyrolysis, 61–66, Copyright (2010) with permission from Elsevier.[22]

	Kinetic parameters		
	E (kJ mol^{-1})	n	lnA (the unit of A is min^{-1})
Alcell lignin (hard wood)	159.5	1.07 ± 0.11	28.50 ± 0.13
Asian lignin (straw and grass)	133.9	1.06 ± 0.11	23.92 ± 0.24
Organosolv lignin (hardwood)	144.2	1.01 ± 0.16	25.68 ± 0.29
ETEK lignin (softwood)	151.9	1.04 ± 0.17	27.37 ± 0.37
Klason lignin (beech, hardwood)	151.1	1.43 ± 0.18	26.80 ± 0.20
Klason lignin (cassava rhizine)	172.2	1.58 ± 0.12	29.87 ± 0.24
Klason lignin (cassava stalk)	171.7	1.53 ± 0.16	29.81 ± 0.26
Klason lignin (mixed softwood)	145.2	1.44 ± 0.08	24.78 ± 0.10
Klason lignin (willow)	156.5	1.53 ± 0.08	27.82 ± 0.17

as renewable materials, including adhesives, automotive brake pads, biodispersants, emulsifiers, epoxides, polymers, polyolefins, resins, wood panel products, and as the principal component of thermoplastic materials.[18] Among the abovementioned applications, depolymerization/deconstruction of lignins to monomeric phenols and aromatic compounds as an alternative to the petrochemical industry is the most promising approach to realize sustainable utilization of lignins.[19] In this chapter, the thermal degradation behavior as well as the pyrolysis features of several kinds of lignins extracted from different biomass sources will be discussed and the physicochemical properties of lignin depolymerized products analyzed using advanced analysis techniques will be summarized.

3.2 Thermal Degradation Behavior of Various Lignins

Thermogravimetric analysis (TGA) is a method of thermal analysis in which changes in the physical and chemical properties of materials are measured as a function of increasing temperature (with constant heating rate), or as a function of time (with constant temperature and/or constant mass loss). It can provide information about physical phenomena, such as second-order phase transitions, including vaporization, sublimation, absorption, adsorption, and desorption. It also provides information on chemical phenomena, including chemisorption, desolvation (especially dehydration), decomposition, and solid–gas reactions (*e.g.* oxidation or reduction).[20] In general, lignin structure is very complex because it is mainly composed of aromatic rings with C_α-C_β-C_γ side chains and has several kinds of functional groups. These structural characteristics result in an extremely wide temperature range of decomposition. From TGA results, several types of structural information, including the frequency of β-O-4 linkage and the condensation degree of lignin, can be obtained.[21]

Several studies have investigated the thermal degradation properties of lignins from various sources. Pyrolysis kinetics of four types of lignins (alkali lignin, hydrolytic lignin, organosolv lignin, and klason lignin) were investigated by Jiang *et al.*[22] using TGA. The kinetic parameters (activation energy, order of reaction and frequency factor) were calculated. As shown in Table 3.2, activation energy and order of reaction depend on the separation method and plant species. Jiang *et al.*[22] reported that differences in activation energy among lignins might be the result of variations in the percentage of the three monolignols. In addition, the order of reaction for klason lignin is 1.5, most probably due to its condensation structure, whereas the other lignins follow first order reaction due to their less modified chemical structure.

Chu *et al.*[23] synthesized a model compound with an average weight molecular weight (M_w) of 1755 Da to represent β-O-4 type oligomeric lignin. They reported that this model compound displayed similar thermal degradation behavior to that of lignin from maple wood (Figure 3.3). The ether linkages, especially β-O-4, were easily broken in the temperature range of 250–350 °C.

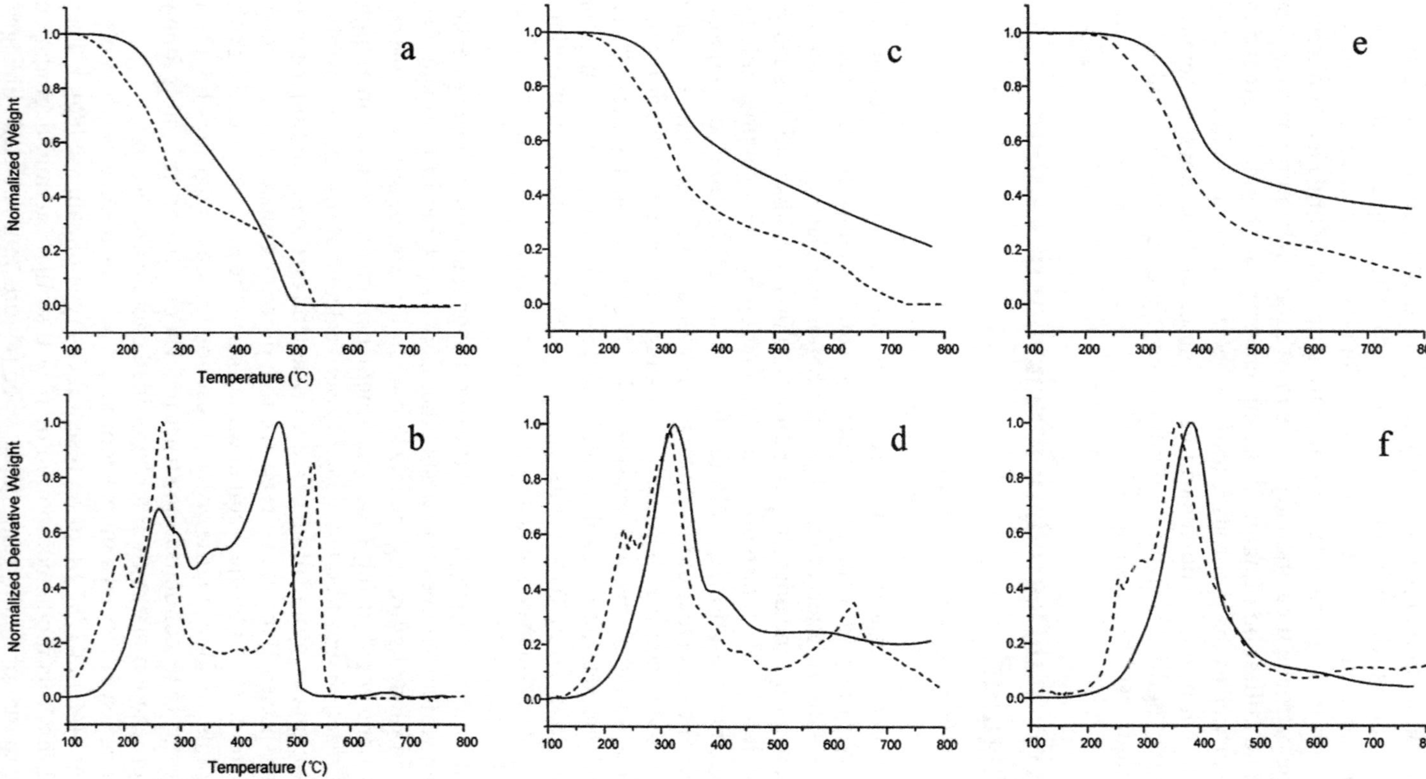

Figure 3.3 Thermogravimetric (a, c and e) and Derivative thermogravimetric (b, d, and f) curves for β-O-4 lignin model compound (dash line) and lignin from maple wood (solid line) at temperature ramps of 1 (a and b) 15 (c and d) and 150 °C min^{-1} (e and f). Reproduced from ref. 23 with permission of the Royal Society of Chemistry.

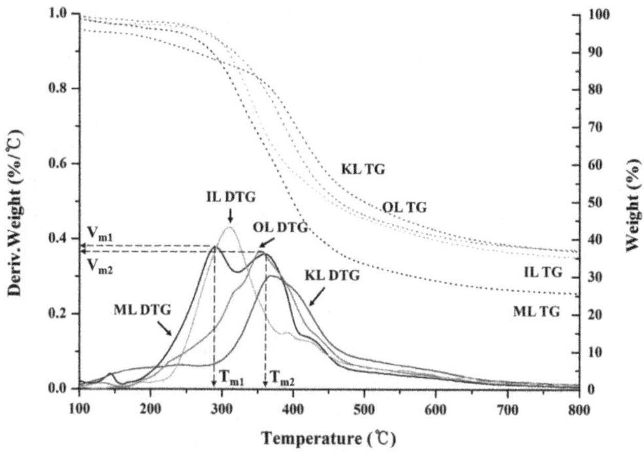

Figure 3.4 TG and DTG curve of lignin samples (Heating rate: 10 °C min^{-1}). Reprinted from Polymer Degradation and Stability, **98**(9), J.-Y. Kim, S. Oh, H. Hwang, U.-J. Kim, J. W. Choi, Structural features and thermal degradation properties of various lignin macromolecules obtained from polar wood (*Populus albaglandulosa*), 1671–1678, Copyright (2013) with permission from Elsevier.[10]

Work by Kim *et al.*[10] also revealed that the frequency of β-O-4 linkage correlated with the thermal degradation behavior of lignin. They prepared four different lignins extracted from poplar wood (milled wood lignin: ML, organosolv lignin: OL, ionic liquid lignin: IL and klason lignin: KL). In the case of ML and IL, which have relatively large amounts of β-O-4 linkages as compared with OL and KL, T_{m1} (maximum decomposition temperature) for ML and T_m for IL were 290.6 °C and 307.8 °C, respectively, indicating that these peaks resulted from the cleavage of β-O-4 linkages (Figure 3.4). OL only had a shoulder peak in this temperature region, due to the relatively low amount of β-O-4 linkages in OL. Meanwhile, T_{m2} for ML and T_m for OL and KL were observed to occur above 350 °C. This was attributed to the cleavage of carbon–carbon bonds.

3.3 Lignin Depolymerization/Deconstruction by Analytical Pyrolysis

During pyrolysis, lignin is thermally decomposed to low molecular weight phenolic compounds by exposure to heat in an inert atmosphere. These pyrolysis products provide information on the structure of lignin.[24] Analytical pyrolysis also has several analytical advantages, including simple preparation with a small sample quantity (μg scale) and rapid analysis times.[25] To date, numerous studies using pyrolysis-gas chromatography-mass spectrophotometry-flame ionization detector (GC-MS-FID) or molecular beam mass spectrometry (MBMS) have been conducted to estimate the lignin potential as a valuable phenol source, as well as to investigate the structural features of various lignins.[26–29] The composition of low molecular products with C_1–C_4 hydrocarbons derived from lignin pyrolysis and their kinetics were investigated using a captive sample reactor.[30,31] Chan and Krieger[32] reported the distribution of lignin pyrolysis products (volatile, char and non-condensable gas) as a function of pyrolysis temperature using a microwave reactor. Recombination mechanisms between the primary lignin pyrolysis products were also investigated using a micro-pyrolyzer coupled with a GC-MS-FID.[33] Figure 3.5 shows the normalized absolute areas of compounds from the three lignins (kraft lignin, prairie cordgrass lignin [PCG lignin], and aspen lignin) reported by Zhang *et al.*[34] Guaiacol, 4-vinylguaiacol, phenol, 4-methylguaiacol, vanillin, acetic acid, and 4-methylphenol were detected as the main pyrolysis products of kraft lignin. In the case of prairie cordgrass lignin, 4-vinylguaiacol, guaiacol, 4-methylguaiacol, 2,3-dihydrobenzofuran, furfural, methyl-isobutyl-ketone, acetic acid, phenol, 4-methyl phenol, and syringol were mainly produced. In addition, syringol, phenol, guaiacol, 4-methylsyringol, 4-methylguaiacol, 4-vinylguaiacol, 3′,5′-dimethoxyacetophenone, and 4-(1-propenyl) guaiacol (trans) were the main compounds derived from aspen lignin.

The reaction chemistry of lignin pyrolysis is very complex. During pyrolysis, lignin is decomposed into monomeric phenols *via* cleavage of both the weaker bonds (*e.g.* hydrogen bonds and aryl ether bonds) and stronger bonds (carbon–carbon) in lignin.[35] Previous research on lignin pyrolysis using various reaction conditions confirmed that lignin is thermally

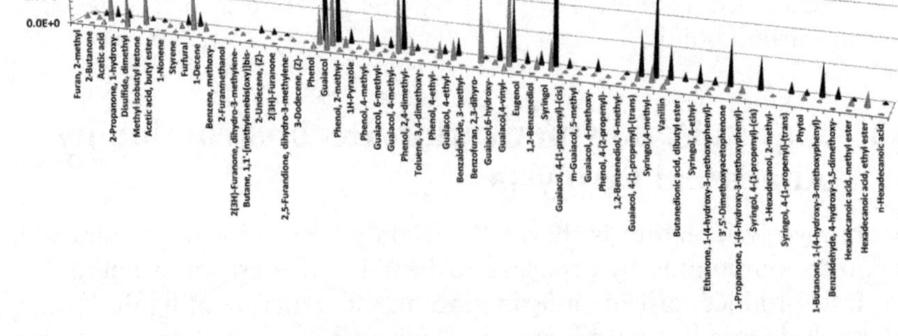

Figure 3.5 Normalized absolute areas of compounds from the three lignins. Reprinted from Journal of Analytical and Applied Pyrolysis, **98**, M. Zhang, F. L. P. Resende, A. Moutsoglou, D. E. Raynie, Pyrolysis of lignin extracted from prairie cordgrass, aspen, and Kraft lignin by Py-GC/MS and TGA/FTIR, 65–71, Copyright (2012) with permission from Elsevier.[34]

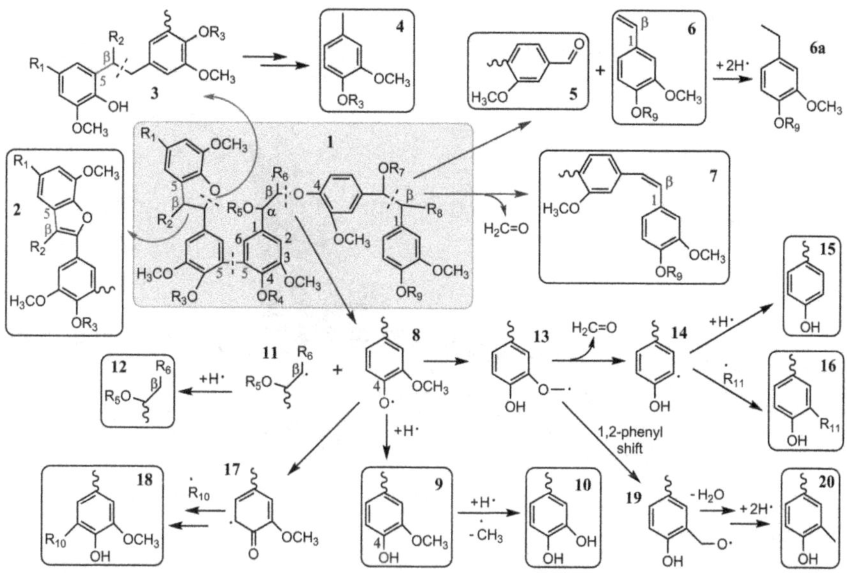

Figure 3.6 Lignin structure dependent pyrolysis reactions: 1 hypothetical lignin source representing the most common bonds in kraft lignin; 2 relatively stable β-5 product in the primary decomposition step; 3 α-O cleavage product; 4 methyl guaiacols; 5 aldehydes; 6 styrenes and 6a ("reduced vinyl-") ethyl-phenols; 7 stilbenes; 8 phenoxy radicals; 9 guaiacols; 10 catechols; 11, 13, 14, 17, 19 transient radicals; 12 aliphatics; 15 phenols; 16, 18 condensation products (*e.g.* 4-O-5 dimers); 20 cresols. Question mark shows possible 5-5 bond breakage, detailed in the 31P NMR section. R_1 cinnamyl group (propanoid unit); R_2, R_6, R_8 γ carbons in cinnamyl groups; R_{3-5}, R_7, R_{9-11} H, CH_3 or whole phenylpropanoid units/macromolecules. Reproduced from ref. 38 with permission from The Royal Society of Chemistry.

decomposed by multiple, parallel radical, and rearrangement pathways.[23,36,37] Figure 3.6 shows several lignin pyrolysis reaction pathways proposed by Kosa *et al.*[38]

Previous investigations led to the generally accepted assumption that lignin degradation follows complex reaction pathways, including the formation of free radicals *via* homolytic dissociation of interunit-linkages between phenolic monomers. The lignin macromolecule can stabilize some of the highly reactive phenolic radicals formed by bond dissociation. Recently published works have reported on the detection of volatile radicals from the gas phase or non-volatile products from lignin pyrolysis using EPR spectroscopy.[39–41] Figure 3.7 indicates that the radical concentration determined by EPR remained at a constant low level up to approximately 300 °C (the starting temperature of the depolymerization process). This work concluded that the stable volatile pyrolytic products could be divided into three main categories: (1) phenols, (2) alkoxy phenols, and (3) phenol alkoxy ketones. These were selectively formed within certain temperature ranges corresponding to

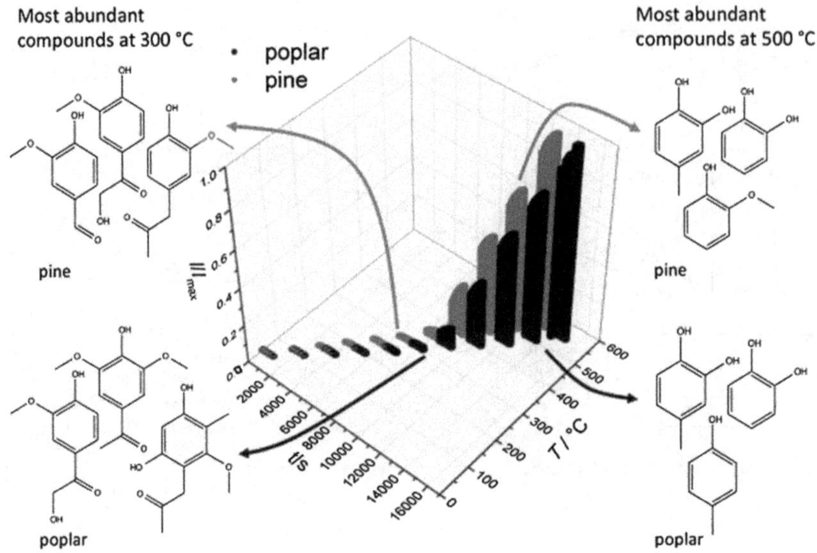

Figure 3.7 Normalized radical concentration of poplar and pine klason lignin during pyrolysis and the most abundant stable organic compounds leaving the sample at 300 and 500 °C. (Reproduced with permission from, ref. 39 John Wiley and Sons, Copyright © 2014 WILEY-VCH Verlag GmbH & Co. KGaA, Weinheim).

different radical concentrations. Furthermore, radicals start to form in significant amounts beyond this temperature.[39]

Pyrolysis-GC-MS-FID is also a powerful analytical instrument that can be applied to investigate lignin structure based on lignin depolymerized products, specifically with regard to the proportion of H-, G-, and S-type monomeric phenols. Ohra-Aho *et al.*[24] investigated the S:G distribution of several Brazilian *Eucalyptus* species and their crossings by pyrolysis-GC-MS and alkaline nitrobenzene oxidation. They revealed that the S:G ratios determined using pyrolysis and nitrobenzene oxidation were close to each other. Nunes *et al.*[42] also reported that pyrolysis-GC-MS was a promising technique for the determination of the S:G ratio in *Eucalyptus* lignin. In addition, Kim *et al.*[10] reported that the S:G ratio determined using the pyrolysis products of lignins varied considerably depending on the pyrolysis temperature.

3.4 Continuous Fast Pyrolysis of Lignin

Fast pyrolysis of extracted lignin has the potential to convert a relatively low value feedstock into valuable phenolic compounds. However, the efforts of several researchers, summarized in Table 3.3, indicate relatively disappointing yields of bio-oil, which range from 11–60 wt%, with most yields around 30–40%.[43–47]

Much of the low bio-oil yield can be ascribed to very high yields of char compared to the pyrolysis of lignocellulosic biomass.[48] Furthermore, the

Table 3.3 Summary of lignin pyrolysis experiments by various researchers.

Lignin	Reactor type	Temperature (°C)	Bio-oil (wt%)	Char (wt%)	Gas (wt%)	Reference
Asian lignin	Fluidized bed, entrained	410–700	22–49.7	34–48.8	5.7–29.7	49
ETEK lignin[a]	flow, batch fluid bed		11.7–72	7–48.8	6.5–39.5	
EMAL[b]	Tubular reactor	400–900	48–60	28–42	6–26	73
Kraft lignin	Fixed bed	500–600	30–38.1	47.1–50.1	14.8–20.0	13
Kraft lignin	Fixed bed	700	44	39	17	43
Corn stover lignin[c]	Fixed bed	800	-	-	-	44
Rice straw lignin[c]						
Kraft lignin						
Kraft lignin	Fluidized bed	450–600	30–36	-	-	45
Organosolv lignin (wheat straw)	Fluidized bed	500	51.9–54.7	30.9–35.6	15.1–17.4	46
Organosolv lignin (hard wood)			47.6	39.0	20.7	
Asian lignin			38.9	43.0	15.2	
Kraft lignin	Fluidized bed	550	32.3–43.0	31.7–36.0	21.5–34.6	47

[a]ETEK lignin was a residue from ethanol production by a 2-stage weak acid hydrolysis of softwood.
[b]EMAL means enzymatic/mild acidolysis lignin.
[c]Residual products derived from the bioethanol process.

Figure 3.8 Char deposit in solid bed (left-up), tar deposits in condenser (left-down) and electrostatic precipitator (right). Reprinted from Journal of Analytical and Applied Pyrolysis, **88**(1), D. J. Nowakowski, A. V. Bridgewater, D. C. Elliott, D. Meier, P. de Wild, Lignin fast pyrolysis: Results from an international collaboration, 53–72, Copyright (2010) with permission from Elsevier.[49]

char appeared as large agglomerates instead of fine powder as is typical in fast pyrolysis of lignocellulosic biomass.[46] A 2010 international collaboration of 14 laboratories[49] on lignin fast pyrolysis reported that continuous pyrolysis of lignin was almost impossible due to melting and subsequent agglomeration and dehydration of lignin particles to form solid char, which clogged reactors and forced shut-downs (see Figure 3.8).

Several researchers have attempted to avoid agglomeration problems by changing the reactor design and process conditions. Oxidative pyrolysis of lignin and its autothermal operation were studied by Li *et al.*[50] Oxygen (air) was injected into the pyrolysis reactor to achieve autothermal operation. They concluded that autothermal pyrolysis of kraft lignin might be practical because it does not require additional external heat input and it also simplifies the equipment requirements. In addition, the bio-oil obtained from this process contained less pyrolytic lignin and more monomeric phenols, although its heating value was slightly lower. Trinh *et al.*[51] used a pyrolysis centrifuge reactor (PCR) system to avoid plugging/agglomeration in the feeder during lignin pyrolysis. Agglomeration disappeared when the reactor temperature was higher than 450 °C, but plugging at the cooling nozzle was still observed after an hour of operation.

Mixing low cost chemicals with extracted lignin as a co-reactant is another approach to overcoming agglomeration. Mukkamala *et al.*[52] added calcium formate to the feed with kraft lignin prior to fast pyrolysis. The calcium formate thermally decomposed into hydrogen, which facilitated *in situ* chemical transformation in the inert atmosphere. They reported that formate-assisted pyrolysis was an effective method for reducing the agglomeration as well as partially deoxygenating lignin-derived bio-oil.

a. b.

Figure 3.9 SEM images of chars obtained from the pyrolysis of untreated (a) and $Ca(OH)_2$ pretreated lignin (b) at 500 °C. Reproduced with from ref. 53 with permission from The Royal Society of Chemistry.

The use of calcium hydroxide pretreatment to overcome the agglomeration issue during lignin fast pyrolysis was developed by Zhou *et al.*[53] They reported that pretreatment with $Ca(OH)_2$ reduced the functional groups in lignin (*e.g.* phenolic hydroxyl, carboxylic acid and aldehyde groups), which are responsible for the melting and agglomeration problems, by forming hydroxylcalcium phenoxides, phenolic alcohols, and phenolic carboxylate salts that inhibited process problems. Figure 3.9 compares surface images between two chars from untreated and pretreated lignin. The char from untreated lignin consisted of agglomerated char that had clearly gone through a melt phase, whereas the char from pretreated lignin formed much smaller particles of char.

3.5 Chemical Properties of Lignin-Derived Oil

Compositional investigation of bio-oil from lignin has been performed by using GC-MS, nuclear magnetic resonance (NMR), and gel permeation chromatography (GPC). Most lignin pyrolysis studies have employed GC-MS to analyze the monomeric phenols in bio-oil.[44–47] Jung *et al.* reported that phenol, 4-ethylphenol, guaiacol, creosol, 4-ethylguaiacol, catechol, and syringol were the main monomeric phenols in bio-oils produced from kraft, constover, and rice straw lignin.[44] A recent published paper by De Wild *et al.*[54] also compared the compositional distribution of bio-oils produced from the pyrolysis of three lignin sources using a 1 kg h^{-1} bubbling fluidized bed reactor at 500 °C (lignin from wheat straw *via* organosolv pulping, kraft lignin and Alcell lignin). Methanol and acetic acid were the main non-aromatic compounds resulting from acetyl and methoxyl groups in lignin. The yield of monomeric phenols ranged from 6 to 7 wt% (based on the dry lignin).

Differences in the yield of monomeric phenols reflected the variety in the composition of lignin sources.

The molecular distribution of lignin-derived bio-oils has been determined by GPC analysis.[38,47,53,55] The average molecular weights (M_w) of bio-oils produced from various lignin sources range from 210 to 350 g mol^{-1} with polydispersity indices ranging from 1.5 to 2.2.[38,47,53,55] This indicates that the phenolic compounds in the bio-oil consist mainly of phenolic monomers and dimers.

Due to the limited detection of high molecular weight compounds in lignin-derived bio-oils, NMR, including quantitative ^{1}H, ^{31}P, and ^{13}C-NMR, and semi-quantitative heteronuclear single-quantum correlation-nuclear magnetic resonance (HSQC-NMR) have also been adapted to further characterization of bio-oils.[38,55,56] Using ^{13}C- and ^{31}P-NMR analysis, Ben *et al.*[55] observed that the aliphatic hydroxyl, carboxyl, and methoxyl groups of kraft lignin were eliminated during pyrolysis. In addition, they confirmed that the ether bonds in lignin were mainly degraded during pyrolysis. This group also investigated the structural characteristics of bio-oils from kraft lignin by HSQC-NMR and found that the methoxyl group in the kraft lignin rearranges to another type and ether bonds were degraded during pyrolysis.[56]

3.6 Catalytic Pyrolysis of Lignin

Catalytic pyrolysis is a promising technology that can convert solid biomass directly into liquid fuels or chemicals such as aromatic hydrocarbons (AH). Benzene, toluene, ethylbenzene, xylenes, and naphthalene are produced as the main AH compounds. In this process, zeolite catalysts play an important role because they affect the chemical composition and yield of the final products.[57] Zeolites are crystalline microporous aluminosilicate materials that can be used in many applications ranging from detergents, ion-exchange agents, and adsorbents to catalysts.[35] Jackson *et al.*[58] found that HZSM-5 was the best catalyst from among the five catalysts (HZSM-5, KZSM-5, Al-MCM-41, solid phosphoric acid, and $CoMo/Al_2O_3$) investigated for producing deoxygenated liquid products. Related to this, prior studies have investigated the effects of various parameters (sample preparation method, pyrolysis temperature, catalyst type, catalyst amount, and lignin source) on the composition of AH using HZSM-5 catalyst.[58–62] Mullen and Boateng[59] investigated catalytic pyrolysis of four lignins from different sources with HZSM-5 zeolite. The aromatic yields increased remarkably from trace amounts in non-catalytic pyrolysis to *ca.* 4–6 wt% in catalytic pyrolysis, but most lignin-derived pyrolysis oxygenates were not converted into AH in their pyrolysis condition. They proposed pathways for lignin depolymerization using catalytic pyrolysis over HZSM-5, as exhibited in Figure 3.10.

The acidity of the zeolite catalyst is one of critical factors for the formation of AH from lignin pyrolysis products. The acidity of HZSM-5 is inversely proportional to the $SiO_2:Al_2O_3$ mole ratio.[63] Ben and Ragauskas[62] studied

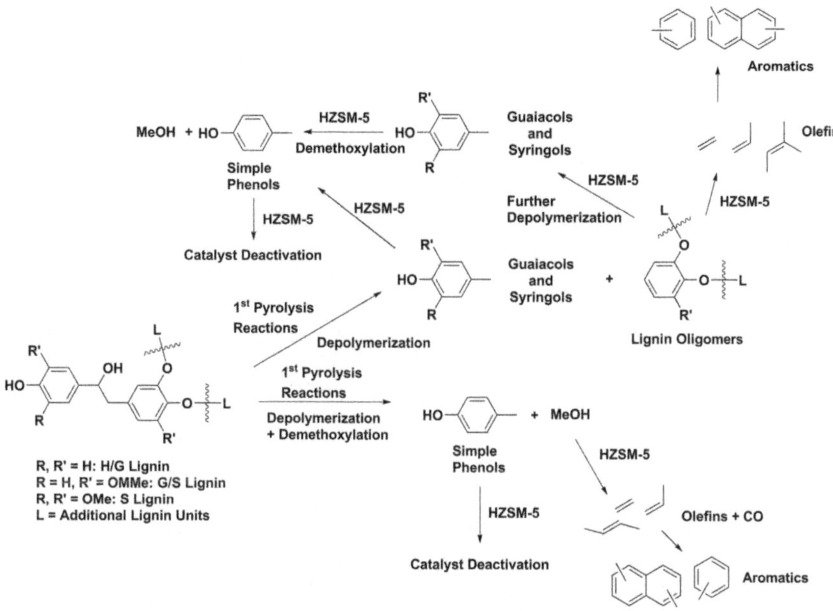

Figure 3.10 Proposed pathways of lignin deconstruction by catalytic pyrolysis over HZSM-5. Reprinted from Fuel Processing Technology, **91**(11), C. A. Mullen, A. A. Boateng, Catalytic pyrolysis-GC/MS of lignin from several sources, 1446–1458, Copyright (2010) with permission from Elsevier.[59]

the influence of the Si : Al mole ratio of the HZSM-5 catalyst on the chemical properties of lignin pyrolysis products. They revealed that the HZSM-5 catalyst with a relatively higher Si : Al mole ratio was more effective at the elimination of ether bonds, aliphatic C–C bonds, and dehydration of aliphatic hydroxyl groups during pyrolysis. Yu *et al.*[57] reported that the capability of HZSM-5 to convert lignin pyrolysis vapors to AH increased with increasing catalyst acidity. Aho *et al.*[64] also found that zeolite with a higher Si : Al mole ratio produced more organic compounds.

The effect of pyrolysis temperature on the formation of AH from lignin pyrolysis products has been also studied in many prior research reports.[65–67] Kim *et al.*[65] reported that the yield of total lignin pyrolysis products increased up to 600 °C, and then gradually decreased at high temperature due to further decomposition of volatiles to non-condensable gas. However, the yield of AH also gradually increases up to 700 °C. This finding is consistent with that of Wang *et al.*[66] who showed that the yield of AH continuously increased with an increment of temperature up to 800 °C. This trend can be explained by the finding that thermal degradation of lignin pyrolysis vapor to low molecular compounds (*e.g.* H_2, C_1–C_3 compound or olefin) is enhanced at high temperature, leading to an increasing conversion possibility of low molecular compounds to AH.[67] Another possible explanation for this finding is that the

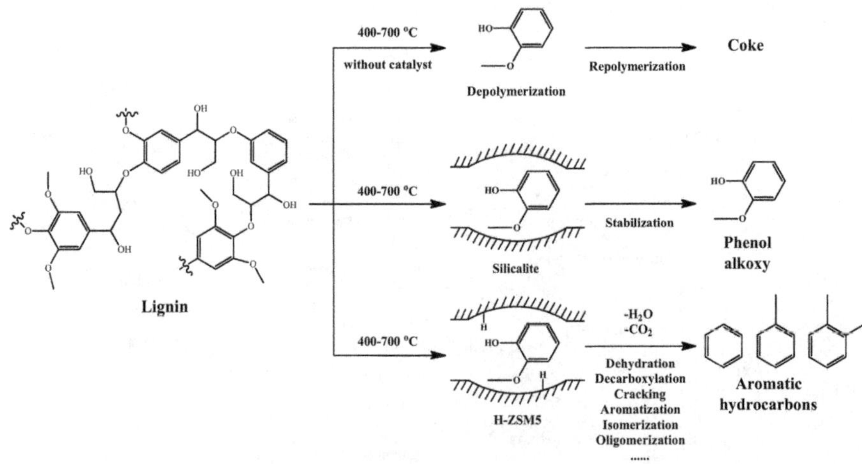

Figure 3.11 The role of porosity to stabilize intermediates and of acid sites to cleave C–O and C–C bonds during catalytic pyrolysis. Reprinted from Applied Catalysis A: General, **423**, Z. Ma, E. Troussard, J. A. van Bokhoven, Controlling the selectivity to chemicals from lignin *via* catalytic fast pyrolysis, 130–136, Copyright (2012) with permission from Elsevier.[69]

pore size of HZSM-5 gets larger with increasing temperature until it is sufficient to allow it to enter lignin pyrolysis vapor.[68]

Similarly, Ma *et al.*[69] suggested that large molecules could possibly enter the pore and then be further converted to AH because the energetic barrier for large molecules to diffuse into the pores would be decreased at high temperatures, as displayed in Figure 3.11. However, some researchers have indicated that lignin pyrolysis vapor can be converted to AH only at the external surface of zeolite catalyst because most of lignin-derived phenols are too large to pass through the inter-pores of the HZSM-5 catalyst.[70] How lignin pyrolysis vapors are converted to AH within the HZSM-5 catalyst framework remains unclear.

Catalyst loading has a significant effect on the formation of AH from lignin pyrolysis products.[65,69,71,72] Zhang *et al.*[72] reported that toluene and *p*-xylene were the two main aromatic hydrocarbons formed in abundance, and to a lesser extent benzene, 2-methylnapthalene, *o*-xylene, along with several others in catalytic pyrolysis of aspen lignin over HZSM-5. They obtained a maximum yield of AH at a catalyst-to-lignin ratio 3 : 1. Lee *et al.*[71] observed that when catalyst (mesoporous Y zeolite) loading increased, a substantial decrease in the production of monomeric phenols and an increase in the production of AH and poly AH occurred. Kim *et al.*[65] also found that the yield of lignin pyrolysis vapor displayed a declining trend with increasing catalyst loading due to an improvement in the catalytic conversion of volatiles to AH or non-condensable gas. This was because the possibility of contact between the lignin pyrolysis vapor and the acid site of catalyst increases with increasing catalyst amount.

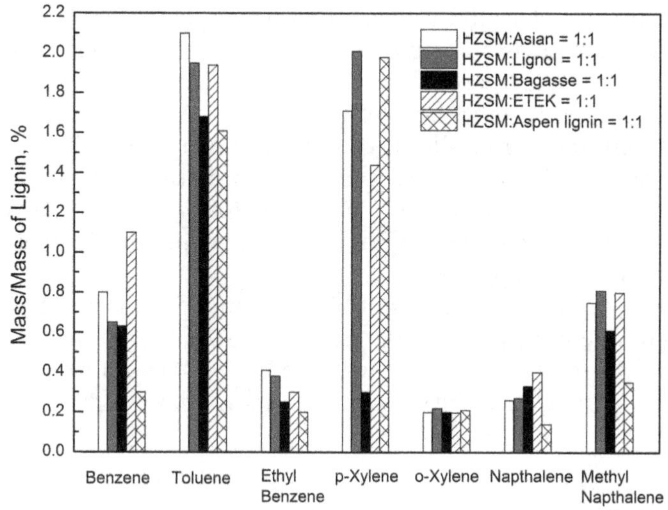

Figure 3.12 Comparison of main AH from catalytic pyrolysis of aspen lignin with the four lignins studied in Mullen and Boateng. Reprinted with permission from Fuel, **116**, M. Zhang, F. L. P. Resende, A. Moutsoglou, Catalytic fast pyrolysis of aspen lignin *via* Py-GC/MS, 358–369, Copyright (2014) with permission from Elsevier.[72]

Lignin source is one of the critical factors for the production of AH and other monomeric phenols. Previous studies revealed that the compositional ratio between pyrolysis products varies significantly depending on lignin source.[59,65,72] Mullen and Boateng[59] used (1) Asian lignin extracted from wheat straw and sarkanda grass *via* the soda pulping process, (2) Lignol lignin (organosolv lignin) supplied by Lignol Innovations, (3) bagasse lignin obtained from the pretreatment of sugarcane bagasse for ethanol production, and (4) ETEK lignin, a mixture of cellulose and lignin (1 : 1) as raw material. They observed that the distribution of individual AH compounds varied based on the structure (H : G : S ratios) of the lignin starting material. Zhang *et al.*[72] compared the yield of the main AH compounds from the catalytic pyrolysis of aspen lignin with the results reported in the Mullen and Boateng study[59] (Figure 3.12). Despite the differences in lignin source, both studies reported yields in the same range (AH). However, the amounts of substantial monomeric phenols, such as phenol, guaiacol, cresols, and syringol, varied significantly among the lignins.

3.7 Conclusion

In this chapter, details of thermal degradation behavior and pyrolysis chemistry of lignin were discussed. Work using a TGA instrument revealed that the frequency of β-O-4 linkage and the condensation degree of lignin correlated with its thermal degradation behavior. Pyrolysis-GC-MS-FID was applied to investigate lignin structure based on lignin pyrolysis products. This instrument was also used to investigate the effects of various process parameters on the production of aromatic hydrocarbon from catalytic pyrolysis. In this process, zeolite catalysts significantly affected the chemical compositions and yields of final products.

The yield of bio-oil from fast pyrolysis of extracted lignin was much lower than produced from lignocellulosic biomass. Furthermore, lignin melted in feed tubes and pyrolysis reactors, causing the formation of large char agglomerates that eventually caused loss of fluidization in fluidized bed reactors. Modified reactors and co-feeding of chemical agents has helped overcome the problems of lignin pyrolysis.

Defining the chemical analysis of bio-oil from lignin is another challenging owing to its compositional distribution. GC-MS analysis has mainly been used to analyze monomeric compounds in bio-oil. In addition, several studies used NMR and GPC to investigate high molecular weight compounds in lignin-derived bio-oils.

References

1. G. Brunow and K. Lundquist, in *Lignin and Lignans: Advances in Chemistry*, ed. C. Heitner, D. R. Dimmel and J. A. Schmidt, CRC Press, 2010, p. 267.
2. R. Chandra, R. Bura, W. Mabee, A. Berlin, X. Pan and J. Saddler, in *Biofuels*, ed. T. Scheper and L. Olssen, Springer, Berlin, 2007, vol. 108, p. 67.
3. F. Yaku, R. Tanaka and T. Koshijima, *Holzforschung*, 1981, **35**, 177.
4. G. Brunow, I. Kilpelainen, J. Sipila, K. Syrjanen, P. Karhunen, H. Setala and P. Rummakko in *Lignin and Lignan Biosynthesis*, ed. N. G. Lewis and S. Sarkanen, American Chemical Society, Washington DC, 1998, vol. 697, p. 131.
5. M. Galbe and G. Zacchi, in *Biofuels*, ed. T. Scheper and L. Olssen, Springer, Berlin, 2007, vol. 108, p. 41.
6. F. van Rantwijk and R. Sheldon, *Chem. Rev.*, 2007, **107**, 2757.
7. S. H. Lee, T. V. Doherty, R. J. Linhardt and J. S. Dordick, *Biotechnol. Bioeng.*, 2008, **102**, 1368.
8. J. C. Cuzens and J. R. Miller, *Renewable Energy*, 1997, **10**, 285.
9. X. Pan, J. F. Kadla, K. Ehara, N. Gilkes and J. N. Saddler, *J. Agric. Food Chem.*, 2006, **54**, 5806.
10. J.-Y. Kim, S. Oh, H. Hwang, U.-J. Kim and J. W. Choi, *Polym. Degrad. Stab.*, 2013, **98**, 1671.
11. J.-Y. Kim, S. Oh, H. Hwang, Y.-H. Moon and J. W. Choi, *Energy*, 2014, **76**, 284.
12. A. Tumbalam Gooty, D. Li, F. Berruti and C. Briens, *J. Anal. Appl. Pyrolysis*, 2014, **106**, 33.
13. H. S. Choi and D. Meier, *J. Anal. Appl. Pyrolysis*, 2013, **100**, 207.
14. D. Meier, R. Ante and O. Faix, *Bioresour. Technol.*, 1992, **40**, 171.
15. H. W. Lee, S. J. Choi, S. H. Park, J.-K. Jeon, S.-C. Jung, S. H. Joo and Y.-K. Park, *Energy*, 2014, **66**, 2.
16. K. Wang and R. C. Brown, *Green Chem.*, 2013, **15**, 675.
17. J. H. Lora and W. G. Glasser, *J. Polym. Environ.*, 2002, **10**, 39.

18. F. G. Calvo-Flores and J. A. Dobado, *ChemSusChem*, 2010, **3**, 1227.
19. C. Li, X. Zhao, A. Wang, G. W. Huber and T. Zhang, *Chem. Rev.*, 2015, **115**, 11559.
20. A. Coats and J. Redfern, *Analyst*, 1963, **88**, 906.
21. C. Wang, J. Pan, J. Li and Z. Yang, *Bioresour. Technol.*, 2008, **99**, 2778.
22. G. Jiang, D. J. Nowakowski and A. V. Bridgwater, *Thermochim. Acta*, 2010, **498**, 61.
23. S. Chu, A. V. Subrahmanyam and G. W. Huber, *Green Chem.*, 2013, **15**, 125.
24. T. Ohra-Aho, F. Gomes, J. Colodette and T. Tamminen, *J. Anal. Appl. Pyrolysis*, 2013, **101**, 166.
25. J. Rodrigues, D. Meier, O. Faix and H. Pereira, *J. Anal. Appl. Pyrolysis*, 1999, **48**, 121.
26. J. C. del Río, A. Gutiérrez, M. Hernando, P. Landín, J. Romero and Á. T. Martínez, *J. Anal. Appl. Pyrolysis*, 2005, **74**, 110.
27. L. Qiang, L. Wen-zhi, Z. Dong and Z. Xi-feng, *J. Anal. Appl. Pyrolysis*, 2009, **84**, 131.
28. J.-Y. Kim, H. Hwang, J. Park, S. Oh and J. W. Choi, *J. Anal. Appl. Pyrolysis*, 2014, **110**, 305.
29. R. J. Evans, T. A. Milne and M. N. Soltys, *J. Anal. Appl. Pyrolysis*, 1986, **9**, 207.
30. B. Iatridis and G. R. Gavalas, *Ind. Eng. Chem. Prod. Res. Dev.*, 1979, **18**, 127.
31. T. R. Nunn, J. B. Howard, J. P. Longwell and W. A. Peters, *Ind. Eng. Chem. Prod. Res. Dev.*, 1985, **24**, 844.
32. R. Chan and B. B. Krieger, *J. Appl. Polym. Sci.*, 1981, **26**, 1533.
33. P. R. Patwardhan, R. C. Brown and B. H. Shanks, *ChemSusChem*, 2011, **4**, 1629.
34. M. Zhang, F. L. Resende, A. Moutsoglou and D. E. Raynie, *J. Anal. Appl. Pyrolysis*, 2012, **98**, 65.
35. L. Xu, Y. Zhang and Y. Fu, *Energy Technol.*, 2016, **5**, 30.
36. V. B. Custodis, P. Hemberger, Z. Ma and J. A. van Bokhoven, *J. Phys. Chem. B*, 2014, **118**, 8524.
37. X. Huang, D.-g. Cheng, F. Chen and X. Zhan, *Ind. Eng. Chem. Res.*, 2013, **52**, 14107.
38. M. Kosa, H. Ben, H. Theliander and A. J. Ragauskas, *Green Chem.*, 2011, **13**, 3196.
39. C. Bährle, V. Custodis, G. Jeschke, J. A. van Bokhoven and F. Vogel, *ChemSusChem*, 2014, **7**, 2022.
40. J. Kibet, L. Khachatryan and B. Dellinger, *Environ. Sci. Technol.*, 2012, **46**, 12994.
41. K. H. Kim, X. Bai, S. Cady, P. Gable and R. C. Brown, *ChemSusChem*, 2015, **8**, 894.
42. C. A. Nunes, C. F. Lima, L. C. Barbosa, J. L. Colodette, A. Gouveia and F. O. Silvério, *Bioresour. Technol.*, 2010, **101**, 4056.
43. H. Ben and A. J. Ragauskas, *Energy Fuels*, 2011, **25**, 4662.
44. K. A. Jung, S. H. Woo, S.-R. Lim and J. M. Park, *Chem. Eng. J.*, 2015, **259**, 107.

45. A. T. Gooty, D. Li, F. Berruti and C. Briens, *J. Anal. Appl. Pyrolysis*, 2014, **106**, 33.
46. P. De Wild, W. Huijgen and H. Heeres, *J. Anal. Appl. Pyrolysis*, 2012, **93**, 95.
47. D. Li, C. Briens and F. Berruti, *Bioresour. Technol.*, 2015, **189**, 7.
48. K. H. Kim, T.-S. Kim, S.-M. Lee, D. Choi, H. Yeo, I.-G. Choi and J. W. Choi, *Renewable Energy*, 2013, **50**, 188.
49. D. J. Nowakowski, A. V. Bridgwater, D. C. Elliott, D. Meier and P. de Wild, *J. Anal. Appl. Pyrolysis*, 2010, **88**, 53.
50. D. Li, C. Briens and F. Berruti, *Biomass Bioenergy*, 2015, **76**, 96.
51. T. N. Trinh, P. A. Jensen, Z. Sárossy, K. Dam-Johansen, N. O. Knudsen, H. R. Sørensen and H. Egsgaard, *Energy Fuels*, 2013, **27**, 3802.
52. S. Mukkamala, M. C. Wheeler, A. R. van Heiningen and W. J. DeSisto, *Energy Fuels*, 2012, **26**, 1380.
53. S. Zhou, R. C. Brown and X. Bai, *Green Chem.*, 2015, **17**, 4748.
54. P. J. De Wild, W. J. Huijgen and R. J. Gosselink, *Biofuels, Bioprod. Biorefin.*, 2014, **8**, 645.
55. H. Ben and A. J. Ragauskas, *Energy Fuels*, 2011, **25**, 2322.
56. H. Ben and A. J. Ragauskas, *Energy Fuels*, 2011, **25**, 5791.
57. Y. Yu, X. Li, L. Su, Y. Zhang, Y. Wang and H. Zhang, *Appl. Catal., A*, 2012, **447**, 115.
58. M. A. Jackson, D. L. Compton and A. A. Boateng, *J. Anal. Appl. Pyrolysis*, 2009, **85**, 226.
59. C. A. Mullen and A. A. Boateng, *Fuel Process. Technol.*, 2010, **91**, 1446.
60. Y. Zhao, L. Deng, B. Liao, Y. Fu and Q.-X. Guo, *Energy Fuels*, 2010, **24**, 5735.
61. X. Li, L. Su, Y. Wang, Y. Yu, C. Wang, X. Li and Z. Wang, *Front. Environ. Sci. Eng.*, 2012, **6**, 295.
62. H. Ben and A. J. Ragauskas, *ACS Sustainable Chem. Eng.*, 2013, **1**, 316.
63. P. L. Benito, A. G. Gayubo, A. T. Aguayo, M. Olazar and J. Bilbao, *J. Chem. Technol. Biotechnol.*, 1996, **66**, 183.
64. A. Aho, N. Kumar, K. Eränen, T. Salmi, M. Hupa and D. Y. Murzin, *Fuel*, 2008, **87**, 2493.
65. J.-Y. Kim, J. H. Lee, J. Park, J. K. Kim, D. An, I. K. Song and J. W. Choi, *J. Anal. Appl. Pyrolysis*, 2015, **114**, 273.
66. K. Wang, K. H. Kim and R. C. Brown, *Green Chem.*, 2014, **16**, 727.
67. H. Yang, R. Yan, H. Chen, D. H. Lee and C. Zheng, *Fuel*, 2007, **86**, 1781.
68. C. E. Webster, R. S. Drago and M. C. Zerner, *J. Phys. Chem. B*, 1999, **103**, 1242.
69. Z. Ma, E. Troussard and J. A. van Bokhoven, *Appl. Catal., A*, 2012, **423**, 130.
70. G. Fogassy, N. Thegarid, Y. Schuurman and C. Mirodatos, *Energy Environ. Sci.*, 2011, **4**, 5068.
71. H. W. Lee, T. H. Kim, S. H. Park, J.-K. Jeon, D. J. Suh and Y.-K. Park, *J. Nanosci. Nanotechnol.*, 2013, **13**, 2640.
72. M. Zhang, F. L. Resende and A. Moutsoglou, *Fuel*, 2014, **116**, 358.
73. R. Lou and S.-b. Wu, *Appl. Energy*, 2011, **88**, 316.

CHAPTER 4

Transport and Secondary Reactions of Depolymerized/ Deconstructed Species

XIANGLAN BAI*[a] AND YUAN XUE[a]

[a]Department of Mechanical Engineering, Iowa State University, Ames, Iowa 50010, USA
*E-mail: bxl9801@iastate.edu

4.1 Introduction

Pyrolysis of biomass is the depolymerization of biomass in an inert environment. In pyrolysis, the decomposition products evolved from biomass escape the biomass matrix as volatiles. Thus, the volatility of the products plays a critical role in biomass pyrolysis. Secondary reactions can occur inside the biomass matrix, in the vapor phase, or in the condensed phase after recovery of vapors as liquid. To mitigate secondary reactions, short vapor residence time and rapid quenching of the vapor products are important. Fast pyrolysis at moderately high temperatures is preferred for maximizing bio-oil yield, not only because the conditions promote biomass depolymerization, but they also encourage evaporation and escape of liquid products from the hot reaction zone. However, excessively high temperatures can thermally crack primary vapor products into non-condensable gases, also reducing bio-oil yield.

Green Chemistry Series No. 50
Fast Pyrolysis of Biomass: Advances in Science and Technology
Edited by Robert C. Brown and Kaige Wang

Published by the Royal Society of Chemistry, www.rsc.org

Lignocellulosic biomass consists of cellulose, hemicellulose and lignin. Due to their distinct molecular structures, each component yields diverse pyrolysis products and exhibits different mass loss profiles during pyrolysis. Differences in volatility and reactivity of primary pyrolysis products can have major influences on yields of final products. For instance, pyrolysis of pure cellulose produces high yields of liquid and low yields of char because levoglucosan (LG), the major product of cellulose pyrolysis, has sufficient volatility to escape the hot pyrolysis zone and condense as bio-oil. Previous studies show that the pathway of product transportation and associated secondary reactions can greatly affect the product recovery during pyrolysis, therefore determining the yield of bio-oil and selectivity of chemicals.[1] In this chapter, we will review the issues of transport and secondary reactions of pyrolysis products and their impact on bio-oil yield and composition.

4.2 Transport and Secondary Reactions of Cellulose Pyrolysis Products

Cellulose depolymerization has been traditionally described as breaking from the ends of the glycosidic chain (unzipping) followed by cyclization to form LG.[2] More recently, Krumm *et al.*[3] suggested an alternative theory in which unzipping from the chain-end occurs only at low temperatures while intra-chain cleavage occurs at high temperatures. The linear chain with non-branched cellulose structure suggests that the yield of LG could reach 100% theoretically, if depolymerization was the sole reaction occurring during pyrolysis. However, LG yields from cellulose pyrolysis are reported to range from a few percent up to 70% (per cellulose) with typical results shown in Table 4.1. Other products include furans, furfurals, carboxylic acids, aldehydes, ketones and sometimes anhydro-oligosaccharides. Light gases (mainly CO_2 and CO) and solid char are also produced.

Table 4.1 Levoglucosan yields reported in previous studies.

Levoglucosan yield	Pyrolysis conditions	Ref.
3.57%	300 °C for 2.5 hr, 1 atm	17
28.1%	300 °C for 2.5 hr, 1.5 MmHg	
58%	400 °C, vacuum, 3 min heating	9
70%	430 °C, reduced pressure (5 kPa), medium heating rate	10
61%	350 °C	73
25.2%	500 °C, fast Py, bench-scale, N_2 gas	74
58.8%	500 °C, fast microPy, He gas	8
42.0%	500 °C fast Py, fluidized bed reactor	
7.5% (plus 23.7% anhydro-oligosaccharides)	1000 °C, flash Py	5
10%	500 °C, flash Py, 10^5 °C s^{-1}	11

In addition to LG and other light oxygenates, the presence of anhydro-oligosaccharides in cellulose bio-oil has been frequently reported by other researchers from bench-scale pyrolysis apparatus. For example, Pouwels *et al.*[4] found anhydro-oligosaccharides up to degree of polymerization (DP) of 6 in the pyrolysis products of cellulose. Piskorz *et al.*[5] reported that yields of anhydro-oligosaccharides in bio-oil were up to 20% in their study. The boiling point of LG is 339 °C and it has sufficient vapor pressure to evaporate during pyrolysis. In comparison, most anhydro-oligosaccharides have low to no volatility under pyrolysis conditions. For example, the boiling points of cellobiosan and cellotriosan are 581 °C and 792 °C, respectively, higher than the usual temperature range of fast pyrolysis (450 to 550 °C). Thus, the origin of these oligomeric sugars sometimes found in bio-oil was postulated by Piskorz *et al.*[6] to be thermo-mechanical ejection from pyrolyzing polysaccharides. They suggested that the large molecular weight (MW) compounds result when partially decomposed fragments are entrained in the product gas stream as aerosols. The authors further speculated that aerosol ejection is due to explosive destruction of the cellulose and high shear stress in the carrier gas stream.

To visualize cellulose pyrolysis at the microscale, Teixeira *et al.*[7] pyrolyzed cellulose on a heated surface at 700 °C and captured time-resolved images using high speed photography. They found that melted cellulose forms bubbles and the bubbles subsequently collapse to form liquid jets, which develop further to aerosols. They showed that the composition of the ejected aerosols was mainly cellobiosan. They argued that if polymerization of LG was responsible for this aerosol, it would have contained similar amounts of cellobiosan and its isomer maltosan. Thus, it was speculated that the cellobiosan was directly formed and ejected from cellulose. The authors further concluded that the non-volatile anhydro-oligosaccharides are also preliminary products of cellulose pyrolysis (*i.e.* primary aerosols), not the results of condensation and reaction of evaporated species.

Patwardhan *et al.*[8] offered an alternative view on the origin of oligosaccharides in bio-oil based on comparisons of cellulose pyrolysis in a micropyrolyzer and a bench-scale fluidized bed reactor system. In the micropyrolyzer, a sample cup containing a few hundred micrograms of biomass was dropped into a preheated reactor oven for pyrolysis and the vapor products were directly swept into an online Gas Chromatography Mass Spectrometry Flame Ionization Detection (GC-MS-FID) system for analysis. Thus, secondary reactions of the primary products in the vapor or condensed phases are thought to be largely limited. The pyrolysis vapors from the fluidized bed reactor, on the other hand, were condensed and analyzed off-line. The authors reported that the LG yield dropped from 58.8% with the micropyrolyzer to 42% with the fluidized bed reactor. The decreased yield of LG from the fluidized bed reactor was compensated by a significant quantity (up to 20%) of oligomeric sugars found in the cellulose bio-oil. They also reported that no oligomeric sugars were detected among the pyrolysis products in the micropyrolyzer experiment, despite the higher gas velocity in the micropyrolyzer compared to in the fluidized bed reactor. Interestingly, the sum of LG and oligomeric sugar yields from the fluidized bed reactor was similar to the LG from the micropyrolyzer. Thus, the authors speculated that the oligomeric sugars are the repolymerization products of LG, either formed during the vapor transportation or condensation of the vapor at the bio-oil recovery system.

As listed in Table 4.1, LG yields from cellulose pyrolysis range from a high of 70% to a low of 3.57% in the literature, depending on the pyrolysis conditions. Shafizadeh *et al.*[9] reported a 58% LG yield during pyrolysis of cotton hydrocellulose at 400 °C for 5 to 9 min under vacuum, a condition that is clearly not within the regime of conventional fast pyrolysis. The highest LG yield to date has been reported by Kwon *et al.*[10] who pyrolyzed cellulose using a rapid cooling, continuous-feed pyrolyzer operated at reduced pressures. The maximum liquid yield was 96%, achieved at a reactor pressure of 1 kPa and a pyrolysis temperature of 430 °C. The maximum yield of LG in the study was 70%, obtained at 430 °C and 5 kPa. LG accounted for as much as 80% of the liquid product at this condition. At the other extreme, Funazukuri *et al.*[11] reported a yield of only 10% LG despite pyrolyzing cellulose in a microfluidized bed reactor with a very high heating rate. Based on the summary of LG yields provided in Table 4.1, it can be concluded that the greater LG yields are often obtained when cellulose was pyrolyzed at moderate temperatures and heating rates under high ventilation rates or reduced system pressure. These results cannot be fully explained without considering secondary reactions of LG.

The secondary reactions of LG during pyrolysis were investigated by several researchers. Suuberg *et al.*[12] reported that the enthalpy of cellulose pyrolysis is very similar to the enthalpy of LG vaporization. Thus, they suggested that cellulose forms liquid LG and LG either evaporates or dehydrates to char. According to a previous study,[13] the GC-MS chromatogram obtained by pyrolyzing LG in a micropyrolyzer was identical to that of cellulose, as might be expected if secondary reactions of LG produced lighter oxygenated products, such as furfurals, furans, acids and aldehydes. It is also known that LG polymerizes at elevated temperatures. Kawamoto *et al.*[14] previously pyrolyzed LG at 250 to 400 °C for up to 5 min and analyzed the products. Based on the Nuclear Magnetic Resonance (NMR) and Gel Permeation Chromatography (GPC) results, the authors reported that the water-soluble fraction of the pyrolysates is made up of randomly linked oligomers without a reducing end-group, formed through ring-opening polymerization of LG. They also pointed out that the same oligomeric sugars were previously isolated from bio-oil of cellulose and cellobiose. Based on this result, the authors proposed a pathway of cellulose pyrolysis *via* LG as an intermediate product. In their model, cellulose depolymerizes to LG and then it simultaneously transforms into low MW products and polysaccharides. The model also suggests that the polysaccharides turn into carbonized solids in the later stage of pyrolysis. In a follow-up study,[15] the same research group pyrolyzed LG in a sealed dual-space ampoule reactor to distinguish the secondary reactions of LG in the vapor from the secondary reactions occurring in the liquid/solid phases.

In the study, a small quartz tube with about 0.1 to ~10 mg of LG was placed inside a larger quartz tube and pyrolyzed at 400 °C for 1 to 20 min. The products were collected from the wall of the larger tube or inside the sample holder. The char observed inside the sample cup is likely to be due to carbonization of the polysaccharide. The condensates collected from the tube wall contained low MW oxygenates, including 5-hydroxymethylfurfural (5-HMF), glycolaldehyde, hydroxylacetone, acetic acid and formic acid in addition to LG. The yields of low MW condensates were dependent on LG sample size, as higher mass loading increased the yields of the products. With the sample loading of 0.1 mg, no char was found and CO and CO_2 were the major secondary products. Thus, the authors speculated that the condensable low MW compounds are somehow closely related to the char formation in the liquid/solid phase. Based on the new findings, an updated model of LG pyrolysis was proposed. According to the new model, liquid LG is subjected to thermal polymerization (to polysaccharides) and volatilization (to LG vapor). The evaporated LG further decomposes to CO and CO_2, due to radical chain initiated vapor phase gasification. The gasification to the non-condensable gases became significant with low sample mass loading. When mass loading is high, the aldehydes among the products inhibit the vapor phase gasification of LG because the aldehydes act as H-donors to stabilize the intermediate radicals.

The physiochemistry of LG during pyrolysis was also investigated by Bai *et al.*[13] In the study, LG was pyrolyzed using a Thermogravimetric Analyzer (TGA) coupled with an online MS to obtain time-resolved information about the evolution of the vapor products. The solid/liquid residues of LG remaining in the sample cup were also collected during the course of pyrolysis to investigate the corresponding compositional change in the liquid/solid products. It was found that LG undergoes two competing processes of evaporation and polymerization, which agrees with the findings by Hosoya *et al.*[15] The study also revealed that polymerized anhydro-oligosaccharides might be subject to two simultaneous reaction pathways: decomposition to low MW oxygenates such as water, carbon dioxide, 5-HMF, furfural, furan and acetic acid; or further condensation to higher DP polysaccharides, which eventually dehydrate to char or decompose to low MW volatiles. Depolymerization of anhydro-oligosaccharides to LG and lower DP anhydro-oligomers was also observed. The schematic diagram of the secondary reactions of LG is given (Scheme 4.1). This finding implies that while the evaporated LG can be recovered in bio-oil as a final product, the polymerized LG would be trapped inside the hot pyrolysis zone and eventually dehydrate to less desired low MW volatiles and char.

Another study also showed that the competing processes of LG evaporation and polymerization are affected by pyrolysis conditions.[16] Increasing LG loading promoted polymerization and inhibited evaporation; pyrolyzing water-dissolved LG inhibited char formation and promoted evaporation. Using helium instead of nitrogen as the sweep gas had a significant impact on LG evaporation, although both of these gases are inert (Figure 4.1).

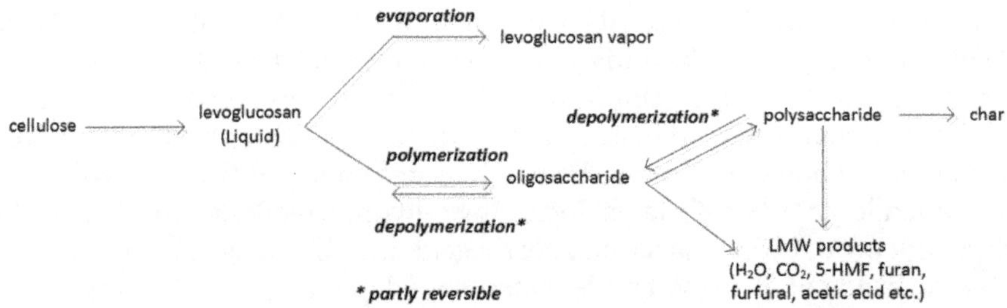

Scheme 4.1 Proposed secondary reaction pathways of LG during pyrolysis. Reprinted from Journal of Analytical and Applied Pyrolysis, **99**, X. Bai, P. Johnston, S. Sadula, R. C. Brown, Role of levoglucosan physiochemistry in cellulose pyrolysis, 58–65, Copyright (2013) with permission from Elsevier.[13]

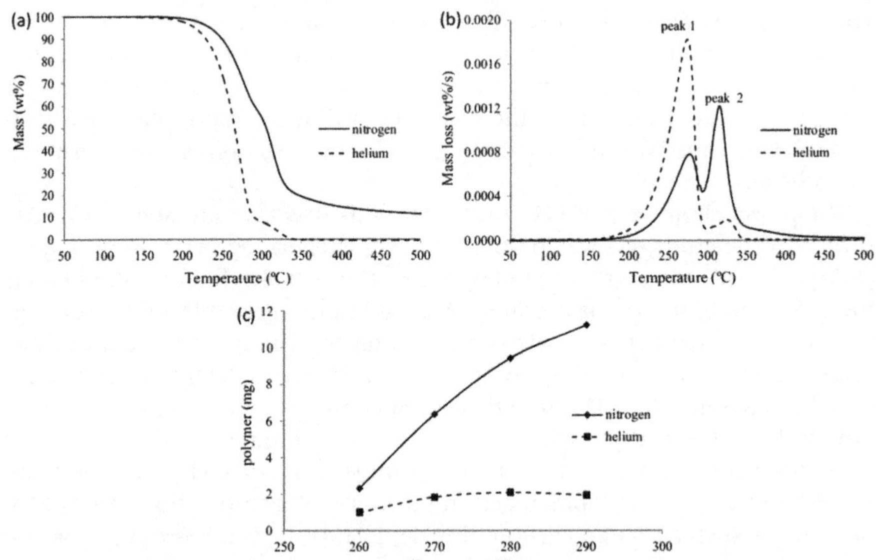

Figure 4.1 Pyrolysis of LG using nitrogen or helium sweep gas: (a) TGA profile; (b) DTG profile: peak 1. LG evaporation; peak 2: LG polymerization and dehydration and (c) Polymerized LG (starting LG weight: 20 mg). Reprinted from Journal of Analytical and Applied Pyrolysis, **99**, X. Bai, P. Johnston, R. C. Brown, An experimental study of the competing processes of evaporation and polymerization of levoglucosan in cellulose pyrolysis, 130–136, Copyright (2013) with permission from Elsevier.[16]

When helium was the sweep gas in TGA experiments, char was not formed and the amount of polymerized LG decreased dramatically, in contrast to experiments with nitrogen as the sweep gas. This is explained from the higher diffusion rate of LG vapor in helium compared to nitrogen, allowing LG to more quickly escape the reactor. As evaporation and polymerization of LG are competing processes, enhancing LG evaporation reduced LG polymerization. Recall in Table 4.1 that cellulose pyrolysis conducted in a vacuum or at reduced pressures overall gave higher LG yields than pyrolysis conducted at atmospheric pressure. For example, Shafizadeh and Fu[17] pyrolyzed cellulose at 300 °C for 2.5 hr at either atmospheric pressure or reduced pressure. The LG yield was only 3.57% for pyrolysis at atmospheric pressure, while the yield increased to 28.1% when reducing the pressure to 1.5 mm Hg. Decomposition of LG vapor is unlikely at such low pyrolysis temperature. Kwon *et al.*[18] also reported that switching the pyrolysis environment from a vacuum to 1 atm nitrogen reduced the yield of LG to less than half of that obtained from vacuum pyrolysis.

On the other hand, increasing pressure and lowering flow rate during pyrolysis increased yields of char and CO_2.[19] Higher pressure inhibits evaporation of LG while encouraging polymerization of LG and dehydration to char. Thus, the yield of LG from cellulose pyrolysis depends on the effectiveness of removing LG from the reaction zone. With reduced pressure, primary pyrolysis products could rapidly diffuse out of cellulose before secondary reactions start. Higher amounts of LG, not anhydro-oligosaccharides, are recovered at reduced pressures, suggesting that LG is the main pyrolysis product.

Heating rate also influences LG yield. Although it was originally thought that LG would quickly evaporate during pyrolysis at higher heating rates, experiments revealed just the opposite: more LG polymerized and dehydrated to char.[16] A model was proposed for the competing processes of LG evaporation and polymerization with kinetic parameters determined from TGA experimental data.[20] The activation energies for evaporation and polymerization of LG were found to be 92 kJ mol^{-1} and 139 kJ mol^{-1}, respectively. The model was then used to predict the effect of heating rate on LG evaporation and polymerization at fast pyrolysis conditions. The model predicted that higher heating rates promote LG polymerization over evaporation. Although both the rates of evaporation and polymerization increase with increasing temperature, polymerization increased faster than evaporation with temperature. It is well known that fast pyrolysis of cellulose yields more LG than slow pyrolysis. However, the detrimental effect of heating rate on LG evaporation suggests that not all the LG produced from cellulose can be evaporated and recovered. For cellulose pyrolysis, depolymerization of cellulose to anhydro-sugars is favored at high temperatures and heating rates, whereas low temperatures and heating rates promote random-cross linking reactions within cellulose to form char and light gases.[21] However, higher heating rates also promote secondary reactions of LG through polymerization, thus preventing efficient evaporation of LG.

When cellulose is pyrolyzed, it is difficult to differentiate between anhydro-oligosaccharides resulting from polymerization of LG and those produced from the depolymerization of cellulose.[22] However, the fact that high heating rates promote LG polymerization helps to explain why flash pyrolysis of cellulose with extremely high heating rates (*e.g.* >10 000 °C s^{-1}) usually yields smaller amounts of LG in comparison to fast pyrolysis.[5] Polymerization of

LG followed by formation of light oxygenates and char from polymerized LG could also be one of the reasons why it is difficult to obtain the theoretical amount of LG from cellulose by pyrolysis. Although higher heating rates/temperatures increase LG production from cellulose, more LG polymerizes under these conditions. High temperature also promotes secondary cracking of LG vapor. On the other hand, although a smaller amount of LG was produced from cellulose with lower heating rates, it more effectively evaporates and the resulting LG vapor is more stable. Thus, the maximum LG yield is probably achieved from cellulose pyrolysis with moderately high heating rates and temperatures.

To increase LG yield, long residence times of LG vapor at high pyrolysis temperatures should be avoided. Stiles and Kandiyoti[23] pyrolyzed cellulose in a fluidized reactor and reported that the secondary reactions of the products in the vapor phase became significant at pyrolysis temperatures above 400 °C, reducing bio-oil yield. At 600 °C, increasing the vapor residence time from 0.25 s to 3.5 s was found to reduce the bio-oil yield by 30%. The decrease was attributed to secondary cracking of LG and other volatiles to light gases. To improve LG yield, LG should be quickly removed from both the hot cellulose matrix and the reactor zone by overcoming mass transfer limitations. Mass transfer limitations could be present even in the micropyrolyzer where high-velocity helium is used as the sweep gas. Proano-Aviles *et al.*[24] found that the formation of stagnation flow inside the sample cup keeps the pyrolysis vapor from being immediately swept away by helium. Their study showed that the time constant of the pyrolysis vapor is as high as 2 s, rather than the 15–20 ms that was assumed by others for the micropyrolyzer.[8] Replacing the standard cup with a perforated shallow cup reduced the time constant to 0.2 s, therefore increasing LG yield by 10% when cellulose was pyrolyzed.

4.3 Transport and Secondary Reactions of Lignin Pyrolysis Products

The mechanism of lignin pyrolysis is more complicated than for cellulose. Continuous pyrolysis of lignin is difficult because of its tendency to melt at about 200 °C, resulting in agglomeration of lignin particles and clogging of feeder tubes and fouling of reactors.[25,26] Although it is commonly considered that the melting of lignin is attributed to its "thermoplastic" property, it is also due to partial depolymerization of lignin[27] and repolymerization of depolymerized lignin. The result is large amounts of char product (up to 70%), depending on lignin type and pyrolysis conditions.[28–30] The yield of lignin bio-oil is usually around 30–40%, much lower than from the pyrolysis of cellulose. Lignin bio-oil is a complex mixture of phenolic compounds, with the yields of most individual compounds less than 1%.[31] The complexity of lignin pyrolysis is attributed to the intrinsic molecular structure of lignin, which is an amorphous macro-polymer randomly cross-linked by its three precursor monomers through biosynthesis. Upon pyrolysis, bond cleavages among lignin result in the phenolic compounds with different functional groups and reactivity. The molecular sizes of the phenolic compounds in bio-oil could range from below 100 Da (in the case of phenol) up to 2000 Da, indicating the products are a mixture of monomers and oligomers. As only phenolic monomers with MW below 220 Da are detectable by GC-MS, a significant fraction of lignin-derived phenolic compounds cannot be detected or identified by GC-MS. Lignin-derived phenolic oligomers have little or no vapor pressure.

Because such a large fraction of these phenolic products are non-volatile, their transport from the reactor is frequently explained by thermal ejection.[32–34] According to this mechanism, uneven heating of polymeric material gives rise to bubbles in the molten phase that carry along some of the melted material when they are ejected. Shear forces from the flow of carrier gas passing the molten material could also elutriate melted material.

Fratini *et al.*[33] conducted small angle neutron scattering analysis of aged pyrolysis oil found the phenolic oligomers occurred as distributed microstructures. They argued that reoligomerization of phenolic monomers during aging would result in aggregated molecular structures. Thus, they concluded that the oligomers must be lignin fragments, rather than repolymerization products. However, several other researchers have shown that the phenolic compounds repolymerize during bio-oil storage, resulting in increased MW of stored bio-oil.[35–38]

Pyrolytic lignin extracted from bio-oil has been used by several investigators to study lignin pyrolysis and secondary reactions. Meier's group previously characterized pyrolytic lignin using various analytical methods, such as GPC, NMR and TOF-MS.[39–42] They proposed the tentative structures of the phenolic oligomers (up to octamers) by including at least one stilbene and one phenyl coumaran structure in each compound, and suggested that pyrolytic lignin contains both thermally ejected phenolic oligomers and recombination products. The authors further speculated that the recombination products are formed through radical reactions or etherification reactions during storage. On the other hand, Mullen and Boateng[43] reported that pyrolytic lignin consists of highly-substituted aromatic rings connected by various aliphatic linkages. Neither phenylcoumaran nor resinol, compounds naturally occurring in lignin, were found by NMR analysis in pyrolytic lignin. Their results suggest that pyrolytic lignin could be a secondary product, rather than partly depolymerized lignin fragments.

The use of pyrolytic lignin from the pyrolysis of whole biomass to investigate the decomposition of lignin is complicated by the presence of carbohydrate. First, the process of water-extracting carbohydrate products from phenolic compounds can separate some of the more water-soluble phenolics.[44] Second, the presence of polysaccharides in biomass can affect lignin depolymerization and subsequent secondary reactions. Thus, some researchers have worked directly with technical lignin isolated through solvent extraction or other means.[45–48]

Micropyrolyzers coupled with online GC-MS (Py-GC-MS) have been frequently used to probe the primary reaction of lignin pyrolysis. Patwardhan *et al.*[47] reported up to 92% of mass closure by pyrolyzing a corn stover organosolv lignin in a Py-GC-MS. The quantified products include 17% of GC-MS-identified phenolic monomers, 37% of char, 19% of non-condensable gases, and other light oxygenates such as water and acids. The remaining 8% of unaccounted products could be the phenolic oligomers that are undetected by the GC-MS. The authors also found that the condensed pyrolysis vapor collected in the reaction zone had lower average MW, in comparison to the phenolic products condensed in the chilled bath downstream. Their study also did not support the thermal ejection theory because altering the gas flow rate in the pyrolyzer to change the shear force did not change the product distribution. The findings suggest that phenolic monomers convert to oligomers, which is contradictory to the pyrolysis mechanism suggested by Zhou *et al.*[34] who claim that phenolic oligomers are the primary products whereas phenolic monomers are the result of the oligomers cracking in the vapor phase.

To extend this research effort, Bai *et al.*[48] also pyrolyzed corn stover organosolv lignin in a micropyrolyzer. In the study, the primary vapor product leaving the reactor was immediately quenched in cold solvent and analyzed by GPC for the MW distribution. The average MW of the condensed primary vapor was only 286 Da, much smaller than the reported MWs of pyrolytic lignin.[40] The main peak in the GPC chromatogram is correspondent to phenolic monomers, and the less intensive peak next to it corresponds to phenolic dimers. The condensed vapor was further analyzed by two different types of high resolution MS spectroscopies (APPI-FTICR and Orbitrap) to determine the accurate MW of each compound. The DPs of the phenolic compounds were then calculated based on the double bond equivalence values of the molecules. As a result, over 500 different compounds with MWs up to 504 Da were identified and determined to be mostly monomers and dimers. The lignin-derived phenolic monomers, such as vinylphenol, guaiacol, ethylphenol and phenol, were also pyrolyzed in the same reactor to investigate the reactivity of the lignin-derived vapors. It was shown that the reactive vapor species, such as vinyl or methoxylphenols, could polymerize to form oligomers in the vapor phase although the residence time of the vapor was extremely short. In the tests conducted in the micropyrolyzer, the secondary reactions of the monomers *via* polymerization was the preferred reaction at temperatures of up to 500 °C, and side-chain cracking and rearrangement became more dominant at higher temperatures. The study also revealed that phenylcoumaran, stilbene and biphenol structures in natural lignin are relatively thermally stable during pyrolysis. Thus, the phenolic dimers with the above structures can also be the primary products upon pyrolyzing lignin.

Zhou *et al.*[49] also investigated secondary reactions of lignin products by pyrolyzing pine wood in a fluidized bed reactor and sending the vapor to a tubular reactor with adjustable temperature and vapor retention time.

Based on analysis of pyrolytic lignin, the authors reported that secondary reactions of pyrolytic lignin in the vapor phase included cracking and polycondensation reactions, which compete with each other. It was also reported that cracking reactions included the cleavage of the ether bonds between the aromatic rings, the removal of carboxylic groups, propanoid side chains, demethoxylation, and the cleavage of $-CH_2-OH$ groups to form aldehyde groups. Ben and Ragauskas[50] pyrolyzed Kraft lignin in a batch reactor and also reported condensation and cracking as the secondary reactions.

Secondary reactions of lignin-derived products were also investigated by Hosoya *et al.*[51] by pyrolyzing lignin in a sealed reactor. It was reported that unsaturated side chains (C=C) and phenolic aromatic rings are the most reactive structures in secondary reactions. The products with these groups first form condensation products and then further decompose *via* side-chain cracking to form low MW aromatics with saturated alkyl groups or –H. Aldehyde groups are also subject to decarbonylation to rearrange the substitution pattern of the aromatic ring during this cracking reaction. It was also found that catechol-type compounds became reactive in the presence of the guaiacol-type compounds, probably because the catechols are activated by radical chain reactions through the H-abstraction from the phenolic hydroxyl groups. Secondary char formation was also investigated by pyrolyzing lignin-derived phenolic monomers.[52] It was found that the extent of secondary char formation is related to the types of functional groups on aromatic side chains, decreasing in the order of (C=C) > (C=O) > ($O-CH_3$). Also, the co-presence of methoxyl groups in addition to unsaturated C=C or carbonyl C=O group in the phenolic compounds significantly enhances char formation. As such, significant secondary char formation was found with the pyrolysis of vinylguaiacol, isoeugenol, eugenol, vanillin and acetovanillone. In comparison, phenol and methylphenols formed nearly no secondary char. Pyrolysis of guaiacol produces p-catechol and ether rearranged products. The authors explained that the latter products are formed *via* an o-quinone methide intermediate, initiated from the H-abstraction of the phenolic hydroxyl groups. O-quinone methide is thought to be an intermediate of lignin char.

Secondary reactions of lignin in the solid/liquid phase also occur, although it has been less frequently explored compared to the vapor phase reactions. The cleavage of β-O-4 linkages starts at temperatures of about 250 °C,[53] while this low temperature is not sufficient for the depolymerized phenolic compounds to evaporate. Reactive free radicals, which could have a lifetime of less than a millisecond, would cross-link and repolymerize in the solid/liquid phase of lignin to increase the MW of the products and eventually dehydrate to form pyrolysis char. Lignin char, the hard-shell carbonized solid, may also create a heat transfer barrier for pyrolysis, and act as a structural barrier for transporting the vapor products out of the lignin matrix. As a result, some intrinsic phenolic molecular structures were found to be remaining in the agglomerated lignin char collected inside a fluidized bed after pyrolysis.[54]

The limitation of using isolated lignin to represent natural lignin in terms of studying lignin pyrolysis and secondary reactions, however, should also be noted. The isolated lignins are structurally modified lignins and they often have lower MWs than corresponding natural lignin due to partial fragmentation/recondensation occurring during chemical or mechanical isolation process of lignin.[55–57] The isolated lignins usually have shorter side-chain lengths and more condensed aromatic structures in comparison to natural lignin. Isolated lignin is more thermally resistant to destruction, producing more char and a lower yield of phenolic oil.[58] It is, thus, expected that the phenolic compounds produced from the isolated lignin have different concentrations of functional groups, in comparison to the products from pyrolysis of natural lignin. For example, natural lignin contains more methoxyl groups and aryl ether linkages, thus producing more reactive phenols toward secondary reactions.

4.4 Interactions Among Cellulose- and Lignin-Derived Pyrolysis Products

When biomass is pyrolyzed, cellulose- and lignin-derived products could interact each other to affect the product composition and selectivity. The influence of biomass-derived compounds on the secondary reaction of LG in the vapor phase was studied by Kawamoto *et al.*[59] In their study, cellulose, xylan and lignin were placed in different sample holders inside of a dual-space reactor to only allow vapor phase contact among the products derived from different biomass compositions. It was found that pyrolysis vapors derived from cellulose or lignin significantly suppress the recovery of LG. In comparison, xylan-derived pyrolysis vapors do not have significant impact on LG. When they are co-present, xylan-derived pyrolysis vapors inhibited the decomposition of LG induced by the cellulose- and lignin-derived vapors.

The effect of pyrolysis char on secondary reactions of LG was studied by Ronsse *et al.*[60] In the study, pyrolysis vapor of cellulose was sent to a heated bed filled with unwashed and acid-washed red oak pyrolysis char. It was found that LG yield reduces from 27.75% without the char bed to 21.84% in the presence of the acid-washed char. The yield of LG reduced even more with the unwashed char bed, down to only 8.15%. The decrease in LG yield was accompanied by increasing yields of light oxygenates, such as formic acid, acetic acid, furan, levoglucosenone and glycoaldehyde. As the acid wash removes most of the inorganic impurities in the biochar, it was concluded that it is the inorganics present in biochar that catalyze the decomposition and dehydration of LG in the vapor, leading to yield reductions.

The solid/liquid- and vapor phase interactions between cellulose and lignin as well as their pyrolysis products were studied by Hosoya *et al.* using the dual-space reactor.[61] They reported that the solid/liquid phase interactions among cellulose and lignin increase bio-oil yield and reduce char yield. Water yield was also decreased by the interactions due to the reduced dehydration.

The yields of most GC-MS- and NMR-detectable compounds derived from cellulose or lignin were increased by the solid/liquid interactions. The yield of LG was decreased, but the yields of furans and C_2–C_3 carbonyls increased. Due to the interactions, the products with unsubstituted (H–) and methyl guaiacols selectively increased and shorter saturated side chains increased. The solid/liquid phase interaction also promoted the formation of the products with lower MWs and increased catechol yield. On the other hand, the yield of CH_4 became lower by co-pyrolysis, possibly because O–CH_3 bond homolysis was inhibited in lignin pyrolysis products when cellulose and lignin were co-pyrolyzed.

The interactions between cellulose- and lignin-derived products were less significant in the vapor phase in comparison to the solid/liquid phase interactions. It was found that vapor phase interactions enhanced the gas formation from cellulose-derived vapors, and reduced the secondary char formation in the lignin-derived vapor products. The yields of methane and catechols also increased greatly due to the cross-interactions between the vapor products of cellulose and lignin. The contents of saturated alkyl and C=O bonds were found to increase, whereas the products with unsaturated C=C bonds on the side chain were reduced by the vapor phase interactions. Nevertheless, the authors also noted that the vapor phase interactions became noticeable only with very long vapor residence times (*e.g.* >80 s).

The mechanism of the cross interaction between lignin- and cellulose-derived products was explained using the hydrogen donor and acceptor theory (Scheme 4.2). Lignin ether cleavages produce high amounts of compounds with unsaturated C=C side chains. These unsaturated bonds are

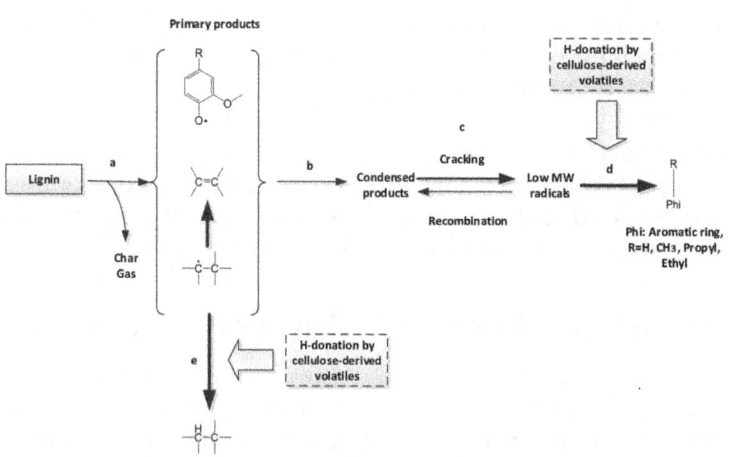

Scheme 4.2 Proposed mechanism for the influences of the cellulose-derived products on the pyrolysis oil formation from lignin. Reprinted from Journal of Analytical and Applied Pyrolysis, **85**(1–2), T. Hosoya, H. Kawamoto, S. Saka, Solid/liquid and vapor phase interactions between cellulose- and lignin-derived pyrolysis products, 237–246, Copyright (2009) with permission from Elsevier.[60]

highly susceptible to polymerization through radical coupling polymerization. When co-pyrolyzed, cellulose-derived products act as H-donors to the radical species to enhance the formation of low MW aromatics with saturated alkyl groups, therefore inhibiting condensation reactions caused by the unsaturated C=C bonds. According to the authors, the effect of interaction by H-donors is effective in the solid/liquid phase because the lower MW products have a much better chance to volatilize and increase bio-oil yield, instead of staying in the solid/liquid matrix to further polymerize to high MW products and eventually char. The concentration of H-donors and the chances for donating or accepting hydrogen are also much higher in solid/liquid interactions than that in diluted vapor products. On the other hand, lignin-derived radicals abstract hydrogens from cellulose-derived products and enhance the gasification of the primary products.

It is worth mentioning that Zhang *et al.*[62] also co-pyrolyzed cellulose with lignin in a micropyrolyzer and stated the interactions between cellulose and lignin were only apparent with herbaceous biomass, but not woody biomass. Thus, the authors speculated that the higher amount of covalent linkages in herbaceous biomass compared to woody biomass caused the variation in the results. Nevertheless, Radlein *et al.*[63] reported that the lignin-derived fraction of bio-oils obtained from fast pyrolysis of poplar wood accounted for 80% of the lignin present in the feedstock biomass. Such a high conversion of lignin to liquid products is usually not achievable with isolated lignin in conventional pyrolysis. Overall, the above results suggest that conversion of lignin to liquid products is efficient when lignin is present in biomass because the secondary reactions among lignin- and cellulose-derived products have positive effects on lignin volatilization.

A negative effect on the product distribution caused by co-pyrolysis was also reported. Pathawardhan *et al.*[47] showed that co-pyrolysis of coumaryl alcohol, coniferyl alcohol, and sinapyl alcohol with acetic acid greatly promotes the formation of phenolic oligomers. Acetic acid is one of the major pyrolysis products of hemicellulose, produced by deacetylation. It is likely that acetic acid catalyzes repolymerization of phenolic compounds. It has also been reported that formic acid derived from polysaccharides has a stronger catalytic effect for polymerization than acetic acid.[64]

4.5 Transport of Products During Biomass Pyrolysis

Secondary reactions of biomass pyrolysis products in the vapor phase include both condensation and thermal cracking, and the vapor residence time and the temperature of the reactor can greatly affect the product distribution. Previously, Stiles and Kandiyoti[23] pyrolyzed silver birch in a fluidized bed reactor, and reported that bio-oil yield maximizes at a pyrolysis temperature of 425 °C and residence time of 0.25 s. Beyond this point, either increasing the temperature or increasing the residence time sharply decreased the bio-oil yield, because secondary cracking of condensable vapors to light gases takes place.

Light pyrolysis gas products include CO_2, CO, CH_4, H_2 and also small amounts of light hydrocarbons. The product gases could influence biomass pyrolysis and the secondary reactions. According to Zhang *et al.*,[65] CO gas reduced bio-oil yield whereas CH_4 improved bio-oil yield. CO and H_2 promoted deoxygenation of the products, therefore increasing the heating values of the bio-oils. It was also found that CO_2 and CO gases increase the yields of acids and ketones in the products, while reducing the content of phenols and sugars, as well as methoxyl-containing products in bio-oil. The decreased methoxyl groups in the products also helped to improve the stability of the bio-oil. It is expected that the effect of the product gases on the biomass pyrolysis through secondary reactions becomes more noticeable in large reactors where the concentration of pyrolysis vapors or gases inside the reactor is relatively high. Based on the reactivity of the pyrolysis gases, Boateng *et al.*[66,67] developed a tail-gas reactive pyrolysis by recycling the non-condensable gas products during biomass pyrolysis. The researchers reported that the resulting bio-oil has significantly lower oxygen content than the bio-oil produced from the same biomass feedstock with nitrogen gas.

In biomass, polysaccharides and lignin are both physically and chemically bonded together to form unique plant structures. When plant biomass is pyrolyzed, the complex plant tissue and cell structure could play a role in biomass heating, as well as how the pyrolysis products are transported out of the plant matrix. Thus, the pyrolysis of biomass particles is governed by both chemical kinetics and physical phenomena. The chemical kinetics include biomass decomposition to form primary products and the secondary reactions of the primary products in the solid/liquid matrix of biomass. The physical phenomena include heat transfer within the biomass particles, and the diffusion of the pyrolysis products out of the solid biomass.

The transport and chemical kinetics of biomass pyrolysis were studied by several research groups. In some previous models, it was described that the primary reaction of biomass pyrolysis forms bio-oil, gas and char and the tar is further subjected to secondary reactions to form char (by repolymerization) and gas (by cracking).[68] It was proposed that the secondary reactions occur in the vapor phase within the pores of solid matrix, and the reaction rates are proportional to the concentration of condensable vapors, strongly coupled with heat and mass transfer in biomass.

Blasi *et al.*[68] suggested that a char layer forms on the surface of large biomass particles, which reduces the heat transfer rate inside the solid biomass. On the contrary, small particles could allow a fast heating rate within the particles. The authors also pointed out that porosity, permeability, thermal conductivity, mass diffusivity, heat capacity and the volume occupied by the solid all have roles in biomass pyrolysis and secondary reactions. The volatile products must be transported out from the unconverted biomass region or the char layer through convection and diffusion. It is possible that the condensation could also occur during the product transportation within the biomass particles. As the particle size of biomass becomes bigger, the secondary reactions are promoted because the temperature gradient inside the particles increases, as well as volatile residence time within the particles. As pyrolytic devolatilization proceeds, the particles also shrink. The shrinking particles and the formation of porous structure due to the escaping volatiles change heat transfer conditions inside the particles, and as a result, could affect both the primary and the secondary reaction paths. Devolatilization becomes faster and volatile residence times decrease as the pyrolysis proceeds. The orientation of heat flux with respect to the biomass grain direction also affects the product yields. A heating perpendicular to biomass grain promotes secondary reactions because a larger temperature gradient along the char layer occurs.

Koufopanos and Papavannakos[69] suggested that pyrolysis of biomass with a particle size of above 1 mm is controlled by both external and internal heat transfer, as well as primary and secondary reactions of the pyrolysis products within the particles. They further stated that the heat transfer effect and secondary reactions become increasingly significant as the pyrolysis temperature and particle size increase. In their reaction model, biomass is pyrolyzed to form (condensable and non-condensable) volatiles and char in two parallel primary reactions, and secondary interaction occurs between the char and the volatiles.

Some experimental observations have provided strong evidence for the unique physical and chemical coupled conversion of biomass pyrolysis. Fraga *et al.*[70] pyrolyzed sugar-cane bagasse and silver birch in a wire-mesh reactor using heating rates ranging from 1 to 1000 °C s^{-1} with a holding time of between 0 and 100 s. They reported that the yields of bio-oil at pyrolysis temperatures of around 500 °C decreased with increasing heating rates when the sample holding time at the reactor was zero (*i.e.* the sample heating time after the sample temperature reached 500 °C). However, when the sample holding time was increased to 30 s, the pyrolysis of the biomass at higher heating rates produced higher yields of bio-oil, in comparison to the pyrolysis at lower heating rates. The results suggest that the pyrolysis does not complete within the short conversion time with fast heating rates, probably related to heat transfer limitation inside the biomass particles. It could also be that the short pyrolysis time and moderately low temperatures are not sufficient to completely devolatilize the pyrolysis products. The authors also stated that secondary reactions of primary products take place inside the particles and such reactions are intensified during pyrolysis at higher heating rates.

A real-time visual observation of biomass pyrolysis has been conducted by Haas *et al.*[71] In their experiment, a wood chip was pyrolyzed on a hot stage and the pyrolysis process was monitored using a microscope. This unique experimental method provided first-hand information about biomass structural changes and product transport at the plant tissue and cell wall level. When the biomass is heated to around 300 °C, a mild expansion was observed within the particle in the plane perpendicular to biomass grain, probably due to gaseous content in biomass. When the temperature reaches about 350 °C, an amber-colored liquid pyrolysis product was found inside of the cell, and the cell wall contraction began. The cell wall (mainly lignin) partly melts and the liquid leaves through the collapsed cell wall and tissue. This phenomenon was observed between temperatures of 350 and 500 °C, which corresponds well to the temperature range where the major mass loss of biomass occurs during pyrolysis. As the cell wall contracts, the wall becomes thinner and weaker and eventually carbonizes. Although it is expected that the increasing porosity of biomass and the structural shrinkage during pyrolysis could increase the extent of liquid devolatilization from the biomass plant matrix, some liquid products were still trapped within the plant matrix after the pyrolysis. Liquid droplets of sizes of <500 nm up to 10 μm and black char-like solid residue were found between the cell walls after pyrolysis (Figure 4.2), although there was enough time for the liquid products to evaporate. This observation implies that the trapped liquid products at elevated temperatures are subject to secondary reactions, thereby reducing bio-oil yield and increasing char yield. It was also found that in the biomass particle, the area closer to the heating source was more severely pyrolyzed, whereas the morphology of the area far from the heating source was similar to unpyrolyzed biomass. The authors examined the biochar produced from a fluidized bed at 700 °C and found that the char also largely retains the morphology of natural biomass. This observation suggests that the complex tissue and cell structure of biomass prevents effective heat and mass transfer within biomass, even during fast pyrolysis at high temperatures.

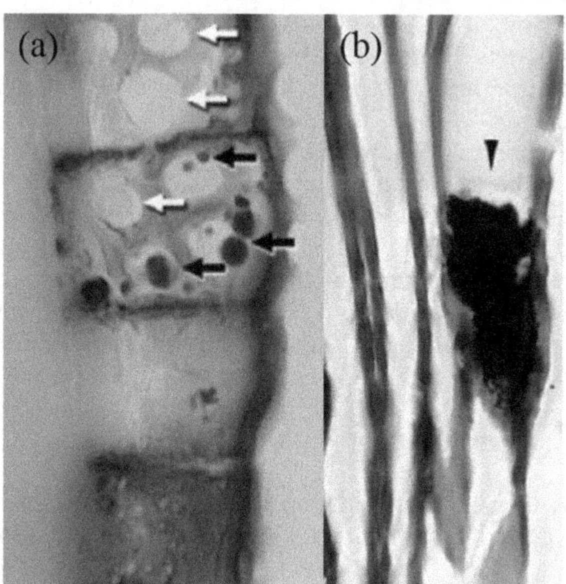

Figure 4.2 Pyrolysis of biomass observed under microscope. (a) white arrow: cell pore; black arrow: liquid pyrolysis products; (b) black arrow: pyrolysis char. Reprinted with permission from ref. 70. Copyright (2009) American Chemical Society.

These results challenge conventional knowledge, which believes that fast pyrolysis is able to quickly decompose biomass and the products are rapidly removed as vapor. The heat and mass transfer limitation within biomass particles and the related secondary reactions could become significant issues in commercial reactors, where larger biomass particle sizes are preferred. For example, heat and mass transfer limitations could be more noticeable in ablative reactors that pyrolyze biomass with large particle sizes or entrained flow reactors where heat transfer rates are low.[72] The fluidized bed reactors are preferred in this regard, as they usually offer high heat transfer rates and also the particle size of biomass is controlled.

4.6 Conclusion

In this chapter, previous studies related to secondary reactions of biomass pyrolysis products and the product transportation were reviewed. Controversial theories related to primary products and secondary products of cellulose and lignin pyrolysis were described and the supporting evidence were also presented. The comparisons of the results strongly suggest that chemical kinetics and physical phenomena of the pyrolysis products must both be taken into consideration when describing biomass pyrolysis and final product distribution. For instance, the physiochemical properties of LG could play a role in the formation of anhydro-oligosaccharides during cellulose pyrolysis. Secondary reactions among cellulose- and lignin-derived products were also reviewed. The solid/liquid phase or the vapor phase interactions among the products have positive effects on lignin volatilization by reducing char formation and increasing lignin-derived bio-oil yield. Due to the complex tissue and cell structure of biomass, heat and mass transfer limitations become important factors in biomass pyrolysis as the biomass particle size increases. The studies also show that faster heating rate/short reaction time are not necessarily always beneficial in the bio-oil production and final product selectivity. The liquid products and char trapped inside the tissue structure of biomass confirm that secondary reactions of the primary liquid products occur at the solid/liquid biomass matrix. Effectively transporting the pyrolysis products out of biomass is as important as reducing the secondary reactions of the products in the vapor phase.

References

1. D. Mohan, C. U. Pittman and P. H. Steele, *Energy Fuels*, 2006, **20**, 848–889.
2. H. B. Mayes and L. J. Broadbelt, *J. Phys. Chem. A*, 2012, **116**, 7098–7106.
3. C. Krumm, J. Pfaendtner and P. J. Dauenhauer, *Chem. Mater.*, 2016, **28**, 3108–3114.
4. A. D. Pouwels, G. B. Eijkel, P. W. Arisz and J. J. Boon, *J. Anal. Appl. Pyrolysis*, 1989, **15**, 71–84.
5. J. Piskorz, P. Majerski, D. Radlein, A. Vladars-Usas and D. Scott, *J. Anal. Appl. Pyrolysis*, 2000, **56**, 145–166.

6. J. Piskorz, P. Majerski and D. Radlein, Biomass: A Growth Opportunity in Green Energy and Value-Added Products, in *Proceedings of the 4th Biomass Conference of the Americas*, ed. R. P. Overend and E. Chornet, Elsevier Science, Amsterdam, 1999, vol. 2, pp. 1153–1159.
7. A. R. Teixeira, K. G. Mooney, J. S. Kruger, C. L. Williams, W. J. Suszynski, L. D. Schmidt, D. P. Schmidt and P. J. Dauenhauer, *Energy Environ. Sci.*, 2011, **4**, 4306–4321.
8. P. R. Patwardhan, D. L. Dalluge, B. H. Shanks and R. C. Brown, *Bioresour. Technol.*, 2011, **102**, 5265–5269.
9. F. Shafizadeh, R. H. Furneaux, T. G. Cochran, J. P. Scholl and Y. Sakai, *J. Appl. Polym. Sci.*, 1979, **23**, 3525–3539.
10. G.-J. Kwon, D.-Y. Kim, S. Kimura and S. Kuga, *J. Anal. Appl. Pyrolysis*, 2007, **80**, 1–5.
11. T. Funazukuri, R. Hudgins and P. Silveston, *J. Anal. Appl. Pyrolysis*, 1986, **9**, 139–158.
12. I. Milosavljevic, V. Oja and E. M. Suuberg, *Ind. Eng. Chem. Res.*, 1996, **35**, 653–662.
13. X. Bai, P. Johnston, S. Sadula and R. C. Brown, *J. Anal. Appl. Pyrolysis*, 2013, **99**, 58–65.
14. H. Kawamoto, M. Murayama and S. Saka, *J. Wood Sci.*, 2003, **49**, 469–473.
15. T. Hosoya, H. Kawamoto and S. Saka, *J. Anal. Appl. Pyrolysis*, 2008, **83**, 64–70.
16. X. Bai, P. Johnston and R. C. Brown, *J. Anal. Appl. Pyrolysis*, 2013, **99**, 130–136.
17. F. Shafizadeh and Y. Fu, *Carbohydr. Res.*, 1973, **29**, 113–122.
18. G.-J. Kwon, S. Kuga, K. Hori, M. Yatagai, K. Ando and N. Hattori, *J. Wood Sci.*, 2006, **52**, 461–465.
19. W. S.-L. Mok and M. J. Antal, *Thermochim. Acta*, 1983, **68**, 155–164.
20. X. Bai and R. C. Brown, *J. Anal. Appl. Pyrolysis*, 2014, **105**, 363–368.
21. D. Shen, S. Gu and A. Bridgwater, *Carbohydr. Polym.*, 2010, **82**, 39–45.
22. J. Lédé, F. Blanchard and O. Boutin, *Fuel*, 2002, **81**, 1269–1279.
23. H. N. Stiles and R. Kandiyoti, *Fuel*, 1989, **68**, 275–282.
24. J. Proano-Aviles, J. K. Lindstrom, P. A. Johnston and R. C. Brown, *Energy Technol.*, 2017, **5**, 189–195.
25. D. J. Nowakowski, A. V. Bridgwater, D. C. Elliott, D. Meier and P. de Wild, *J. Anal. Appl. Pyrolysis*, 2010, **88**, 53–72.
26. S. Beis, S. Mukkamala, N. Hill, J. Joseph, C. Baker, B. Jensen, E. Stemmler, C. Wheeler, B. Frederick and A. Van Heiningen, *BioResources*, 2010, **5**, 1408–1424.
27. Q. Sun, R. Khunsupat, K. Akato, J. Tao, N. Labbé, N. C. Gallego, J. J. Bozell, T. G. Rials, G. A. Tuskan and T. J. Tschaplinski, *Green Chem.*, 2016, **18**, 5015–5024.
28. R. K. Sharma, J. B. Wooten, V. L. Baliga, X. Lin, W. G. Chan and M. R. Hajaligol, *Fuel*, 2004, **83**, 1469–1482.
29. M. P. Pandey and C. S. Kim, *Chem. Eng. Technol.*, 2011, **34**, 29–41.

30. W. Mu, H. Ben, A. Ragauskas and Y. Deng, *BioEnergy Res.*, 2013, **6**, 1183–1204.
31. G. Jiang, D. J. Nowakowski and A. V. Bridgwater, *Energy Fuels*, 2010, **24**, 4470–4475.
32. S. Wang, H. Lin, B. Ru, W. Sun, Y. Wang and Z. Luo, *J. Anal. Appl. Pyrolysis*, 2014, **108**, 78–85.
33. E. Fratini, M. Bonini, A. Oasmaa, Y. Solantausta, J. Teixeira and P. Baglioni, *Langmuir*, 2006, **22**, 306–312.
34. S. Zhou, M. Garcia-Perez, B. Pecha, S. R. Kersten, A. G. McDonald and R. J. Westerhof, *Energy Fuels*, 2013, **27**, 5867–5877.
35. A. Oasmaa and E. Kuoppala, *Energy Fuels*, 2003, **17**, 1075–1084.
36. E. Alsbou and B. Helleur, *Energy Fuels*, 2014, **28**, 3224–3235.
37. L. Ingram, D. Mohan, M. Bricka, P. Steele, D. Strobel, D. Crocker, B. Mitchell, J. Mohammad, K. Cantrell and C. U. Pittman Jr, *Energy Fuels*, 2007, **22**, 614–625.
38. D. Chen, J. Zhou, Q. Zhang and X. Zhu, *Renewable Sustainable Energy Rev.*, 2014, **40**, 69–79.
39. B. Scholze and D. Meier, *J. Anal. Appl. Pyrolysis*, 2001, **60**, 41–54.
40. B. Scholze, C. Hanser and D. Meier, *J. Anal. Appl. Pyrolysis*, 2001, **58**, 387–400.
41. R. Bayerbach, V. D. Nguyen, U. Schurr and D. Meier, *J. Anal. Appl. Pyrolysis*, 2006, **77**, 95–101.
42. R. Bayerbach and D. Meier, *J. Anal. Appl. Pyrolysis*, 2009, **85**, 98–107.
43. C. A. Mullen and A. A. Boateng, *J. Anal. Appl. Pyrolysis*, 2011, **90**, 197–203.
44. A. Oasmaa, T. Sundqvist, E. Kuoppala, M. Garcia-Perez, Y. Solantausta, C. Lindfors and V. Paasikallio, *Energy Fuels*, 2015, **29**, 4373–4381.
45. J. Rencoret, G. Marques, A. Gutiérrez, L. Nieto, J. Jiménez-Barbero, Á. T. Martínez and C. José, *Ind. Crops Prod.*, 2009, **30**, 137–143.
46. G. Jiang, D. J. Nowakowski and A. V. Bridgwater, *Thermochim. Acta*, 2010, **498**, 61–66.
47. P. R. Patwardhan, R. C. Brown and B. H. Shanks, *ChemSusChem*, 2011, **4**, 1629–1636.
48. X. Bai, K. H. Kim, R. C. Brown, E. Dalluge, C. Hutchinson, Y. J. Lee and D. Dalluge, *Fuel*, 2014, **128**, 170–179.
49. S. Zhou, M. Garcia-Perez, B. Pecha, A. G. McDonald, S. R. Kersten and R. J. Westerhof, *Energy Fuels*, 2013, **27**, 1428–1438.
50. H. Ben and A. J. Ragauskas, *Energy Fuels*, 2011, **25**, 2322–2332.
51. T. Hosoya, H. Kawamoto and S. Saka, *J. Anal. Appl. Pyrolysis*, 2008, **83**, 78–87.
52. T. Hosoya, H. Kawamoto and S. Saka, *J. Anal. Appl. Pyrolysis*, 2009, **84**, 79–83.
53. S. Chu, A. V. Subrahmanyam and G. W. Huber, *Green Chem.*, 2013, **15**, 125–136.
54. S. Zhou, R. C. Brown and X. Bai, *Green Chem.*, 2015, **17**, 4748–4759.
55. S. Wang, B. Ru, H. Lin, W. Sun and Z. Luo, *Bioresour. Technol.*, 2015, **182**, 120–127.

56. T. Ikeda, K. Holtman, J. F. Kadla, H.-m. Chang and H. Jameel, *J. Agric. Food Chem.*, 2002, **50**, 129–135.
57. A. Jääskeläinen, Y. Sun, D. Argyropoulos, T. Tamminen and B. Hortling, *Wood Sci. Technol.*, 2003, **37**, 91–102.
58. S. Zhou, Y. Xue, A. Sharma and X. L. Bai, *ACS Sustainable Chem. Eng.*, 2016, **4**, 6608–6617.
59. H. Kawamoto, H. Morisaki and S. Saka, *J. Anal. Appl. Pyrolysis*, 2009, **85**, 247–251.
60. F. Ronsse, X. Bai, W. Prins and R. C. Brown, *Environ. Prog. Sustainable Energy*, 2012, **31**, 256–260.
61. T. Hosoya, H. Kawamoto and S. Saka, *J. Anal. Appl. Pyrolysis*, 2009, **85**, 237–246.
62. J. Zhang, Y. S. Choi, C. G. Yoo, T. H. Kim, R. C. Brown and B. H. Shanks, *ACS Sustainable Chem. Eng.*, 2015, **3**, 293–301.
63. D. S. Radlein, J. Piskorz and D. S. Scott, *J. Anal. Appl. Pyrolysis*, 1987, **12**, 51–59.
64. J. Meng, A. Moore, D. Tilotta, S. Kelley and S. Park, *ACS Sustainable Chem. Eng.*, 2014, **2**, 2011–2018.
65. H. Zhang, R. Xiao, D. Wang, G. He, S. Shao, J. Zhang and Z. Zhong, *Bioresour. Technol.*, 2011, **102**, 4258–4264.
66. Y. Elkasabi, C. A. Mullen and A. A. Boateng, *ACS Sustainable Chem. Eng.*, 2014, **2**, 2042–2052.
67. C. A. Mullen, A. A. Boateng and N. M. Goldberg, *Energy Fuels*, 2013, **27**, 3867–3874.
68. C. Di Blasi, *Chem. Eng. Sci.*, 1996, **51**, 1121–1132.
69. C. Koufopanos, N. Papayannakos, G. Maschio and A. Lucchesi, *Can. J. Chem. Eng.*, 1991, **69**, 907–915.
70. A.-R. Fraga, A. F. Gaines and R. Kandiyoti, *Fuel*, 1991, **70**, 803–809.
71. T. J. Haas, M. R. Nimlos and B. S. Donohoe, *Energy Fuels*, 2009, **23**, 3810–3817.
72. A. Bridgwater, D. Meier and D. Radlein, *Org. Geochem.*, 1999, **30**, 1479–1493.
73. M. Essig, *Cellulose and Wood*, 1989, p. 841.
74. J. Piskorz, D. Radlein and D. S. Scott, *J. Anal. Appl. Pyrolysis*, 1986, **9**, 121–137.

CHAPTER 5

Catalytic Biomass Pyrolysis with Reactive Gases

DAVID C. DAYTON*[a], KAIGE WANG[a], JONATHAN E. PETERS[a] AND OFEI D. MANTE[a]

[a]Energy Technology Division, RTI International, 3040 East Cornwallis Road, Research Triangle Park, NC, USA 27709
*E-mail: ddayton@rti.org

5.1 Introduction

Thermochemical conversion pathways can be classified as direct and indirect liquefaction processes that convert biomass into intermediates that are upgraded to biofuels. Direct liquefaction pathways for predominantly dry feedstocks include biomass fast pyrolysis, catalytic fast pyrolysis (CFP) (*in situ* and *ex situ*), and hydropyrolysis (HYP) that yield a liquid referred to as bio-oil or bio-crude suitable for upgrading to final fuels. An attractive strategy has been to leverage years of technical development and capital expenditures in the petrochemical industry by adapting crude oil processing technologies for upgrading these liquid intermediates into advanced biofuels. These advanced biofuels are expected to be drop-in hydrocarbon-based intermediates or fuels to replace petroleum liquids in the existing refining and distribution infrastructure.[1–3]

Although bio-oil has prospects for being upgraded in a manner similar to petroleum, the similarities between the two kinds of oil end there. The elemental composition of bio-oil more closely resembles that of biomass than

Green Chemistry Series No. 50
Fast Pyrolysis of Biomass: Advances in Science and Technology
Edited by Robert C. Brown and Kaige Wang

Published by the Royal Society of Chemistry, www.rsc.org

petroleum, and the chemical and physical properties of bio-oil, such as high oxygen content, low pH, and poor thermal stability, are not directly translatable to petroleum processing technologies. Therefore, recent research and development efforts are focused on (1) developing catalytic processes to modify the chemical and physical properties of bio-oil to better match petroleum processes for biofuel production and (2) developing new processes and catalysts to upgrade bio-oils into biofuels. Techno-economic analyses of these pathways highlight the importance of overall product yield for economic viability of biofuel process options.

Biomass pyrolysis is the thermal depolymerization of biomass at modest temperatures in the absence of added oxygen. The slate of products from biomass pyrolysis depends on the feedstock, process temperature, pressure, and residence time of the liberated pyrolysis vapors. The poor fuel properties of biomass fast pyrolysis oils make it unsuitable for direct integration into the current petroleum refining infrastructure.[4] Upgrading bio-oil into transportation fuels requires the removal of oxygen and an increase in the hydrogen-to-carbon ratio.[5] Therefore, one of the challenges in developing biofuels technologies is efficient utilization of hydrogen while maximizing carbon (energy) conversion efficiency and minimizing oxygen content in the product.

Over the past several years, CFP has been explored to improve the quality of liquid intermediates. The role of the catalyst in direct biomass liquefaction processes is to promote deoxygenation of the pyrolysis vapors while minimizing carbon loss to char, light gases, and coke. Oxygen removal during direct biomass liquefaction can occur by dehydration (loss of H_2O), decarboxylation (loss of CO_2), and decarbonylation (loss of CO). Dehydration of the cellulose and hemicellulose fractions during biomass pyrolysis (with or without a catalyst) produces water, referred to as water of pyrolysis, that is the most abundant single component of the liquid phase product. Biomass is inherently oxygen rich and hydrogen deficient, and the catalytic pyrolysis products become even more hydrogen deficient as dehydration occurs, which increases the tendency for aromatic formation and ultimately leads to char/coke production. Deoxygenation by CO and CO_2 removal (decarboxylation and decarbonylation) plus any carbon losses from coke formation on the catalyst lead to lower hydrocarbon liquid yields and lower energy recovery in the bio-crude intermediate.

Numerous researchers have investigated zeolite catalysts for biomass CFP.[5–13] However, hydrocarbon yields from CFP are typically below 20%, while coke yields can be as high as 42% even at optimal reaction conditions in a fluidized bed reactor.[12,13] Such low carbon efficiency and excessive coke formation have challenged the development of CFP technology for advanced biofuels production.

Many industrial refining processes like hydrocracking, isomerization, and naphtha reforming use hydrogen to control coke formation on the catalyst surface. Consequently, biomass HYP has emerged as a potential solution for maximizing carbon efficiency with improved hydrogen utilization.

Biomass HYP involves reacting biomass with hydrogen at elevated temperature and pressure in the presence of a hydrodeoxygenation (HDO) catalyst. Several groups have demonstrated the technical feasibility of producing hydrocarbon-rich liquids from biomass at HYP conditions.[14–21]

One of the technical challenges that faces the continued development of biomass HYP technology is feeding biomass into a high-pressure reactor. Several recent studies have explored the potential of combining the robustness of CFP with the carbon efficiency of biomass HYP by using catalysts optimized for HDO of biomass pyrolysis vapors in hydrogen at atmospheric pressure.[20–23]

The objective of this chapter is to provide an overview of recent progress in the basic scientific understanding of and technical developments in direct biomass liquefaction processes that use reactive gases, like hydrogen and CO, in combination with selective HDO catalysts to produce low-oxygen-containing bio-crude. A review of HDO using model compounds with a variety of catalysts at a range of process conditions provides a mechanistic understanding of deoxygenation pathways. Similar studies using real biomass as a feedstock in micropyrolyzers or small laboratory reactors investigate the impact of reactive gases on bio-crude yields and quality.

5.2 Model Compound Studies

There has been significant effort in recent years to develop catalysts for deoxygenation of bio-oil.[24–26] Many of these studies used model compounds to measure deoxygenation and HDO activity in high-pressure hydrogen.[27–29] Rogers and Zheng[24] have also recently provided a review of deoxygenation studies in low-pressure hydrogen environments covering a wide range of model compounds and catalyst types.

5.2.1 Deoxygenation Reaction Pathways

Many studies typically report on conversion of aromatic oxygenates such as lignin-derived phenolic compounds and holocellulose-derived furanic compounds.[25] One compound that is often studied is the methoxy-substituted phenol; guaiacol. It has both phenolic and methoxy functional groups that are commonly found in lignin and lignin derivatives. In general, two main pathways have been identified for the interaction of hydrogen with aromatic oxygenates. The first pathway is hydrogenation (HYD) in which the aromatic ring is first saturated and then oxygen is removed. The second pathway is called direct deoxygenation (DDO) in which oxygen is removed first, followed by aromatic ring saturation. Catalyst functionality is the main factor that drives the reaction toward one pathway over the other.

Petroleum hydrotreating catalysts like bimetallic molybdenum sulfides (NiMo and CoMo) have been tested for bio-oil model compound deoxygenation. In general, molybdenum-based sulfided catalysts tend to favor the

DDO pathway.[30,31] On the contrary, noble metals have been shown to promote the HYD pathway.[27,28,32] Catalysts that can selectively promote the DDO pathway without ring saturation are preferred because hydrogen consumption can be minimized.

Another reaction pathway that is observed when catalytically converting methoxy-substituted aromatics is transalkylation and demethlyation by cleavage of the O–CH_3 bond. This pathway has been observed in studies with anisole and guaiacol reacted over various catalysts and is attributed, in part, to the activity of the acidic supports.[33–36] This reaction pathway could be very beneficial for maintaining carbon in the liquid phase instead of losing it as methane in the gas phase. Figure 5.1 shows how guaiacol may undergo DDO and transalkylation to produce methyl-substituted aromatics.

5.2.2 Catalyst Development

Conventional sulfided catalysts have typically been evaluated under high-than-atmospheric pressure conditions for bio-oil hydrotreating and usually support the DDO pathway;[32,37] however, some recent studies have looked at their application under low-pressure conditions to limit hydrogen consumption. Bui *et al.* measured the promoting effect of cobalt on MoS_2 and reported that it strongly increased the DDO pathway while reducing

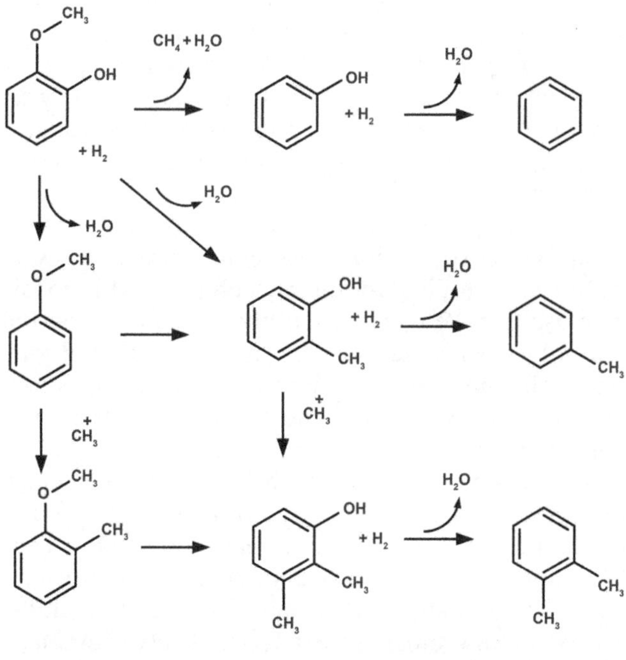

Figure 5.1 Reaction pathways for catalytic hydrodeoxygenation of guaiacol near atmospheric pressure.

hydrogenated products at 300 °C and 40 bar.[38] Also, there was a significant increase in methylated products with alumina-supported CoMoS *versus* the unsupported catalyst.[38] Loricera *et al.* evaluated the effect of sulfided CoMoW supported on mesoporous aluminas SBA-15 and SBA-16 modified with varying amounts of phosphate. It was reported in this case that anisole was converted to mostly phenol and methyl phenols by demethylation and transmethylation reactions promoted by the acidic function of the support; however, very little deoxygenation was observed.[39] Both of these cases indicated significant activity for the transalkyation pathway, but the latter did not suggest significant deoxygenation. The deoxygenation activity may significantly decrease because a sulfur source was not included in the feed to maintain the activity of the metal-sulfide catalysts. The presence of sulfur in the feed stream to maintain catalyst activity presents a drawback to using sulfided catalysts in catalytic pyrolysis applications because the bio-oil feedstocks typically contain very low sulfur compared with the petroleum feedstocks for which these catalysts were designed.

Transition metal oxides have received significant attention for bio-oil deoxygenation. Shimming *et al.* recently reported that oxygen vacancies in ceria-zirconia catalysts promote DDO of guaiacol to primary products phenol and catechol and secondary products cresol and benzene without significant deactivation after 72 hours.[40] Prasomsri *et al.* screened several metal oxide catalysts, including V_2O_5, Fe_2O_3, CuO, WO_3, and MoO_3, for HDO near atmosphere pressure.[35] All of these metal oxides investigated were active for HDO to convert biomass-derived oxygenates into unsaturated hydrocarbons but MoO_3 had the highest selectivity to unsaturated hydrocarbons.[41] Shetty *et al.* also showed that MoO_3 supported on ZrO_2 and TiO_2 has slower deactivation rates caused by coke deposition and could be regenerated by calcining with oxygen at lower temperatures compared to a γ-alumina-supported catalyst.[42] The ability to regenerate with oxygen has made supported metal oxides suitable for catalytic pyrolysis.

Transition metal phosphides are a new class of HDO catalysts that show promise for bio-oil deoxygenation. Zhao *et al.* compared several transition metal phosphides, including nickel phosphide (Ni_2P), cobalt phosphice (Co_2P), iron phosphide (Fe_2P), tungsten phosphide (WP), and molybdenum phosphide (MoP), on silica support for gas-phase HDO of guaiacol.[43] Ni_2P was found to be the most active catalyst forming mostly benzene and phenol, some anisole but no catechol, and no ring saturation at long contact times. Cresol, which is usually a secondary product, was detected at short contact times. Methylated phenols such as cresol are usually formed *via* acidic supports, but in this case, Ni_2P could have promoted the transalkylation pathway because metal phosphides can exhibit both Lewis acid sites and metallic functionality.[44] Ni_2P also showed greater deoxygenation activity and stability when compared with commercial Pd/Al_2O_3 and conventional $CoMoS/Al_2O_3$. Wu *et al.* evaluated the effect of silica, alumina, and zirconia supports for Ni_2P and determined that zirconia promoted the greatest overall deoxygenation activity.[45] Zirconia also increased the demethylation and transalkylation pathways compared with the other supports. Metal phosphides, Ni_2P in particular, are promising catalysts because of their bifunctional nature and their resistance to deactivation.

Recent studies conducted at RTI International investigated the effect of temperature on guaiacol HDO conversion and product selectivity with 2.5 wt% Ni_2P supported on Al_2O_3. Guaiacol vapor in 1 bar of hydrogen was fed over a fixed bed of catalyst at a range of reaction temperatures. Continuous product monitoring during catalyst reduction, guaiacol conversion, and catalyst regeneration provides insights into deoxygenation pathways. The reactor system, analytical methods, and experimental protocols are described in detail in the literature.[46]

Model compound studies for catalytic HDO are typically performed at temperatures between 300 and 400 °C, but catalytic biomass pyrolysis occurs at higher temperatures. Therefore, it is of interest to see how these catalysts perform at temperatures up to 500 °C. Guaiacol conversion and product selectivity from recent experiments conducted at RTI International are provided as a function of temperature in Table 5.1.

Guaiacol conversion increased monotonically from 40 to 99% with increasing temperature. The higher concentration of products formed by direct deoxygenation correlates with increased water formation, indicating that hydrogen consumption increased with temperature. Cresol (9 to 20% sel.), xylenol (1 to 23% sel.), and methyl-anisole (6 to 13.5% sel.) were major products, indicating that the transalkylation pathway was also significant. However, there was not a clear trend in methylated products with temperature. Cresol production increased with temperature, xylenol production decreased with temperature, and methyl-anisole production had a minimum at 400 °C. Because both metal phosphide and the alumina support can have

Table 5.1 Guaiacol conversion and product selectivity for catalytic hydrodeoxygenation of guaiacol over 2.5 wt% Ni_2P on Al_2O_3 at various temperatures under 1 bar hydrogen.

Catalyst	2.5 wt% Ni_2P/Al_2O_3				
Temperature (°C)	300	350	400	450	500
Guaiacol conversion (%)	40.1	49.9	71.8	80.7	99.0
Product selectivity (wt% of total products measured)					
Water	3.7	3.0	6.7	9.9	13.7
CO_2	0.0	0.0	0.0	0.0	0.1
Benzene	8.2	8.9	7.0	4.1	4.6
Toluene	6.3	5.9	4.7	2.3	2.4
Phenol	13.8	15.0	24.0	26.1	32.0
Cresol	8.8	10.6	11.4	20.2	20.2
Anisole	7.7	8.8	7.8	4.2	0.1
Methyl-anisole	10.8	13.5	5.7	8.6	9.7
Xylenol	21.3	21.6	22.7	15.1	1.2
CO	2.2	2.4	0.5	0.3	0.0
Methane	0.3	0.4	0.9	2.2	6.5
Coke	16.9	9.9	8.5	7.1	9.5

active catalytic sites, the formation of different methylated products corresponding to the different acid sites may be favored at different temperatures. Overall, the concentration of products from transalkylation reactions is maximized between 400 and 450 °C, while the sequential demethylation reaction is favored at 450 °C and above.

Organic liquid yields from biomass pyrolysis are typically maximized at 500 °C; however, that temperature may not be ideal for catalytic pyrolysis in hydrogen with nickel phosphide supported on alumina if a significant amount of carbon is lost to the gas phase by methane production. Methane selectivity increased from 0.3% at 300 °C to 6.5% at 500 °C. Biomass experiments are required to find the optimum temperature to maximize liquid yield and deoxygenation while minimizing methane production.

Transition metal carbides are also relatively new catalysts that have emerged as potential HDO catalysts for biomass pyrolysis.[47,48] Wang *et al.* showed that molybdenum carbide/oxycarbides supported on carbon have higher DDO selectivity than MoP, MoO_2, and MoS_2. However, transition metal carbides are susceptible to irreversible deactivation and are not suitable for catalytic pyrolysis.[49]

Bimetallic catalysts that include an oxophilic metal such as Cu, Fe, and Sn and a noble metal have also been investigated for increased selectivity for the DDO reaction pathway.[34,50,51] Sun *et al.* reported that the bimetallic PdFe catalyst supported on carbon substantially enhanced HDO activity compared with just Pd or Fe alone while also avoiding ring saturation.[51] The Fe is believed to be the active deoxygenation site, while the noble metal activates hydrogen and facilitates deoxygenation as well as Fe reduction. This synergistic effect of bimetallic catalysts is promising; however, noble metals add significant cost to catalyst formulations.

5.3 Biomass Pyrolysis in Reactive Gases

A range of techniques have been applied to understand the impact of adding reactive gases during biomass pyrolysis with and without catalysts. This includes fundamental studies with micropyrolyzers connected directly to the inlet of a gas chromatograph with a mass spectrometric detector (GC-MS) and applied studies that use laboratory-scale reactors (fixed and fluidized beds) to produce limited quantities of bio-crude for analysis and testing.

5.3.1 Batch Pyrolysis: Micro-Scale Py-GC-MS Studies

Micropyrolzyers coupled with GC/MS (Py-GC-MS) have been used extensively in fundamental research of biomass pyrolysis. Inert gases, including He, Ar, and nitrogen, are typically used as the carrier gas in Py-GC-MS systems. Relatively few Py-GC-MS studies are available that report the influence of ambient and high-pressure reactive gases, especially hydrogen, on biomass pyrolysis and catalytic biomass pyrolysis.

The effect of hydrogen on non-catalytic biomass pyrolysis is minimal, especially at ambient pressure. Studies have shown that the yield of detectable products decreased when hydrogen was added during pyrolysis.[52–54] In addition, the relative concentration of aromatic hydrocarbons increased substantially compared with the results for pyrolysis in an inert atmosphere (helium) and was enhanced with increasing hydrogen pressure.[52–54] Work by Melligan *et al.*[52,53] on Py-GC-MS using miscanthus showed that the overall yield of detectable pyrolysis vapors decreased significantly with increasing hydrogen pressure up to 30 bar. Furthermore, the composition of the pyrolysis vapors changed as a function of hydrogen pressure. The most noticeable effect is on the concentration of phenolics and aromatics suggesting that hydrogen plays an important role in lignin thermal decomposition but has a smaller effect on cellulose and hemicellulose thermal decomposition.[52,53] Another Py-GC-MS study conducted by Resende and co-workers using poplar lignin reported that no aromatic hydrocarbons were observed from Py-GC-MS experiments in He.[55] Helium pressures up to 26 bar had no observable influence on lignin pyrolysis. A slight increase in phenolic yield was observed at hydrogen partial pressures up to 17 bar. However, phenolic yields tended to decrease as the pressure was increased to 26 bar. These results suggest that hydrogen may enhance hydrogenolysis, but the effects are minimal. Adding a catalyst to promote hydrogen radical production would clearly enhance demethoxylation, decarbonylation, and dehydration reactions.

CFP is being developed as a promising pathway to improve the quality of pyrolysis oils. When high-pressure hydrogen is present, the process is referred to as HYP. Zeolite catalysts, especially HZSM-5, have been frequently used in CFP based on their activity and shape selectivity toward gasoline-range aromatic hydrocarbons.[5] However, excessive coke formation and low yield of desirable hydrocarbons have limited the application of zeolites. Using catalyst promoters and hydrogen at optimized process conditions has been a strategy explored to increase the yield of desirable hydrocarbons and decrease coke formation.[54–56] Thangalazhy-Gopakumar *et al.*[54,56] compared catalytic pyrolysis with HZSM-5 under helium and hydrogen on a Py-GC-MS under HYP conditions. They concluded that pressurized hydrogen or helium (up to 27.6 bar) had no influence on the yield of aromatics. Even with the presence of high-pressure hydrogen, HZSM-5 acted as a cracking catalyst only with no indication that it promoted HYD. However, the aromatic yield increased when ZSM-5 catalysts doped with Ni, Co, Mo, and Pt were used under 27.6 bar hydrogen pressure. Those metals impregnated on a ZSM-5 catalyst can activate hydrogen for HDO and HYD. Formation of a low concentration of cycloalkanes using Pt-impregnated ZSM-5 was reported in another study by Jan *et al.*,[55] which suggested that the noble metal dopant facilitated HDO reactions at elevated hydrogen pressure (17 bar). HYD in high-pressure hydrogen was also observed with Ni-impregnated HZSM-5 catalysts, but at atmospheric pressure, no noticeable effect on the product distribution was observed.[57] These Py-GC-MS studies suggested that catalytic pyrolysis using a pure zeolite catalyst was not affected by the presence of hydrogen. Only metal-impregnated zeolite catalysts can activate hydrogen and enhance HDO and HYD under HYP reaction conditions.

The successful work with model compounds discussed in the last section has led researchers to the conclusion that biomass CFP with atmospheric pressure reactive gas (hydrogen, in particular) is a promising pathway to produce high-quality bio-oil from biomass. This process is referred to as reactive CFP, or RCFP. Although numerous catalysts have been tested with bio-oil model compounds, only a small subset of them have been tested with real biomass using a Py-GC-MS system. Nolte *et al.*[58] performed *ex situ* HDO of pyrolysis vapor in a tandem micropyrolyzer under RCFP conditions. A pure MoO_3 catalyst was found to effectively hydrodeoxygenate pyrolysis vapor to produce mostly linear alkanes and aromatics from lignin, cellulose, and corn stover. The carbon yield of larger hydrocarbons (C_{4+}) was 44 to 53% from cellulose, while the respective yields from lignin and corn stover were slightly lower at 16 to 23% and 15 to 26%. Murugappan *et al.*[59] performed RCFP experiments using a Py-GC-MS to evaluate supported MoO_3 (10 wt% MoO_3/TiO_2 and MoO_3/ZrO_2) catalyst performance. Both supported MoO_3 catalysts showed similar product distributions of *ca.* 7 C% aromatic hydrocarbons, 17 to 19 C% alkenes, 2 C% alkanes, 3 to 5 C% CO_2, 7 C% coke, and 39 C% char. In general, previous Py-GC-MS studies suggested that molybdenum-based catalysts were the most effective HDO catalysts under RCFP conditions.

5.3.2 Continuous Biomass Feed: Laboratory-Scale Reactor Studies

As discussed above, researchers have conducted a large number of model compound studies under various reaction conditions. In contrast, relatively few studies have reported real biomass conversion in continuous laboratory reactor systems. Bubbling fluidized bed reactors are frequently used for laboratory-scale biomass pyrolysis to produce bio-oil and char for comprehensive characterization to support technology development.

Work by Rocha *et al.*[60,61] on cellulose pyrolysis in a fixed bed reactor under a 100 bar hydrogen atmosphere reported a higher yield of bio-oil with a slightly lower oxygen content. However, another cellulose HYP study[19] on a continuous-flow reactor in the absence of a catalyst reported varied results. Addition of 25 bar hydrogen did not have a significant effect, within experimental error, on the yields of solids, liquids, and gases from cellulose pyrolysis at 480 °C compared with an experiment at 25 bar helium. Also, the composition of the liquid product did not differ significantly in the experiments using 25 bar He or hydrogen partial pressure. However, hydrogen appeared to suppress the formation of reactive light oxygenated species like glycolaldehyde and formic acid at higher temperature (580 °C). The authors concluded that hydrogen activation was minimal without a catalyst, so little effect on biomass pyrolysis was observed.

Several groups have also demonstrated the technical feasibility of producing hydrocarbon-rich liquids at HYP conditions in continuously fed reactors using a variety of catalysts.[14,15,17,18,20,22,23,62] This pathway has shown the potential to produce very low oxygen-containing intermediates (below 5 wt% dry basis) with attractive carbon yields.

Agrawal *et al.*[19,20] reported a process that integrated non-catalytic HYP with vapor-phase catalytic HDO. A 5 wt% Pt and 2.5 wt% Mo impregnated multi-walled carbon nanotube catalyst was found to be the optimal HDO catalyst. Carbon yield of C_{4+} hydrocarbons from wood and cellulose were 32% and 55%, respectively. However, the majority of the C_{4+} products are C_4–C_6 light hydrocarbons. Marker and co-workers integrated fast catalytic HYP with HDO (known as the IH^2 process) to produce a liquid fuel with extremely low oxygen and rich in gasoline- and diesel-range hydrocarbons.[21,23] Different proprietary catalysts and reaction temperatures were used in the HYP and HDO reactors. At optimal reaction conditions, a C_{4+} liquid yield of 46.3 wt% from microalgae and 26.4 wt% from wood chips with less than 1 wt% oxygen content was produced from the IH^2 process. The IH^2 process consumed 39 to 61 g hydrogen per kg biomass. Theoretically, sufficient hydrogen can be produced by reforming CO and light C_1–C_3 hydrocarbon gas generated from IH^2 to maintain hydrogen balance.

RTI has also conducted HYP experiments in a fluidized bed reactor. A biocrude yield of 23.7 wt% with oxygen content below 1 wt% was obtained at the most severe condition. Catalyst activity and stability were evaluated over a 10 day period during which biomass HYP was also performed. Catalyst activity did not appear to decrease significantly over the duration of the 10 day experiment. HDO is thought to be the dominant oxygen rejection pathway during HYP. In 20% H_2 at 20.7 bar total pressure, the hydrogen consumption measured during HYP with the commercial hydrotreating catalyst ranged from 20 to 25 g per kg of biomass fed. At higher hydrogen concentrations, the hydrogen consumption increased from 35 to 38 g per kg biomass fed.

Overall, biomass HYP can be used to produce a liquid intermediate with low oxygen content, which will lower the H_2 demand for any further downstream processing to produce a finished transportation fuel. However, feeding solid biomass into a pressurized HYP reactor remains a significant technical challenge that poses considerable risk as the technology is demonstrated at larger scale. Consequently, developing an atmospheric pressure process with similar technical performance to HYP could be an attractive alternative. Several recent studies have shown encouraging results for catalytic biomass pyrolysis in hydrogen at atmospheric pressure.

Zhang *et al.*[63] report results for biomass fast pyrolysis in a fluidized bed reactor with N_2 and reactive gases (H_2, CH_4, and CO). Their results showed that reactive gases affected both product yield and composition. Addition of CO and H_2 enhanced deoxygenation of pyrolysis vapor by producing CO_2 and H_2O, respectively. Consequently, pyrolysis in CO and H_2 increased the higher heating value of the bio-oil (23.7 MJ kg^{-1} and 24.4 MJ kg^{-1}, respectively) compared with bio-oil produced in a N_2 atmosphere (17.8 MJ kg^{-1}), but the bio-oil yield was lower. Lower char yield was measured for biomass pyrolysis in a CO_2 atmosphere, suggesting that char was reacting with CO_2 *via* the Boudouard reaction.[63]

More recently, Boateng and co-workers have conducted a series of studies on fast pyrolysis with tail-gas recycling.[64–67] The gas stream produced from biomass fast pyrolysis, consisting of primarily CO, CO_2, H_2, and light hydrocarbons, was used as a fluidizing gas and reaction atmosphere for the tail-gas reactive pyrolysis (TGRP). Significant deoxygenation was measured, and the TGRP bio-oil had a higher hydrocarbon content compared with pyrolysis in N_2. However, the TGRP bio-oil is highly aromatic and hydrogen deficient. The oxygen content of the bio-oil was reduced with a concurrent decrease of hydrogen content, suggesting that dehydration and aromatization occurred instead of HYD.

Mante *et al.*[7] also explored the influence of hydrogen on the pyrolysis process with a fluid catalytic cracking (FCC) catalysts. Even with the presence of 5 vol% hydrogen, bio-oil yield and its properties were improved significantly. Compared with CFP in nitrogen, RCFP with a minimal amount of hydrogen produced bio-oil with a higher heating value and pH and a lower viscosity and density. Other gases, including CO, CO_2, and tail gas from the pyrolysis process, were also found to be beneficial for obtaining a higher yield of bio-oil with improved quality.

Murugappan *et al.*[59] performed RCFP experiments in a horizontal quartz tube reactor using two types of supported MoO_3 (10 wt% MoO_3/TiO_2 and MoO_3/ZrO_2) and bulk MoO_3. Their study showed that the supported MoO_3 catalysts are active over a longer operating time than bulk MoO_3, which is in agreement with the study using bio-oil model compounds.[68] All of the catalysts became less effective for HDO as biomass feeding increased, suggesting rapid catalyst deactivation. Regeneration of catalysts is suggested for long-term operation of the RCFP process.

Zhou *et al.*[69] studied *ex situ* RCFP of lignin and beech wood using MoO_3 catalysts in atmospheric pressure hydrogen. RCFP bio-crude was collected in a series of condensers. Temperatures over 400 °C were required to obtain a bio-crude with 24.8 wt% oxygen content; however, the organic liquid yield was only 4.6 wt%. This low yield suggests over-cracking of the pyrolysis vapors at the *ex situ* RCFP conditions. In contrast, the yield of organic liquid from lignin RCFP at 450 °C catalyst temperature and 89 vol% hydrogen concentration was 16.2 wt% with 11.5 wt% oxygen content.

RTI has also explored RCFP with proprietary catalysts in a fluidized bed reactor. Biomass is continuously fed at a rate of 1 g min^{-1} into a nominal 2.54 cm diameter bubbling fluidized bed reactor maintained at a temperature between 400 and 600 °C. The multiphase product stream passes through a cyclone separator to remove solids (char and catalyst fines) and then through a condensation system that includes a shell-in-tube heater exchanger, cooled impingers, and an electrostatic precipitator. Process gas is monitored continuously with online GC, and all solid and liquid samples are analysed to determine elemental (C, H, N, O, S) and chemical (GC-MS) composition. A detailed description of the reactor system can be found in the literature.[70,71]

Table 5.2 Effect of hydrogen concentration on RCFP with RTI-A9P catalyst and loblolly pine as feedstock.

Hydrogen concentration/vol%	0%	25%	60%	93%
Temperature/°C	500	500	500	500
Mass balance (wt% of biomass)				
Gas	24.7	26.2	21.6	25.4
Solid	22.4	26.9	18.5	17.1
Aqueous	31.7	28.8	35.7	36.8
Organics	14.1	17.8	18.8	21.1
Wt% oxygen in organics	22	16.5	11.1	4.24
Carbon efficiency in organics	19.8	24.8	31.1	36.9
Carbon efficiency in aqueous	6.2	4.0	3.5	2.5

A range of different process conditions (temperature, residence time, and hydrogen concentration) was studied with a number of commercially available and developing catalysts. Table 5.2 shows the mass balances and carbon efficiency *versus* oxygen content for bio-crudes from RCFP using a platinum-doped solid acid catalyst (designated RTI-A9P) with varying H_2 concentrations. With the addition of hydrogen in the reactor gas at atmospheric pressure, coke production decreased and the carbon yield of organics increased substantially from 19.8% in nitrogen to 36.9% in 93 vol% hydrogen. Meanwhile, the oxygen content in the organics was reduced from 22 wt% to 4.2 wt%. These results indicate that the presence of hydrogen enhanced HDO of pyrolysis vapor. The hydrocarbon concentration in the aqueous fraction also decreased as more hydrogen was added because deoxygenated hydrocarbons more easily phase separated into the organic fraction. At 93 vol% hydrogen, the water content of the aqueous fraction was 93% compared with 80% in the aqueous phase without added hydrogen. Better hydrocarbon separation equates to easier downstream processing and potentially easier aqueous fraction cleanup.

5.4 Discussion

Reactive catalytic biomass pyrolysis is a promising process for using hydrogen during *in situ* catalytic biomass pyrolysis to maximize the biomass carbon and energy recovery in a low oxygen content, thermally stable bio-crude intermediate that can be efficiently upgraded into a finished biofuel. The key to this novel process is developing a robust catalyst that efficiently uses hydrogen for HDO and increases the hydrogen-to-carbon ratio in volatile products to limit char and coke formation. The amount of oxygen removed during the RCFP process is the primary metric for a successful catalyst, but the use of hydrogen also increases carbon efficiency in the liquid bio-crude product. A wide array of parameters ranging from catalyst properties to process conditions can affect the observed HDO activity and coke formation rate measured in model compound screening experiments. A wide variety of catalysts have been screened for HDO activity with model compounds and with real biomass. Figure 5.2 shows the measured carbon efficiency (C_{4+} products)

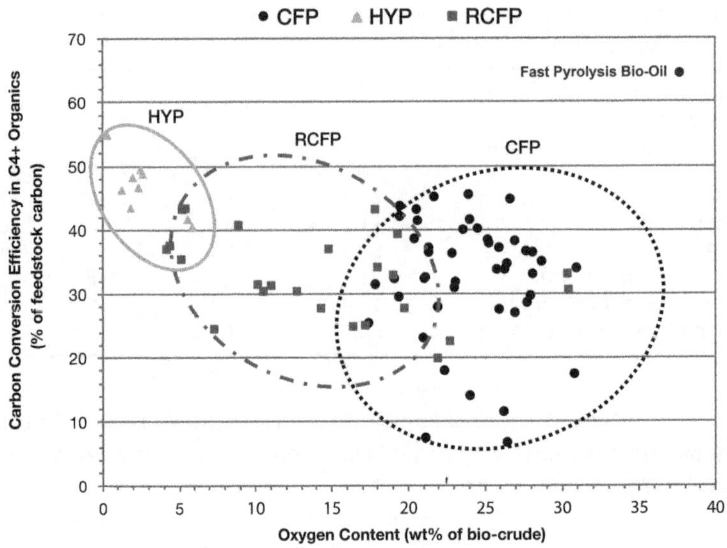

Figure 5.2 Bio-crude oxygen content as a function of carbon efficiency for CFP, RCFP, and HYP.

as a function of oxygen content measured for bio-crude intermediates produced from CFP, HYP, and RCFP. While HYP clearly demonstrates the highest yields at the lowest oxygen content, the RCFP has been successfully developed to the point where greater than 42% of the carbon input from biomass can be recovered in a bio-crude intermediate with less than 8 wt% oxygen.

The commercial viability of these direct biomass liquefaction pathways can be assessed by comparing how the relative yields, hydrogen demand, and carbon recovery potential affect the preliminary techno-economics of each process. The biofuels yield and hydrogen demand are estimated from the laboratory material balances and measured bio-crude oxygen content. The calculated fuel yield (light gray bars) presented in Figure 5.3 is based on upgrading the bio-crudes produced from each process in a hydroprocessing step. The calculation assumes no carbon losses during hydroprocessing. The hydrogen demand determined for each conversion process (dark gray bars) shown in Figure 5.3 was experimentally measured, while the hydrogen demand for upgrading (medium gray bars) was calculated by assuming the remaining oxygen in the bio-crude was removed as water and the hydrogen-to-carbon ratio in the finished biofuel is two. The volumetric yield of the finished biofuels is based on a density of 0.8 g ml^{-1}, similar to diesel.

The total hydrogen demand (sum of the medium and dark gray bars) for each process effectively correlates with the bio-crude yield because more hydrogen is required to deoxygenate and upgrade the intermediate bio-crude produced. The CFP and RCFP processes have about the same hydrogen demand for upgrading, but the RCFP process uses hydrogen in the conversion step, so the total hydrogen demand is higher. Less hydrogen

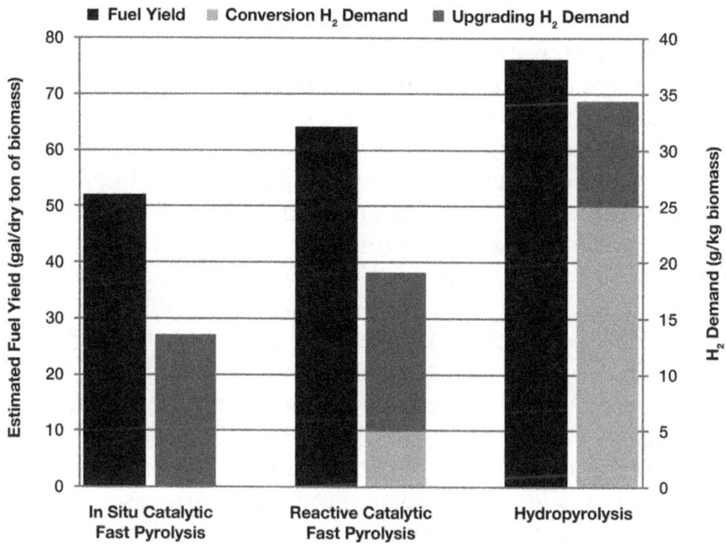

Figure 5.3 Summary of the biofuel yield and hydrogen demand for three bio-oil pathways.

is required for deoxygenating the HYP bio-crude because it has low oxygen content, but the hydrogen demand during HYP is comparatively high because carbon efficiency is higher and there is near complete HDO in the conversion step.

5.5 Summary and Future Work

In summary, reactive gases, especially hydrogen, have led to improved bio-oil quality by slightly reducing oxygen content even in the absence of catalysts. Although free radical mechanisms are proposed, researchers generally do not have an in-depth understanding of how the chemistry of biomass fast pyrolysis changes by adding reactive gases. The resulting bio-oil from non-catalytic pyrolysis with reactive gases is highly aromatic and hydrogen deficient. Catalysts to promote the HYD reaction are recommended to add to the reaction network to further improve the quality of the bio-oil.

Many of these studies have been conducted with model compounds or in small laboratory reactors with biomass. The technology is promising but requires scale-up and demonstration to become a commercial reality. One key activity for technology development includes testing the RCFP concept in pilot-scale reactors with continuous biomass feeding and integrated reaction/regeneration to optimize the process and evaluate the long-term performance of catalysts. Pilot-scale testing will require parallel development and scale-up of catalysts with suitable HDO activity and physical strength to maximize performance and minimize catalyst loss by attrition. Removing coke deposits during regeneration will be required to maintain catalyst activity, but irreversible deactivation by sintering or catalyst poisoning will require substantial investigation to guarantee long-term performance.

Upgrading RCFP bio-crude into refinery blendstock or finished biofuel will require additional research and development. To date, large enough reactor systems to produce enough bio-crude for upgrading studies have not been available. A working hypothesis assumes that RCFP bio-crude will be easier to hydroprocess than CFP or fast pyrolysis intermediates. The true test of the RCFP process will be to evaluate the final biofuel yield and hydrogen demand and compare it to existing advanced biofuels processes.

A clear logistical and economic challenge for this technology is the need for hydrogen in the pyrolysis reactor. Co-locating future plants with a source of hydrogen, like a petroleum refinery, methanol synthesis process, or integrated gasification combined-cycle plant, could be an option to meet the hydrogen demand of the RCFP process. Another option is to leverage the development of small-scale hydrogen production processes that are being developed as a strategy for hydrogen fueling stations of the future. Regardless, integrated process concepts that minimize capital costs to maximize the economic feasibility of the RCFP process will need to be developed, demonstrated, and deployed.

References

1. J. C. Serrano-Ruiz, E. V. Ramos-Fernandez and A. Sepulveda-Escribano, *Energy Environ. Sci.*, 2012, **5**, 5638.
2. J. J. Cheng and G. R. Timilsina, *Renewable Energy*, 2011, **36**, 3541.
3. F. de Miguel Mercader, M. J. Groeneveld, S. R. A. Kersten, N. W. J. Way, C. J. Schaverien and J. A. Hogendoorn, *Appl. Catal., B*, 2010, **96**, 57.
4. D. Mohan, C. U. Pittman and P. H. Steele, *Energy Fuels*, 2006, **20**, 848.
5. C. Liu, H. Wang, A. M. Karim, J. Sun and Y. Wang, *Chem. Soc. Rev.*, 2014, **43**, 7594.
6. D. C. Dayton, J. R. Carpenter, A. Kataria, J. E. Peters, D. Barbee, O. D. Mante and R. Gupta, *Green Chem.*, 2015, **17**, 4680.
7. O. D. Mante, J. A. Rodriguez, S. D. Senanayake and S. P. Babu, *Green Chem.*, 2015, **17**, 2362.
8. H. Zhang, R. Xiao, J. Nie, B. Jin, S. Shao and G. Xiao, *Bioresour. Technol.*, 2015, **192**, 68.
9. C. A. Mullen, A. A. Boateng, D. J. Mihalcik and N. M. Goldberg, *Energy Fuels*, 2011, **25**, 5444.
10. S. Du, Y. Sun, D. P. Gamliel, J. A. Valla and G. M. Bollas, *Bioresour. Technol.*, 2014, **169**, 188.
11. C. Mukarakate, M. J. Watson, J. ten Dam, X. Baucherel, S. Budhi, M. M. Yung, H. Ben, K. Iisa, R. M. Baldwin and M. R. Nimlos, *Green Chem.*, 2014, **16**, 4891.
12. T. R. Carlson, Y.-T. Cheng, J. Jae and G. W. Huber, *Energy Environ. Sci.*, 2011, **4**, 145.

13. Y. T. Cheng, J. Jae, J. Shi, W. Fan and G. W. Huber, *Angew. Chem.*, 2012, **124**, 1416.
14. R. Agrawal and N. R. Singh, *AIChE J.*, 2009, **55**, 1898.
15. A. J. Guell, C. Z. Li, A. A. Herod, B. J. Stokes, P. Hancock and R. Kandiyot, *Biomass Bioenergy*, 1993, **5**, 155.
16. T. L. Marker, L. G. Felix, M. B. Linck and M. J. Roberts, *Environ. Prog. Sustainable Energy*, 2012, **31**, 191.
17. R. V. Pindoria, A. Megaritis, A. A. Herod and R. Kandiyoti, *Fuel*, 1998, **77**, 1715.
18. J. D. Rocha, C. A. Luengo and C. E. Snape, *Org. Geochem.*, 1999, **30**, 1527.
19. V. K. Venkatakrishnan, J. C. Degenstein, A. D. Smeltz, W. N. Delgass, R. Agrawal and F. H. Ribeiro, *Green Chem.*, 2014, **16**, 792.
20. V. K. Venkatakrishnan, W. N. Delgass, F. H. Ribeiro and R. Agrawal, *Green Chem.*, 2015, **17**, 178.
21. T. L. Marker, L. G. Felix, M. B. Linck, M. J. Roberts, P. Ortiz-Toral and J. Wangerow, *Environ. Prog. Sustainable Energy*, 2014, **33**, 762.
22. D. C. Dayton, J. Hlebak, J. R. Carpenter, K. Wang, O. D. Mante and J. E. Peters, *Energy Fuels*, 2016, **30**, 4879.
23. T. L. Marker, L. G. Felix, M. B. Linck and M. J. Roberts, *Environ. Prog. Sustainable Energy*, 2012, **31**, 191.
24. K. A. Rogers and Y. Zheng, *ChemSusChem*, 2016, **9**, 1750.
25. Q. Bu, H. W. Lei, A. H. Zacher, L. Wang, S. J. Ren, J. Liang, Y. Wei, Y. P. Liu, J. M. Tang, Q. Zhang and R. Ruan, *Bioresour. Technol.*, 2012, **124**, 470.
26. M. Saidi, F. Samimi, D. Karimipourfard, T. Nimmanwudipong, B. C. Gates and M. R. Rahimpour, *Energy Environ. Sci.*, 2014, **7**, 103.
27. C. R. Lee, J. S. Yoon, Y. W. Suh, J. W. Choi, J. M. Ha, D. J. Suh and Y. K. Park, *Catal. Commun.*, 2012, **17**, 54.
28. A. Gutierrez, R. K. Kaila, M. L. Honkela, R. Slioor and A. O. I. Krause, *Catal. Today*, 2009, **147**, 239.
29. M. V. Bykova, D. Y. Ermakov, S. A. Khromova, A. A. Smirnov, M. Y. Lebedev and V. A. Yakovlev, *Catal. Today*, 2014, **220**, 21.
30. E. Laurent and B. Delmon, *Appl. Catal., A*, 1994, **109**, 77.
31. A. Centeno, E. Laurent and B. Delmon, *J. Catal.*, 1995, **154**, 288.
32. Y. C. Lin, C. L. Li, H. P. Wan, H. T. Lee and C. F. Liu, *Energy Fuels*, 2011, **25**, 890.
33. S. Boonyasuwat, T. Omotoso, D. E. Resasco and S. P. Crossley, *Catal. Lett.*, 2013, **143**, 783.
34. X. W. Xu, E. C. Jiang, Y. H. Du and B. S. Li, *Renewable Energy*, 2016, **96**, 458.
35. T. Prasomsri, T. Nimmanwudipong and Y. Roman-Leshkov, *Energy Environ. Sci.*, 2013, **6**, 1732.
36. R. C. Runnebaum, T. Nimmanwudipong, D. E. Block and B. C. Gates, *Catal. Sci. Technol.*, 2012, **2**, 113.
37. H. M. Wang, J. Male and Y. Wang, *ACS Catal.*, 2013, **3**, 1047.
38. V. N. Bui, D. Laurenti, P. Afanasiev and C. Geantet, *Appl. Catal., B*, 2011, **101**, 239.

39. C. V. Loricera, B. Pawelec, A. Infantes-Molina, M. C. Alvarez-Galvan, R. Huirache-Acuna, R. Nava and J. L. G. Fierro, *Catal. Today*, 2011, **172**, 103.
40. S. M. Schimming, O. D. LaMont, M. König, A. K. Rogers, A. D. D'Amico, M. M. Yung and C. Sievers, *ChemSusChem*, 2015, **8**, 2073–2083.
41. T. Prasomsri, T. Nimmanwudipong and Y. Román-Leshkov, *Energy Environ. Sci.*, 2013, **6**, 1732.
42. M. Shetty, K. Murugappan, T. Prasomsri, W. H. Green and Y. Román-Leshkov, *J. Catal.*, 2015, **331**, 86.
43. H. Y. Zhao, D. Li, P. Bui and S. T. Oyama, *Appl. Catal., A*, 2011, **391**, 305.
44. K. L. Li, R. J. Wang and J. X. Chen, *Energy Fuels*, 2011, **25**, 854.
45. S. K. Wu, P. C. Lai, Y. C. Lin, H. P. Wan, H. T. Lee and Y. H. Chang, *ACS Sustainable Chem. Eng.*, 2013, **1**, 349.
46. J. E. Peters, J. R. Carpenter and D. C. Dayton, *Energy Fuels*, 2015, **29**, 909.
47. M. M. Sullivan, C.-J. Chen and A. Bhan, *Catal. Sci. Technol.*, 2016, **6**, 602.
48. W.-S. Lee, Z. Wang, W. Zheng, D. G. Vlachos and A. Bhan, *Catal. Sci. Technol.*, 2014, **4**, 2340.
49. H. Y. Wang, S. D. Liu and K. J. Smith, *Energy Fuels*, 2016, **30**, 6039.
50. M. A. Gonzalez-Borja and D. E. Resasco, *Energy Fuels*, 2011, **25**, 4155.
51. J. M. Sun, A. M. Karim, H. Zhang, L. Kovarik, X. H. S. Li, A. J. Hensley, J. S. McEwen and Y. Wang, *J. Catal.*, 2013, **306**, 47.
52. F. Melligan, M. H. B. Hayes, W. Kwapinski and J. J. Leahy, *J. Anal. Appl. Pyrolysis*, 2013, **103**, 369.
53. F. Melligan, M. Hayes, W. Kwapinski and J. Leahy, *Energy Fuels*, 2012, **26**, 6080.
54. S. Thangalazhy-Gopakumar, S. Adhikari, R. B. Gupta, M. Tu and S. Taylor, *Bioresour. Technol.*, 2011, **102**, 6742.
55. O. Jan, R. Marchand, L. C. Anjos, G. V. Seufitelli, E. Nikolla and F. L. Resende, *Energy Fuels*, 2015, **29**, 1793.
56. S. Thangalazhy-Gopakumar, S. Adhikari and R. B. Gupta, *Energy Fuels*, 2012, **26**, 5300.
57. D. P. Gamliel, G. M. Bollas and J. A. Valla, *Energy Technol.*, 2016, **5**, 172–182.
58. M. W. Nolte, J. Zhang and B. H. Shanks, *Green Chem.*, 2016, **18**, 134.
59. K. Murugappan, C. Mukarakate, S. Budhi, M. Shetty, M. Nimlos and Y. Roman-Leshkov, *Green Chem.*, 2016, **18**, 5548–5557.
60. J. Rocha, S. Brown, G. Love and C. Snape, *J. Anal. Appl. Pyrolysis*, 1997, **40**, 91.
61. J. Dilcio Rocha, C. A. Luengo and C. E. Snape, *Org. Geochem.*, 1999, **30**, 1527.
62. D. C. Dayton, J. Carpenter, J. Farmer, B. Turk and R. Gupta, *Energy Fuels*, 2013, **27**, 3778.
63. H. Zhang, R. Xiao, D. Wang, G. He, S. Shao, J. Zhang and Z. Zhong, *Bioresour. Technol.*, 2011, **102**, 4258.
64. C. A. Mullen, A. A. Boateng and N. M. Goldberg, *Energy Fuels*, 2013, **27**, 3867.

65. C. A. Mullen, A. A. Boateng and N. M. Goldberg, Methods For Producing Bio-oil, US20140238835 A1, 2013.
66. P. C. Tarves, C. A. Mullen and A. A. Boateng, *ACS Sustainable Chem. Eng.*, 2016, **4**, 930.
67. A. L. M. T. Pighinelli, A. A. Boateng, C. A. Mullen and Y. Elkasabi, *Energy Sustainable Dev.*, 2014, **21**, 42.
68. M. Shetty, K. Murugappan, T. Prasomsri, W. H. Green and Y. Roman-Leshkov, *J. Catal.*, 2015, **331**, 86.
69. G. Zhou, P. A. Jensen, D. M. Le, N. O. Knudsen and A. D. Jensen, *ACS Sustainable Chem. Eng.*, 2016, **4**, 5432–5440.
70. K. Wang, O. D. Mante, J. E. Peters and D. C. Dayton, *Energy Technol.*, 2016, **5**, 183–188.
71. M. Von Holle, J. R. Carpenter and D. C. Dayton, Reactive catalytic fast pyrolysis process and system, WO2015038754 A1, 2015.

CHAPTER 6

Characterization and Separation of Bio-Oil

SHURONG WANG*[a], JUNHAO CHEN[a], FAN ZHANG[a] AND YURONG WANG[a]

[a]State Key Laboratory of Clean Energy Utilization, Zhejiang University, Hangzhou 310027, China
*E-mail: srwang@zju.edu.cn

6.1 Introduction

Bio-oil, the liquid product of fast pyrolysis, can be used as both an energy source and feedstock for chemical productions. Generally, bio-oil is a dark brown, viscous organic liquid comprised of several complex oxygenated compounds. Raw bio-oil contains viscous, water-insoluble tar, mainly the oligomeric lignin-containing fraction, which tends to spontaneously precipitate in the presence of a large amount of water.[1] In addition, due to the large amount of oxygen from various functional groups, bio-oil is thermally unstable.[2] Bio-oil that has been stored for longer than two months is more difficult to ignite and burn than fresh bio-oil.[3] Bio-oil is mainly comprised of acids, alcohols, aldehydes, esters, ketones, sugars, phenolic compounds, furans and extractable terpene with multi-functional groups. The content of some compounds is too low to be analyzed precisely.[4] Besides, there are still a certain number of oligomeric species of which the molecular weights vary from 900 to around 2500,[5] which are mainly derived from the pyrolysis of lignin.

Green Chemistry Series No. 50
Fast Pyrolysis of Biomass: Advances in Science and Technology
Edited by Robert C. Brown and Kaige Wang

Published by the Royal Society of Chemistry, www.rsc.org

High-molecular-weight species have high boiling points and viscosity, which could greatly affect the whole bio-oil properties. Apart from the high molecular weights of oxygenated compounds, the wide range of polarities also makes the separation of bio-oil quite difficult by conventional column chromatography.[1] All of these properties make the characterization, separation and utilization of bio-oil quite challenging.

An exhaustive understanding of bio-oil's physicochemical properties has not been achieved, though numerous studies were conducted and great progress has been made in revealing the corresponding properties. Knowing the physical properties of bio-oil helps to determine how it behaves in a processing facility; properties including density, water content, acidity, viscosity, stability index, *etc.* Physical characterization techniques usually follow those of the petroleum industry, such as ultimate analysis and the Karl-Fischer titration for water content determination. Chemical characterization can provide detailed information on the component distribution in bio-oil. It has been reported that analyzing bio-oil compounds distribution in chemical families is very useful, and necessary since bio-oil can be regarded as a mixture of chemical compounds with distinct functional groups.[6] This could benefit modeling bio-oil evaporation, combustion and subsequent upgrading. The common chemical characterization techniques include thermogravimetry (TG), gel permeation chromatography (GPC), gas chromatography-mass spectrometry (GC-MS), high performance liquid chromatography (HPLC), Fourier transform infrared spectroscopy (FTIR), nuclear magnetic resonance analysis (NMR), *etc.* Considering the complex composition and the targeted characterizations of bio-oil, it is often required to undergo pretreatment such as separation or derivatization.

6.2 Bio-Oil Properties

The basic physical properties of bio-oil include density, water content, pH value, total acid number, viscosity, gross calorific value, molecular weight, ultimate analysis (C, H, O, N, S), ash and solid content, thermal sensibility, combustion characteristics (flashing point, pour point, ignition feature), *etc.*

Previous analysis results showed that many common physical analysis techniques are applicable for bio-oil, such as rotational viscometry, capillary electrophoresis, *etc.*, which are still commonly used and trusted today.[7]

The water of bio-oil mainly comes from the inherent water in biomass and the subsequent reactions during pyrolysis such as dehydration. The water content of bio-oil is an important parameter which can greatly influence physical and chemical properties such as viscosity, polarity and subsequent reaction activity. It is frequently determined using chemical titrations, for example, the Karl-Fischer titration. The standard of ASTM D 1744 is often adopted. Bio-oil has a high water content varying from 10 wt% to 37 wt%.[8,9] For slow pyrolysis with a heating ramp of 15 °C min^{-1}, the water content of bio-oil could be as high as 50 wt% from the pyrolysis of softwoods, 57 wt% from hardwoods and more than 80 wt% from wheat shell.[2]

The pH value of bio-oil can be conveniently measured by a pH meter. As reported, bio-oil from lignocellulosic materials has low pH values, typically from 2.0 to 3.7,[10,11] which was attributed to a certain amount of carboxyl acids and phenolic compounds. The carboxyl acids content varies from 1.0–23.4 wt% according to the bio-oil collection system.[11] It was also found that the pH value of bio-oil increased with pyrolysis temperature.[12] In vacuum pyrolysis, the aqueous phase was separated from the bio-oil, which contributes to lower acid content.[13,14] Low pH can cause severe corrosion problems during storage, transportation and subsequent combustion. Total acid number (acid value or acidity) is determined by titrating bio-oil with potassium hydroxide (KOH), represented as milligrams of KOH to neutralize one gram of oil. There are standard methods such as ASTM D 974 and DIN 51558 (for mineral oils, biodiesel). Bio-oil from softwood bark is reported to have a total acid number (TAN) of 35 mg KOH g^{-1}[11] while bio-oil from oak bark and oak wood is 120 mg KOH g^{-1}.[10]

The viscosity of bio-oil can be measured using ASTM-445-88. Depending upon feedstock and pyrolysis operating conditions, viscosity of bio-oil ranges from 25 $mm^2\ s^{-1}$ to 1000 $mm^2\ s^{-1}$ at 40 °C, compared to 1.3–4.1 $mm^2\ s^{-1}$ of diesel.[15] The viscosity of bio-oil can increase over time due to the formation of larger-molecule compounds, and is accelerated by increasing temperature and oxygen and UV light exposure.[16]

Some compounds in bio-oil remain highly reactive, which leads to the deterioration of bio-oil after long-term storage. Polymerization of bio-oil at ambient conditions is referred to as "aging". Diebold[17] found that ten main types of chemical reactions occur during aging, including: esterification between acids and alcohols; esterification between acids and alkene; aldehydes and water forming aquo-complexes; aldehydes and alcohols forming hemiacetal or acetal compounds; self-condensation of aldehydes; polymerization of aldehydes and phenolic compounds; polymerization of organics containing sulfur; polymerization of other unsaturated compounds; and oxidation by air. Among these reactions, some are reversible reactions, during which the species distribution in bio-oil can be influenced through changing reaction conditions, while some reactions generating polymers are irreversible. Bio-oil viscosity has been reported to increase as much as 43.8% after 6 months' storage at room temperature.[18] The aging rate depends on feedstock composition, and pyrolysis conditions. However, the most important factor is temperature as it affects the rates of chemical reactions exponentially.[19] For example, viscosity of a hardwood pyrolysis oil doubled after one year of storage at room temperature, or in a week when stored at 60 °C, or in a day at 80 °C.[20] The aging rate for a hot-filtered poplar pyrolysis oil was found to be $2.317 \times 10^{13} \exp(-9659\ T^{-1})$ cP day^{-1}.[21] Furthermore, severe storage conditions can lead to an increase in water content and molecular weight in the bio-oil, followed by phase separation, generating a low-viscosity water-rich upper layer and a thick tarry bottom layer. In other studies, the gain in molecular weight was about the same for different storage regimes, whether one week at 80 °C or one year at ambient temperature.[22]

The gross calorific value of bio-oil is 13–27 MJ kg^{-1}, which is only 30–50% of that of hydrocarbon fuels[19] and is largely influenced by its water content.[11,12,23] Although water reduces its gross calorific value, it decreases its viscosity and enhances its atomization in engines.[24] Generally, bio-oil from lipid-rich feedstocks has a higher gross calorific value compared to feedstocks such as straw, wood or agricultural residues. For example, safflower seed and rapeseed cake were found to produce bio-oil with gross calorific values of 41.0 and 36.4 MJ kg^{-1} at a maximum yield of 44% and 60%, respectively.[25,26]

Bio-oil contains a large amount of carbon and oxygen at 35–63 wt% and 29–58 wt% respectively, while hydrogen, sulfur and nitrogen occupy merely 6–7 wt%, 0.02–0.3 wt% and 0.1–1 wt% respectively.[10,11,23,27] For many lignocellulosic feedstocks, the chemical composition of bio-oil can be expressed as $CH_{1.3}O_{0.47}$[21] or, including nitrogen, $CH_{1.38}O_{0.37}N_{0.002}$.[28] The ash content of bio-oil is typically about 0.04–0.5 wt%; thus, bio-oil needs to be filtered before further processing.[10] The molecular weight of bio-oil can be determined by GPC, and is found to range from 370 g mol^{-1} to 1000 g mol^{-1}.[19] The molecular weight distribution in bio-oil can be optimized by increasing the cracking severity (time–temperature relationship) so that the vaporization characteristic during combustion is improved.[24,28]

The pour point of bio-oil can be determined using ASTM D 97, the standard manual method for crude oils, or ASTM D 5949, the automatic method. The pour point of bio-oil is between −36 °C and −9 °C, rather lower than that of light fuel oil or heavy fuel oil. The flash point of bio-oil is between 50–100 °C.[19] The higher water content of bio-oil increases its ignition delay and decreases combustion rate compared to diesel fuels. In addition, the oligomer compounds are also difficult to ignite due to the high molecular weight. Bio-oil with lower average molecular weight were reported to have higher combustion rate and lower ignition delay in diesel engine tests.[29] Better combustion performance is observed for bio-oil with lower water content and lower molecular weight.[24]

6.3 Characterization of the Bio-Oil

For lignocellulosic biomass, the components of bio-oil are mainly from pyrolysis fragments of cellulose, hemicellulose and lignin. Bio-oil from cellulose pyrolysis consists mainly of levoglucosan, levoglucosenone, hydroxyacetone, glycolaldehyde, aldehyde, ketone, 5-hydroxymethyl furfural, furfural *etc.*, while hemicellulose pyrolysis can produce primarily hydroxyacetone, acids, furfural, 5-methylfurfural, levoglucosenone, and cyclopentenone. Lignin pyrolysis predominantly generates phenolic compounds such as guaiacols, syringols, phenolic compounds, benzenediols. The chemical compositions of bio-oil obtained from several different biomass feedstocks are presented in Table 6.1.[2]

Bio-oil is a very complex mixture containing hundreds of oxygen-containing organic compounds with a wide distribution of boiling points, molecular weights, polarities and solubilities. It is therefore hard to characterize bio-oil

Table 6.1 Chemical composition of the bio-oil from different biomass. (Reproduced from Fuel, **95**, M. Bertero, G. de la Puente, U. Sedran, Fuels from bio-oils: Bio-oil production from different residual sources, characterization and thermal conditioning, 263–271, Copyright (2012) with permission from Elsevier[2]).

	Pine sawdust	Mesquite sawdust	Wheat shell
Acids	18.71	16.99	18.91
Esters	8.29	7.62	8.62
Linear aldehydes and ketones	17.52	9.12	7.34
Cyclic ketones	6.11	5.77	6.06
Furans	5.95	6.25	3.44
Alcohols and sugars	9.81	8.06	6.04
Ethers	0.31	1.22	1.38
Phenolic compounds	15.74	28.23	19.02
Other oxygenated cyclic compounds	2.91	3.96	2.64
Hydrocarbons and derived	0.80	2.15	3.81
Nitrogen compounds	1.53	1.18	1.48
Unknown	11.96	11.32	21.75

by a single analytical method. One-dimensional gas chromatography coupled with mass spectroscopy is commonly used to identify volatile, low-boiling-point and low-molecular-weight compounds. According to the results of mass spectrometry identification, some compounds could be quantitatively analyzed using the internal standard method and the external standard method. The external standard method is able to identify a sample with better accuracy by drawing the concentration curve of different standards, but it is always limited by the lack of standard samples. The internal standard method is to take a pure substance as an internal standard to calculate the content of each compound, which is suitable for the detection of compounds in the absence of standard samples. There remains a series of challenges in GC analysis of complex bio-oil, such as insufficient chromatographic resolution, peak co-elution, unavailability of mass spectra of some bio-oil components in MS libraries and the lack of analytical standards. Research indicates that about only 40 wt% of bio-oil can be detected by gas chromatography, while nearly 25 wt% of high-molecular-weight compounds cannot be detected by chromatographic methods.[28]

Some improvement over one-dimensional GC is attained with two-dimensional GC × GC. In comparing the two techniques, Almeida[30] identified 81 compounds in the bio-oil and 42 in the aqueous phase using GC-MS, but 95 in the bio-oil and 68 in the aqueous phase GC × GC-MS. This improvement could be ascribed to resolving some co-eluted peaks.

Although the resolution ratio of two-dimensional GC × GC-MS has been improved, by which the number of peaks identified increases 4 to 7 times over one-dimensional GC,[31] it is still not suitable for detecting compounds with high molecular weight and low volatility, such as sugars and phenolic oligomers. It is worth noting that some compounds in bio-oil may undergo secondary reactions in packed columns leading to uncertainty of quantification in the process of temperature programming.[32] Fourier transform ion cyclotron resonance-mass spectrometry (FT-ICR-MS), which has been developed in recent years, can not only give information on molecular weight and elemental composition of compounds but also classify the detected compounds by equivalent parameters such as the number of heteroatoms, carbon number and double bonds. It is also able to identify high-molecular-weight compounds and has better resolution ratios for nitrogen- and sulfur-containing compounds in trace amounts. Smith *et al.*[33] performed the analysis of bio-oil by FT-ICR-MS, detecting over 800 bio-oil components, only 40 of which were identified by GC-MS. Liu *et al.*[34] used negative-ion electrospray ionization (ESI) FT-ICR-MS to analyze the bio-oil and its solvent fractions. The components with 4–39 carbon atoms and 2–17 oxygen atoms were identified in the molecular weight range of 150–700 g mol^{-1}. However, detailed chemical structure of bio-oil components could not be identified precisely with this method due to the lack of analytical standards.

NMR and FTIR are two common characterization methods to detect the composition of "whole" bio-oil, but they are mainly used for qualitative and semi-quantitative analysis of macro functional groups. Lu *et al.*[35] used FTIR to analyze pyrolysis bio-oil produced from rice husk and various functional groups were detected, which consisted of C=C bonds, C=C bonds, C–O bonds and even N–H bonds. FTIR was employed by Lievens *et al.*[36] to characterize carbonyl groups in bio-oil and NMR can estimate the types of functional groups and linkages. Commonly adopted one-dimensional spectra include ^{1}H, ^{13}C, ^{31}P NMR. The samples detected by ^{31}P NMR require a pre-derivatization process. For liquid NMR technology, identification needs a suitable solvent for complete dissolution, and the corresponding structure information is obtained through different signal assignment. Appropriate solvents include dimethyl sulfoxide, methanol, ethanol, and THF. However, the solvent and sample can interact with each other, and some functional groups have wide signal spans, resulting in the overlapping of certain signals. Therefore, solvent selection should be carefully taken into the consideration. Acetone and dimethyl sulfoxide are most commonly used as low cost solvents. The corresponding peaks of H and C spectra observed using the two solvents are shown in Tables 6.2 and 6.3,[37,38] respectively. The peak signal of the H spectrum is stronger, but the display window is narrow and only several typical functional groups can be identified, especially hydroxyl, which can be easily affected by deuterated solvent. The C spectrum can provide more detailed structural information, but it takes longer to obtain satisfactory results due to low abundance. The attribution of carbon in typical compounds of bio-oil can be identified through the partition of the C spectrum chemical shift, and semi-quantitative analysis of linkages and functional groups can then be conducted through the integration of peak area corresponding to different chemical shift, so as to further understand the composition of bio-oil. Sugars and phenolic compounds in bio-oil can be semi-quantitatively analyzed by combining

Table 6.2 ^{1}H NMR analysis of bio-oil. (Reproduced with permission from ref. 37, Copyright (2009) American Chemical Society, and ref. 38. Copyright (2010) American Chemical Society).

Chemical shift (ppm)[a]	Assignment	Chemical shift (ppm)[b]	Assignment
0.5–1.5	Alkanes	0–2	Aliphatic
1.5–3.0	Aliphatics α-to hetero-atom or unsaturation	2–3	$CH_2C{=}O$, aliphatic
3.0–4.4	Alcohols, methylene-dibenzene	3–4.2	Ether, methyoxyl
4.4–6.0	Methoxy, carbohydrates	4.2–6	Aliphatic OH, –C=C–, Ar–CH_2–O–
6.0–8.5	(Hetero-) aromatics	6–8.25	Aromatic, conjugated H
9.5–10.1	Aldehydes	8.25–11	CHO, ArOH
		11–12.5	COOH

[a]based on acetone solvent.[37]
[b]based on dimethylsulfoxide solvent.[38]

Table 6.3 ^{13}C NMR analysis of bio-oil. (Reproduced with permission from ref. 37, Copyright (2009) American Chemical Society, and ref. 38. Copyright (2010) American Chemical Society).

Chemical shift (ppm)[a]	Assignment	Chemical shift (ppm)[b]	Assignment
0–28	Short aliphatics	0–54	Alkyl
28–55	Long and branched aliphatics	54–70	Methoxy/hydroxy
55–95	Alcohols, ethers, phenolic-methoxys, carbohydrate sugars	70–103	Carbohydrate
95–165	Aromatics, olefins	103–163	Aromatic
165–180	Esters, carboxylic acids	163–215	Carbonyl
180–215	Ketones, aldehydes		

[a]based on acetone solvent.[37]
[b]based on dimethylsulfoxide solvent.[38]

C spectrum and element analysis to identify typical functional groups such as carbonyl and phenolic hydroxyl (combining with the NMR structure of sample after derivatization).[39,40]

One-dimensional NMR can be used to semi-quantitatively analyze bio-oil composition. Due to the complexity of bio-oil, the peaks of compounds with similar structures are close to each other, which may lead to peak overlapping. High-field NMR instruments with high signal resolution can be adapted to two-dimensional NMR analysis, which can confirm signal assignments precisely from the coupling signal chemical shift of C and H. There are many types of two-dimensional spectra, including homonuclear H–H correlation spectroscopy (COSY), heteronuclear H–C COSY (heteronuclear multiple quantum correlation (HMQC), heteronuclear singular quantum correlation (HSQC), heteronuclear multiple bond correlation (HMBC)), total correlation spectroscopy (TOCSY), *etc.* COSY directly detects and associates

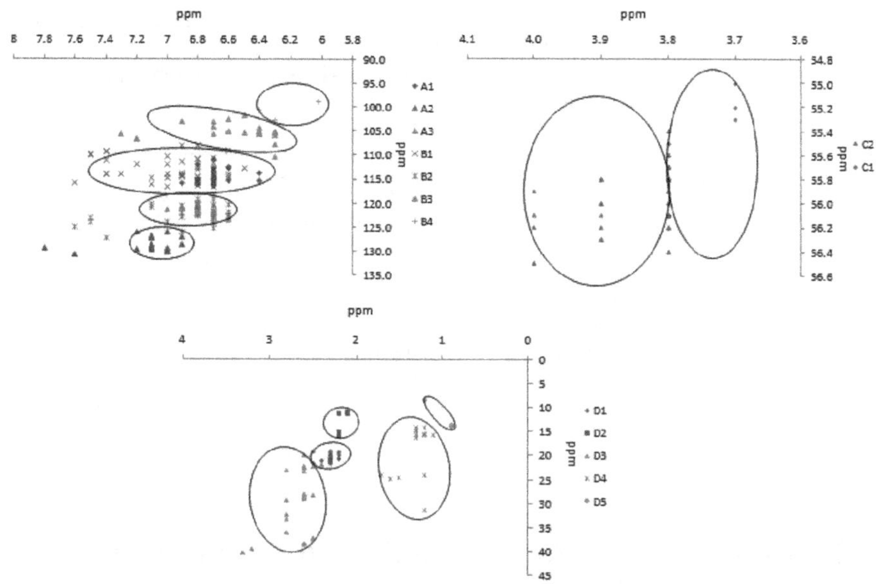

Figure 6.1 Various functional groups in bio-oil detected by 2D HSQC NMR. Reproduced with permission from ref. 41. Copyright (2011) American Chemical Society.

coupled protons, but in TOCSY, the detected protons are affected by other protons in the detection system. HMQC and HSQC simply associate protons linked directly with carbon atoms. HMBC experiments correlate protons and carbons that are related by long-range coupling interactions over 2–3 bonds, providing extremely valuable connectivity information. HMQC and HSQC are usually combined to accurately characterize functional groups and linkages. Ben *et al.*[41] used HSQC to characterize bio-oil, and presented the typical functional groups in spot distribution, especially functional groups with similar natures, as shown in Figure 6.1.[41]

Because the same type of functional groups in bio-oil exhibit in the same range of FTIR, semi-quantitative analysis can be achieved by calculating the relative content of functional groups with Gaussian peak separation algorithms according to the peaks in FTIR. Bio-oil is in the liquid phase, and therefore it is diluted using the liquid-film method combined with isopropanol, so as to obtain a linear relation between the concentration and peak area. Semi-quantitative results can be used to compare and analyze functional groups in different bio-oils, especially carbonyl functional groups in the same range. As is shown in Table 6.4, the types of carbonyl groups include aldehyde carbonyl, ketone carbonyl, carboxyl, ester groups, and carbonyl linked with a benzene ring, which are mainly derived from aldehydes, ketones, acids, esters, and lactones. High wave number is assigned to hydroxyl vibration, illustrating the existence of alcohols, phenolic compounds and water.

Table 6.4 The functional groups of bio-oil determined by the FTIR analysis. (Reproduced from Biomass and Bioenergy, 27(3), P. Das, T. Sreelatha, A. Ganesh, Bio-oil from pyrolysis of cashew nut shell-characterisation and related properties, 265–275, Copyright (2004) with permission from Elsevier[42]).

Wavenumber (cm^{-1})	Functional group	Class of compound
3600–3300	O–H stretching	Polymeric O–H, water impurities
3050–2800	C–H stretching	Alkanes
1750–1650	C=O stretching	Ketones, aldehydes, carboxylic acids
1650–1580	C=C stretching	Alkenes
1550–1490	$-NO_2$ stretching	Nitrogenous compounds
1470–1350	C–H bending	Alkanes
1300–950	C–O stretching	Primary, secondary and tertiary alcohol
915–650	O–H bending	Phenolic compounds, esters, ethers, aromatic compounds

6.4 Characterization of Bio-Oil After Separation

6.4.1 Importance of Separation in the Bio-Oil Characterization

At present, suitable technology is still not available to completely characterize bio-oil composition, and the concentrations of some compounds in bio-oil are too low to be detected. The pre-separation of bio-oil can simplify the composition of bio-oil and enrich some compounds that are suitable for certain characterization technologies, and it facilitates subsequent bio-oil upgrading and chemicals extraction. The main separation technologies include membrane separation, centrifugal separation, extraction, column chromatography, distillation, fractional condensation, *etc.*

Due to different condensation temperatures of the components in bio-oil, substances with similar condensation temperatures can be enriched by fractional condensation. Fractional condensation is mainly performed in two ways. One is recuperative direct condensation that controls the temperature of coolant. This type of fractional condensation has been extensively explored by Ellens *et al.*[43] and Brown *et al.*[44] The other is to use a temperature-controlled quench of a liquid that does not dissolve the bio-oil; the bio-oil can then be easily separated from the quench liquid by distillation.[45] In addition, bio-oil in the form of aerosols can be collected with electrostatic precipitators. For laboratory experiments, bio-oil can also be obtained by adopting dry ice or liquid nitrogen.[46] A multiple-stage condensing system has been developed to collect distinct fractions of bio-oil.[43,44,47] In the five-stage collection device designed by Pollard *et al.*,[47] the condensation temperature was reduced from 85 °C to 18 °C, together with a two-stage electrostatic precipitator (ESP) trapping device. The water content in bio-oil was reduced to only 6–15 wt% in the first four-stages while it was up to 63.3 wt% in the last stage. Levoglucosan, 5-hydroxymethyl furfural, alkyl phenol, guaiacols, syringyls and benzenediols could be collected at a condensation stage with high temperature, while water-soluble substances such as acetic acid, furfural and small-molecular-weight ketones could be enriched at a condensation stage with low temperature.

6.4.2 Membrane Separation and Centrifugation

Bio-oil is known to contain char particles with different sizes. Hydrophobic char particles are surrounded by organic compounds in the bio-oil. Agglomeration may occur, forming larger particles.[48]

Membrane separation technology can separate, purify, concentrate and refine compounds according to the permselectivity properties of the membranes. Membranes can be used under room temperature with less constituent loss, no phase change or chemical change and better selectivity. They are very applicable for the separation of thermosensitive substances. They can also be used to remove solid particles from bio-oil. Large molecular and solid particles can be removed with micron-grade membrane. When using membrane with a size of 0.5 μm for filtration, most char particles would be excluded, and ash content decreased by about 60%, while chemical compounds were retained.[49]

Centrifugal separation is often used to separate bio-oil into water and organic phases. Although bio-oil is homogenous when water content is low, stratification will occur during centrifugal separation.[50] As shown in Table 6.5,[51] the upper layer consists of an acidic water phase containing ash with a waxy surface layer. The bottom layer is rich in lignin-derived compounds.

6.4.3 Extraction and Column Chromatography

It is difficult to characterize high-molecular-weight compounds in bio-oil, such as pyrolytic lignin, especially when limited to GC. High-molecular-weight compounds are hard to vaporize and readily deposit in the GC column. Analysis of bio-oil can often be facilitated by first separating it into an aqueous phase and an organic phase. Water-soluble compounds such as carboxylic

Table 6.5 Physicochemical properties of crude bio-oil and its fractions from centrifugation. (Reproduced with permission from ref. 51, Copyright (2004) American Chemical Society).

Property	Crude bio-oil	Upper layer	Bottom layer
Water content (wt%)	13.0	3.5	14.6
Density at 28 °C (kg m^{-3})	1188	1089	1222
Kinematic viscosity (cSt)	62	88	66
pH	3.00	3.03	2.98
Ash content (wt%)	0.3	0.1	0.3
Heating value (MJ kg^{-1})	27.9	34.3	26.4
Low-molecular-weight acids (wt%)	3.53	1.77	3.96

acids, ketones, aldehydes, monophenolic compounds and sugars readily dissolve in the aqueous phase of bio-oil. High-molecular-weight and less polar compounds, such as pyrolytic lignin, that are poorly water soluble appear in the organic phase. Scholze *et al.*[52,53] and Bayerbach *et al.*[54,55] who extensively investigated the analysis of the organic phase, sometimes referred to as pyrolytic lignin, obtained by water extraction. They concluded that adding bio-oil dropwise into water under constant vigorous stirring favored the extraction of pyrolytic lignin, which accounted for 13.5–27.7 wt% of bio-oil. Subsequently, they analyzed the structure and functional groups of pyrolytic lignin with a series of chromatography and spectrum characterization methods to infer the reactions that might occur in the pyrolysis process of lignin. Due to the heterogeneity of pyrolytic lignin and the high consumption of water during the extraction process, pyrolytic lignin could be further separated by combining water extraction and organic solvent or acid and alkali solution to obtain pyrolytic lignin with different activities convenient for further analysis.[56–60]

Water extraction combined with other organic solvents with low boiling points can achieve high efficiency separation of compounds. Low polarity organic solvents, including *n*-hexane, cyclohexane, petroleum ether and toluene, are commonly used to extract hydrocarbons and benzene series with low polarity. Medium polarity organic solvents, such as dichloromethane, ethyl ether and ethyl acetate, have better extraction efficiency on monophenolic compounds; solvents with strong polarity, such as methanol, ethanol and tetrahydrofuran (THF) are preferred for their ability to completely dissolve bio-oil. Oasmaa *et al.*[57] used *n*-hexane to extract crude bio-oil and obtained 5.5 wt% of soluble substances. This was followed by water extraction to produce a water-soluble phase and insoluble phase. The water-soluble phase was further extracted using dichloromethane and ethyl ether. It was found that the extractives by dichloromethane and ethyl ether included 50 wt% aldehydes, 25 wt% ketones and 25 wt% lignin degradation products. The water-insoluble phase mainly consisted of pyrolytic lignin which accounts for 20 wt% of crude bio-oil. Dichloromethane is frequently used in bio-oil extracts as it has higher selectivity to phenolic compounds, especially when combined with other pre-separation methods. A detailed quantitative or semi-quantitative analysis of phenolic compounds can be achieved using acid and alkali solvents to control pH. Wang *et al.*[60] combined pH control with dichloromethane extraction, and obtained a phenolic compounds content of 94.33 wt%. Remarkably, the content of guaiacol compounds reached 48.27 wt%, and the concentrations of 2-ethyl-5-methyl-phenol and 2-methoxy-3-(2-propenyl)-phenol, which could not be detected in the crude bio-oil, were enriched at high pH with dichloromethane as the extraction solvent.

Supercritical fluid extraction has drawn attention because of the ability to tune its dissolution power through adjustment of pressure and temperature, allowing selective extraction of compounds. Rout *et al.*[61] and Naik *et al.*[62] explored supercritical fluid extraction of bio-oil. Oxygenated benzenoids, hydrocarbons and fatty acids with low molecular weights were mainly gathered in three extracts obtained at 45 °C and 25 MPa. The fraction obtained at 30 MPa included more hexadecanoic acid (44 wt%) and oxygenated benzenoids (16 wt%) than that obtained at 25 MPa. In consequence, valuable compounds like furans, pyran and benzene compounds could be enriched in the extraction phase at high concentrations free of water. In subsequent studies, it was found that oil fractions obtained at 10 MPa and 25 MPa contained more furans, pyrans and benzene compounds, while the oil fraction obtained at 30 MPa included more high-molecular-weight fatty acids and alcohols.

Comparing supercritical CO_2 extraction, liquid CO_2 extraction and solvent extraction of bio-oil absorbed onto SiO_2, researchers found that the components extracted by supercritical CO_2 and liquid CO_2 were similar.[63] Compounds without benzene rings, like acids, ketones, furans and esters, had better enrichment effects in CO_2 extraction fractions, while almost all the aromatic compounds in pyrolysis oil could be extracted by hexane. In contrast with the traditional extraction and chromatography methods, supercritical extraction could reach a higher extraction rate, and it had a better selectivity to low polarity compounds, whereas the water and organics with high polarity were reserved in the residue components.

Column chromatography separates substances based on different adsorption capacities on the stationary phase. Its extraction principle is similar to that of solvent extraction, and it plays an important role in the separation of medicine, food, natural products and petroleum chemical products. Column chromatography can separate compounds with similar structures or properties. There are many factors which can affect this separation method including adsorbent, samples, drench lotion, and separation analysis conditions. Commonly used absorbents include silica gel, alumina (neutral, acidic, alkaline), active carbon, macroporous resin, ion exchange resin and octadecylsilane (C18). When choosing absorbent and operating conditions, the characteristics of the target compounds should be taken into consideration. Positive chromatographic separation is usually used to realize raw or refined separation of bio-oil by selecting different solvents or solvent deployment system. Solvent types range from low polarity solvents like alkane to high polarity solvents such as alcohols and water. Five kinds of compositions can be obtained by eluting bio-oil samples on silica gel with different solvents, and the total recycle rate was up to 98.35 wt%.[64] According to the characterization results *via* one-dimensional and two-dimensional GC, it was found that acetone and methanol eluted mainly polar compounds such as sugars, ketones and phenolic compounds at very high yields. Methylbenzene and dichloromethane eluted medium polarity compounds such as hydrocarbons and alcohols with low yield. A small amount of nonpolar fatty hydrocarbons (3.5 wt%) were eluted by pentane.

As the principles of column chromatography and solvent extraction are similar, they are frequently combined to achieve high resolution of sample components. Firstly, the compounds with similar polarity and structure are obtained by solvent extraction. Then based on the polarity distribution, various solvents are adjusted into different proportions so as to further separate the extraction components. More purified compounds can be obtained with this method. Ma *et al.*[65] firstly used solvents with different polarities to extract poultry litter pyrolysis oils. As expected, viscosities declined for the low and medium polarity fractions. The main substances in *n*-hexane extractives were oleic acid and sterol. Toluene extractives contained octadecanoic acid and linoletic acid, and chloroform extractives contained levoglucosan and 1,4-dehydration xylose, while substances soluble in methanol included levoglucosan, 1,4-dehydration xylose and glucose. In the subsequent chromatography, high-purity guaiacol, syringol, levoglucosan, xylose, glucose, linoleic acid and other compounds were obtained, which greatly simplified the composition of bio-oil. The physical properties of bio-oil and its fractions are shown in Table 6.6.[65]

Solid phase microextraction is another pretreatment method of bio-oil. Bio-oil sample is initially adsorbed on certain adsorbents and then desorbed with solvents or at high temperature. Afterwards, the sample can be directly introduced into detection devices such as GC, GC-MS and liquid chromatography (LC). This method just needs a small amount of raw material and no extraction process, which is appropriate for analyzing both volatile and non-volatile substances with good reproducibility. It can effectively separate the bio-oil preliminarily to facilitate subsequent detection. The key to solid phase microextraction is selecting coatings on quartz fibers (adsorbent) which adsorb the target compounds on the coating. In order to prevent the adsorption of irrelevant compounds and solvents, the coating is selected to have similar polarity of target compounds. Combining pyrolysis and solid phase microextraction, Torri and Fabbri[66] employed solid phase column to absorb the products derived from biomass pyrolysis. Then the desorption products occurred under high temperature, which would be subsequently analyzed *via* GC-MS. As a result, quantitative analysis of volatile compounds (C1–C4) and qualitative analysis of the semi-volatile compounds were achieved.

Table 6.6 Physical properties of bio-oil and its fractions. (Reproduced from Biomass and Bioenergy, **64**, B. Ma, F. A. Agblevor, Polarity-based separation and chemical characterization of fast pyrolysis bio-oil from poultry litter, 337–247, Copyright (2014) with permission from Elsevier[65]).

Sample	Mass fraction (%)	Viscosity (Pa s at 60 °C)	Heating value (MJ kg^{-1})
Crude bio-oil	100	11.929	28.34
Hexane fraction	32.8	0.137	40.94
Toluene fraction	20.7	0.764	35.59
Chloroform fraction	34.3	1.365	26.32
Methanol fraction	10.0	Hard gel	23.31
Ethyl acetate fraction	81.8	0.514	33.25

6.4.4 Distillation

Because the boiling point span of bio-oil compounds is wide, distillation might be considered a strategy for separating components prior to analysis as is done in the petrochemical industry. However, unlike petroleum, bio-oil is thermally sensitive, thus the selection of temperature is important in traditional distillation methods.

The distillation curve presents as 'S style' at atmospheric pressure, and it could be divided into low and high temperature boiling sections, as shown in Figure 6.2.[67] The low temperature boiling section indicates the removal of water, which could effectively improve the organics concentrations. At 150 °C, the curve suddenly enters the high temperature boiling zone, and polymerization began to occur and coke formation was observed. It has been suggested that the residual coke from heating bio-oil above 240 °C could be co-pyrolyzed with biomass to enhance bio-oil yield.[68] The recovery rate of low-molecular-weight compounds achieved could be more than 80 wt%, including acetic acid, propionic acid and furfural, but newly observed compounds in distilled fractions also suggested that chemical reactions happened during the distillation process, which were not in favor of further characterization and identification.

Separation of bio-oil at a relatively low temperature can be achieved *via* vacuum distillation, which reduces the boiling point of compounds. In this case, polymerization reactions as well as the formation of high-molecular-weight compounds are suppressed to a certain degree. Similar to atmospheric

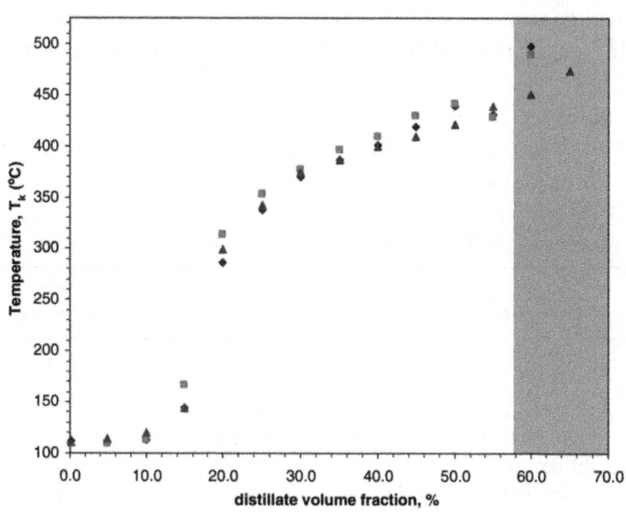

Figure 6.2 Distillation curves of bio-oil at atmospheric pressure. Reproduced from Fuel, 87(15–16), L. Starkey Ott, B. L. Smith, T. J. Bruno, Advanced distillation curve measurement: Application to a bio-derived crude oil prepared from swine manure, 3379–3387, Copyright (2008) with permission from Elsevier.[67]

distillation, benzene derivatives (about 20%) and other oxygenated compounds (about 18%) were detected in low temperature distilled fractions *via* vacuum distillation. However, the composition in medium temperature distilled fractions were quite different to the fraction obtained by atmospheric distillation, including more phenolic compounds.[69]

Molecular distillation achieves separation based on differences in mean free path length among chemical species. The mean free path length of molecular motion mainly relates to environment pressure, temperature and effective molecular diameter. At certain temperatures and pressures, molecules with enough energy can escape from the liquid surface when the liquid mixture is heated. Low-molecular-weight compounds would constantly escape from the liquid surface and then condense on the cooling surface when the corresponding mean free path length is longer than the distance between cooling and heating surfaces. High-molecular-weight compounds have shorter mean free path lengths than the distance between cooling and heating surface, so they cannot reach the cooling surface and just return to the liquid phase. In this way, effective separation of different molecules is achieved. Molecular distillation equipment has a unique structure and exhibits a series of advantages such as high vacuum degree, low operating temperature, short heating time, high separation degree and high product yield. It is quite appropriate for the separation and characterization of bio-oil with high boiling point and poor thermal stability. Refined division of bio-oil species can be realized over molecular distillation by changing the temperature, pressure and separation grades. The physical characteristics of bio-oil fractions obtained through common distillation methods are shown in Table 6.7. It is found that molecular distillation can obtain a higher yield of distillate at low temperature and high vacuum degree compared with traditional distillation.

Wang *et al.*[9] first introduced molecular distillation for bio-oil separation. Experimental results showed that low-molecular-weight compounds with strong volatility and high reactivity, such as acids, ketones and some

Table 6.7 Distillation of bio-oil with different methods.

Distillation	Feedstock	Pressure	Temperature (°C)	Yield (wt%)	Reference
Ambient distillation	Soft wood bark	0.1 MPa	140	11.7	75
	Corn stover	0.1 MPa	100 + 180[a]	65.0	69
Vacuum distillation	Corn stover	0.05 MPa	80 + 160[a]	57.1	69
	Rice husk	2 kPa	80	61.0	76
Molecular distillation	Sawdust	60 Pa	70	57.0	9
		60 Pa	100	62.4	9
		60 Pa	130	83.0	9
		1600 + 340 Pa[a]	80	43.0	70

[a]two-stage distillation.

monophenolic compounds along with water, were mainly enriched in the light distillation fraction due to their long molecular free path length, while high-molecular-weight compounds such as sugars and pyrolytic lignin were highly enriched in the heavy distillation fraction, which contained almost no water. Since molecular free path length is related not only to the diameters of compound molecules but also the corresponding environment, increasing evaporation temperature and decreasing pressure can further enrich the high-molecular-weight compounds in the light distillation fraction.[70] Therefore, light distillation fraction can be better characterized by GC, while heavy distillation fraction requires further separation to remove high-molecular-weight pyrolytic lignin, which is then characterized separately. Sugars in the pyrolytic lignin can be identified by LC and NMR.[59,71] Molecular distillation is one of the most promising bio-oil separation methods because it can achieve efficient separation on the premise of maintaining the original characteristics of bio-oil.[72–74]

6.5 Characterization of Bio-Oil After Derivatization

The characterization of bio-oil is difficult due to its complex composition as well as the inferior elution behavior of many components, resulting in poor detector response. Many researchers have explored derivatization methods to improve the detector response, peak symmetry, and peak resolution to achieve the precise qualitative and quantitative characterization of bio-oil.[77] Common derivatization methods include acetylation of phenolic compounds, trimethylsilylation of anhydrosugars, catalytic methanololysis of aldehydes, alkylation of volatile carboxylic acids, *etc.*[78]

6.5.1 Acetylation of Phenolic Compounds

In the GC analysis of bio-oil, phenolic compounds tend to give broad peaks with long tails due to their high polarity, leading to high detection limits,[79] which severely restricts the accuracy of phenol determination. Therefore, prior derivatization on bio-oil samples is required, among which acetylation is the most frequently used method in the pretreatment of samples for phenolic determination.

Acetylation of phenolic compounds is to substitute hydrogen atoms in hydroxyls with acetyl groups (CH_3–CO–) to produce corresponding esters or acetates. The most common acetylating reagent is acetic anhydride.

In a typical acetylation process, starting with the mixture of bio-oil with pyridine and acetic anhydride, the mixed solution heated in a sealed water bath followed by extraction and enrichment for further analysis. Murwanashyaka *et al.*[80,81] identified the bio-oil after acetylation with the volume proportion of acetic anhydride to pyridine at 1 : 1, and more than 30 phenolic compounds (C6–C11) were detected by GC-MS.

6.5.2 Trimethylsilylation of Sugars

Sugars have high boiling points and poor thermal stability complicating their analysis *via* chromatography. Trimethylsilylation is a widely utilized derivation method for GC-MS analysis of sugars in bio-oil. The trimethylsilyl group (–$Si(CH_3)_3$) takes the place of active hydrogen atoms in hydroxyl or carboxyl groups. This replacement reduces the restriction of hydrogen bond and the polarity of sugars, increasing their volatility and thermal stability. Furthermore, the characteristic fragments generated from the trimethylsilylation derivatives in the mass spectra help to identify the carbohydrate precursor.[82] Bistrimethylsilylacetamide (BSA), Bistrimethylsilyltrifluoroacetamide (BSTFA) Hexamethyldisilzane (HMDS), and trimethylchlorosilane (TMCS) are commonly adopted as trimethylsilylation reagents.[77]

Busetto *et al.*[40] identified sugars such as glucose and levoglucosan in crude bio-oil and hydrogenated bio-oil using GC-MS after trimethylsilylation. As developed by Fabbri *et al.*,[83] cellulose pyrolysis products were eluted with acetonitrile dissolved into the internal standard solution, trimethylchlorosilane and pyridine added, the mixture heated at 60 °C for 30 min, then diluted with acetonitrile and finally analyzed by GC-MS. The result showed that the anhydrosugars yields, including levoglucosan, levoglucosenone as well as other carbonyl compounds, were precisely quantified.

6.5.3 Other Derivatization Methods

Sipilä *et al.*[84] used catalytic alkylation to determine the content of volatile acids (C1–C6) in the water-soluble fraction of bio-oil. The acids were converted into corresponding benzylesters, which were subsequently analyzed by GC-MS. From the analysis results, formic and acetic acids were shown as the main detected acids, constituting about 80–90 wt% of the total amount of acids.

Busetto *et al.*[40] performed catalytic methanololysis of some polar compounds in the crude and hydrogenated bio-oil such as acetaldehyde, hydroxyacetaldehyde and methylglyoxal. The corresponding generated dimethyl acetals were then analyzed by GC-MS and the proportion of hydroxyacetaldehyde was determined to be as low as 0.01 wt% in the hydrogenated bio-oil (4.2 wt% in the raw bio-oil).

Some other researchers utilized derivatization with special reagents such as 2,4-DNPH (2,4-dinitrophenylhydrazine) and PFBHA (*o*-(2,3,4,5,6-pentafluorobenzyl) hydroxylamine hydrochloride), followed by GC-MS and HPLC to identify carbonyl compounds. They achieved the detection and quantification of carbonyl compounds such as formaldehyde, acetaldehyde, propionaldehyde and acetone.[85,86]

6.6 Conclusion

This chapter introduces the basic physical properties of bio-oil and characterization methods. These properties of bio-oil include density, water content, pH value, viscosity, gross calorific value, ultimate analysis, ash and solid content, thermal sensibility, and combustion characteristics. Standard detection methods and pretreatments such as fractional condensation, membrane separation, centrifugal separation, extraction, column chromatography and distillation, which can realize rational division of species and can effectively improve the identification of compounds in bio-oil, were described. Moreover, conventional characterization methods, such as GC-MS, HPLC, FTIR and NMR were discussed. In summary, most components in bio-oil could be detected by GC-MS and LC, while some complex compounds like phenolic compounds, sugars and aldehydes require appropriate separation and derivatization methods.

Acknowledgements

The authors are grateful for the financial support from the National Natural Science Foundation of China (51476142, 50676085) and the National Science and Technology Supporting Plan Through Contract (2015BAD15B06). The authors also sincerely appreciate Furong Leng, Yuan Zhao, Xiaoliu Wang, Liang Lu and Yixi Liu for their great help on literature collection, illustration arrangement and text proofreading.

References

1. Q. Song, J. Nie, M. Ren and Q. Guo, *Energy Fuels*, 2009, **23**, 3307.
2. M. Bertero, G. de la Puente and U. Sedran, *Fuel*, 2012, **95**, 263.
3. J. Zheng and Y. Kong, *Energy Convers. Manage.*, 2010, **51**, 182.
4. Y. Guo, Y. Wang, F. Wei and Y. Jin, *Chem. Ind. Eng. Prog.*, 2000, 13.
5. D. Meier and B. Scholtze, *Biomass Gasification and Pyrolysis, State of the Art and Future Prospects*, CPL Scientific, 1997, pp. 431–441.
6. T. Ba, A. Chaala, M. Garcia-Perez, D. Rodrigue and C. Roy, *Energy Fuels*, 2004, **18**, 704.
7. A. Oasmaa and C. Peacocke, *A Guide to Physical Property Characterisation of Biomass-Derived Fast Pyrolysis Liquids*, Technical Research Centre of Finland Espoo, 2001.
8. A. V. Bridgwater, S. Czernik, J. Diebold, D. Meier, A. Oasmaa, C. Peacocke, J. Piskorz and D. Radlein, *Fast Pyrolysis of Biomass: A Handbook*, CPL Scientific Publishing Services Limited, Newbury, U.K., 1999.
9. S. Wang, Y. Gu, Q. Liu, Y. Yao, Z. Guo, Z. Luo and K. Cen, *Fuel Process. Technol.*, 2009, **90**, 738.
10. L. Ingram, D. Mohan, M. Bricka, P. Steele, D. Strobel, D. Crocker, B. Mitchell, J. Mohammad, K. Cantrell and C. U. Pittman, *Energy Fuels*, 2008, **22**, 614.
11. M. Garcìa-Pérez, A. Chaala, H. Pakdel, D. Kretschmer and C. Roy, *J. Anal. Appl. Pyrolysis*, 2007, **78**, 104.
12. S. Thangalazhy-Gopakumar, S. Adhikari, H. Ravindran, R. B. Gupta, O. Fasina, M. Tu and S. D. Fernando, *Bioresour. Technol.*, 2010, **101**, 8389.
13. C. Branca, P. Giudicianni and C. Di Blasi, *Ind. Eng. Chem. Res.*, 2003, **42**, 3190.

14. C. Branca, C. D. Blasi and R. Elefante, *Ind. Eng. Chem. Res.*, 2005, **44**, 799.
15. A. Demirbas, *Energy Convers. Manage.*, 2009, **50**, 2782.
16. G. W. Huber and A. Corma, *Angew. Chem., Int. Ed.*, 2007, **46**, 7184.
17. J. P. Diebold, *A Review of the Chemical and Physical Mechanisms of the Storage Stability of Fast Pyrolysis Bio-Oils*, Citeseer, Golden, Colorado, 2000.
18. A. Chaala, T. Ba, M. Garcia-Perez and C. Roy, *Energy Fuels*, 2004, **18**, 1535.
19. A. Oasmaa and S. Czernik, *Energy Fuels*, 1999, **13**, 914.
20. A. Oasmaa, E. Leppamaki, P. Koponen, J. Levander and E. Tapola, *Physical characterisation of biomass-based pyrolysis liquids: Application of standard fuel oil analyses*, VTT Publications, 1997.
21. J. P. Diebold and S. Czernik, *Energy Fuels*, 1997, **11**, 1081.
22. M. E. Boucher, A. Chaala, H. Pakdel and C. Roy, *Biomass Bioenergy*, 2000, **19**, 351.
23. X. Guo, S. Wang, Q. Wang, Z. Guo and Z. Luo, *Chin. J. Chem. Eng.*, 2011, **19**, 116.
24. Q. Zhang, J. Chang, T. Wang and Y. Xu, *Energy Convers. Manage.*, 2007, **48**, 87.
25. S. H. Beis, Ö. Onay and Ö. M. Koçkar, *Renewable Energy*, 2002, **26**, 21.
26. D. Özçimen and F. Karaosmanoğlu, *Renewable Energy*, 2004, **29**, 779.
27. J. Piskorz, D. S. Scott and D. Radlein, *Pyrolysis Oils from Biomass*, American Chemical Society, 1988, vol. 376, pp. 167–178.
28. D. Mohan, C. U. Pittman and P. H. Steele, *Energy Fuels*, 2006, **20**, 848.
29. A. L. Shihadheh, Rural electrification from local resources: biomass pyrolysis oil combustion in a direct injection diesel engine, PhD Dissertation, Massachusetts Institute of Technology, 1999.
30. T. M. Almeida, M. D. Bispo, A. R. T. Cardoso, M. V. Migliorini, T. Schena, M. C. V. D. Campos, M. E. Machado, J. A. López, L. C. Krause and E. B. Caramão, *J. Agric. Food Chem.*, 2013, **61**, 6812.
31. N. S. Tessarolo, L. R. M. dos Santos, R. S. F. Silva and D. A. Azevedo, *J. Chromatogr. A*, 2013, **1279**, 68.
32. R. Venderbosch and W. Prins, *Biofuels, Bioprod. Biorefin.*, 2010, **4**, 178.
33. E. A. Smith, S. Park, A. T. Klein and Y. J. Lee, *Energy Fuels*, 2012, **26**, 3796.
34. Y. Liu, Q. Shi, Y. Zhang, Y. He, K. H. Chung, S. Zhao and C. Xu, *Energy Fuels*, 2012, **26**, 4532.
35. Q. Lu, X. Yang and X. Zhu, *J. Anal. Appl. Pyrolysis*, 2008, **82**, 191.
36. C. Lievens, D. Mourant, M. He, R. Gunawan, X. Li and C.-Z. Li, *Fuel*, 2011, **90**, 3417.
37. C. A. Mullen, G. D. Strahan and A. A. Boateng, *Energy Fuels*, 2009, **23**, 2707.
38. J. Joseph, C. Baker, S. Mukkamala, S. H. Beis, M. C. Wheeler, W. J. DeSisto, B. L. Jensen and B. G. Frederick, *Energy Fuels*, 2010, **24**, 5153.
39. G. Gellerstedt, J. Li, I. Eide, M. Kleinert and T. Barth, *Energy Fuels*, 2008, **22**, 4240.
40. L. Busetto, D. Fabbri, R. Mazzoni, M. Salmi, C. Torri and V. Zanotti, *Fuel*, 2011, **90**, 1197.

41. H. Ben and A. J. Ragauskas, *Energy Fuels*, 2011, **25**, 5791.
42. P. Das, T. Sreelatha and A. Ganesh, *Biomass Bioenergy*, 2004, **27**(3), 265.
43. C. J. Ellens and R. C. Brown, *Bioresour. Technol.*, 2012, **103**(1), 374.
44. R. C. Brown, S. T. Jones and A. Pollard, U.S. Patent No. 8,476,480, 2013.
45. R. J. M. Westerhof, D. W. F. Brilman, M. Garcia-Perez, Z. H. Wang, S. R. G. Oudenhoven, W. P. M. van Swaaij and S. R. A. Kersten, *Energy Fuels*, 2011, **25**, 1817.
46. A. Pattiya, *Bioresour. Technol.*, 2011, **102**, 1959.
47. A. S. Pollard, M. R. Rover and R. C. Brown, *J. Anal. Appl. Pyrolysis*, 2012, **93**, 129.
48. T. Tzanetakis, N. Ashgriz, D. James and M. Thomson, *Energy Fuels*, 2008, **22**, 2725.
49. A. Javaid, T. Ryan, G. Berg, X. Pan, T. Vispute, S. R. Bhatia, G. W. Huber and D. M. Ford, *J. Membr. Sci.*, 2010, **363**, 120.
50. N. M. Bennett, S. S. Helle and S. J. Duff, *Bioresour. Technol.*, 2009, **100**, 6059.
51. T. Ba, A. Chaala, M. Garcia-Perez and C. Roy, *Energy Fuels*, 2004, **18**, 188.
52. B. Scholze and D. Meier, *J. Anal. Appl. Pyrolysis*, 2001, **60**, 41.
53. B. Scholze, C. Hanser and D. Meier, *J. Anal. Appl. Pyrolysis*, 2001, **58**, 387.
54. R. Bayerbach, V. D. Nguyen, U. Schurr and D. Meier, *J. Anal. Appl. Pyrolysis*, 2006, **77**, 95.
55. R. Bayerbach and D. Meier, *J. Anal. Appl. Pyrolysis*, 2009, **85**, 98.
56. A. Oasmaa, E. Kuoppala, S. Gust and Y. Solantausta, *Energy Fuels*, 2003, **17**, 1.
57. A. Oasmaa, E. Kuoppala and Y. Solantausta, *Energy Fuels*, 2003, **17**, 433.
58. A. Oasmaa and E. Kuoppala, *Energy Fuels*, 2003, **17**, 1075.
59. Y. Wang, S. Wang, F. Leng, J. Chen, L. Zhu and Z. Luo, *Sep. Purif. Technol.*, 2015, **152**, 123.
60. S. Wang, Y. Wang, Q. Cai, X. Wang, H. Jin and Z. Luo, *Sep. Purif. Technol.*, 2014, **122**, 248.
61. P. Rout, M. Naik, S. Naik, V. V. Goud, L. Das and A. K. Dalai, *Energy Fuels*, 2009, **23**, 6181.
62. S. Naik, V. V. Goud, P. K. Rout and A. K. Dalai, *Bioresour. Technol.*, 2010, **101**, 7605.
63. Y. Feng and D. Meier, *Biomass Bioenergy*, 2016, **85**, 346.
64. M. E. da Cunha, J. K. Schneider, M. C. Brasil, C. A. Cardoso, L. R. Monteiro, F. L. Mendes, A. Pinho, R. A. Jacques, M. E. Machado and L. S. Freitas, *Microchem. J.*, 2013, **110**, 113.
65. B. Ma and F. A. Agblevor, *Biomass Bioenergy*, 2014, **64**, 337.
66. C. Torri and D. Fabbri, *Microchem. J.*, 2009, **93**, 133.
67. L. S. Ott, B. L. Smith and T. J. Bruno, *Fuel*, 2008, **87**, 3379.
68. X. S. Zhang, G. X. Yang, H. Jiang, W. J. Liu and H. S. Ding, *Sci. Rep.*, 2013, **3**, 1120.
69. J. A. Capunitan and S. C. Capareda, *Fuel*, 2013, **112**, 60.
70. Z. Guo, S. Wang, Y. Gu, G. Xu, X. Li and Z. Luo, *Sep. Purif. Technol.*, 2010, **76**, 52.

71. S. Wang, Y. Wang, F. Leng and J. Chen, *Energy Fuels*, 2016, **30**, 2233.
72. J. S. Kim, *Bioresour. Technol.*, 2015, **178**, 90.
73. L. Zhang, R. H. Liu, R. Z. Yin and Y. F. Mei, *Renewable Sustainable Energy Rev.*, 2013, **24**, 66.
74. S. M. Kang, X. L. Li, J. Fan and J. Chang, *Renewable Sustainable Energy Rev.*, 2013, **27**, 546.
75. M. Boucher, A. Chaala and C. Roy, *Biomass Bioenergy*, 2000, **19**, 337.
76. J. Zheng and Q. Wei, *Biomass Bioenergy*, 2011, **35**, 1804.
77. F. Orata, *Derivatization Reactions and Reagents for Gas Chromatography Analysis*, INTECH Open Access Publisher, Rijeka, 2012.
78. M. Staš, D. Kubička, J. Chudoba and M. Pospíšil, *Energy Fuels*, 2014, **28**, 385.
79. M. Saraji and M. Bakhshi, *J. Chromatogr. A*, 2005, **1098**, 30.
80. J. N. Murwanashyaka, H. Pakdel and C. Roy, *Sep. Purif. Technol.*, 2001, **24**, 155.
81. J. N. Murwanashyaka, H. Pakdel and C. Roy, *J. Anal. Appl. Pyrolysis*, 2001, **60**, 219.
82. D. Fabbri, G. Chiavari, S. Prati, I. Vassura and M. Vangelista, *Rapid Commun. Mass Spectrom.*, 2002, **16**, 2349.
83. D. Fabbri, C. Torri and V. Baravelli, *J. Anal. Appl. Pyrolysis*, 2007, **80**, 24.
84. K. Sipilä, E. Kuoppala, L. Fagernäs and A. Oasmaa, *Biomass and Bioenergy*, 1998, **14**(2), 103.
85. C. Tessini, N. Müller, C. Mardones, D. Meier, A. Berg and D. von Baer, *J. Chromatogr. A*, 2012, **1219**, 154.
86. E. D. Christensen, G. M. Chupka, J. Luecke, T. Smurthwaite, T. L. Alleman, K. Iisa, J. A. Franz, D. C. Elliott and R. L. McCormick, *Energy Fuels*, 2011, **25**, 5462.

CHAPTER 7

Role of Free Radicals in Fast Pyrolysis

KWANG HO KIM*[a], SEEMA SINGH[a], VICTORIA CUSTODIS[b] AND JEROEN VAN BOKHOVEN[b]

[a]Sandia National Laboratories, Biomass Science and Conversion Technology Department, P.O. Box 969, Livermore, California, CA 94551, USA; [b]ETH Zurich, Institute for Chemical and Bioengineering, Zurich, Switzerland
*E-mail: kwakim@sandia.gov

7.1 Introduction

Fast pyrolysis of biomass involves a complex set of concurrent parallel and series chemical reactions and ends up with a large number of pyrolysis products. Because biomass is a mixture of cellulose, hemicellulose, lignin and minor amounts of other organic substances, each component thermally decomposes at different rates and by different mechanisms and pathways.[1] Although it depends on the process conditions of the pyrolysis reaction, cellulose and hemicellulose typically decompose over narrower temperature ranges whereas lignin decomposes over wider temperature ranges.[1] It is important to study pyrolysis fundamentals because they allow for accurate process optimization as well as feedstock flexibility, both of which are critical to any recent high-throughput processes.[2] For these reasons, kinetics and reaction mechanisms for the pyrolysis of biomass have been extensively studied.[3] However, the pyrolysis of biomass proceeds *via* a very complex set of competitive and concurrent reactions and therefore, the exact mechanisms

Green Chemistry Series No. 50
Fast Pyrolysis of Biomass: Advances in Science and Technology
Edited by Robert C. Brown and Kaige Wang

Published by the Royal Society of Chemistry, www.rsc.org

for pyrolysis still remain unclear. Due to the complexity originating from structural heterogeneity of biomass, the fundamentals of biomass pyrolysis have been studied using the individual components of biomass or model compounds.

Thermal decomposition of cellulose occurs at 315–400 °C.[4] Upon pyrolysis reaction, cellulose is mostly converted into levoglucosan. The glycosidic linkages in cellulose are first cleaved to glucose and, in the next, levoglucosan is formed by the elimination of water.[5] The pyrolysis behavior of hemicellulose was found to be considerably different to cellulose.[6] First of all, the hemicellulose thermally decomposes more readily than cellulose upon pyrolysis. By thermogravimetric analysis (TGA) the weight loss of hemicellulose was found to mainly occur at 220–315 °C.[4] Decarboxylation and deacetylation appeared to be one of the preliminary steps of hemicellulose pyrolysis forming acetic acid, 2-furaldehyde and some anhydrosaccharides.[6] Unlike the relatively homogeneous structure of cellulose and hemicellulose, lignin is an amorphous heteropolymer consisting of phenylpropane units with hydroxyl, methoxyl functionality at the aromatic rings.[7,8] Different types of lignin units, in the native lignin, are linked by β-O-4, α-O-4, 4-O-5, β-5 and β-β linkages. Lignin pyrolysis is found to occur from 160–900 °C and produces mono-, di-, tri- and oligomeric phenolic compounds along with substantial amount of char. Lignin pyrolysis starts mainly from a cleavage of aryl-ether linkages, at which the bond dissociation energy is the lowest, forming monomeric phenolic compounds such as 2-methoxyphenol, 4-vinylphenol, 2-methoxy-4-vinylphenol and 2,6-dimethoxyphenol as major products. When these primary volatiles undergo secondary repolymerization, they form large molecular weight products or char.[9,10]

Previous studies have provided mechanistic insights towards biomass pyrolysis. Particularly, there have been studies on detecting free radicals in the vapor phase or condensed phase, which clearly show the evidence of homolytic cleavage reactions that generate free radicals. The free radicals likely play a crucial role in subsequent reactions. Recently, *in situ* high-temperature electron paramagnetic resonance (EPR) spectroscopy was used to detect free radicals from lignin pyrolysis.[11] Also, volatile intermediate radicals such as methoxyl, phenoxy, and substituted phenoxy radicals that evolved during lignin pyrolysis were observed.[12] EPR analysis of bio-oils produced from cellulose, lignin and whole biomass showed the presence of free radicals during pyrolysis.[13] Free radicals were found to be mainly present in the lignin fraction, which could be a combination of carbon-centered (benzyl radicals) and oxygen-centered (phenoxy radicals) organic species.[13] Interestingly, radicals observed in the condensed phase were stable in aromatic environments due to the mesomeric or resonance effect. Clearly, free radicals are generated during pyrolysis of biomass (both carbohydrates and lignin), and transformed into a stable form upon condensation to liquid.[13,14]

This chapter will review the reaction mechanisms of free radicals occurring during biomass pyrolysis. Also, the stability of radicals in bio-oil and biochar will be briefly discussed.

7.2 Free Radicals in General Organic Chemistry

7.2.1 Stability of Free Radicals

Free radical species in organic chemistry is typically classified into the following three categories:[15]

- *reactive radicals* with half-lives of less than 10^{-3} s,
- *stable radicals* with half-lives of greater than 10^{-3} s,
- *persistent radicals* mainly resulting from steric hindrance

The stability of radical species depends mainly on the environment. Electron-poor species are stabilized by the adjacent atoms that can donate electron density. For example, an increase in the number of alkyl groups on the radical center increases the stability of radical, which is called the hyperconjugation effect, as illustrated in Figure 7.1.

Another factor that contributes to the thermodynamic stability of radicals is the mesomeric effect. This mesomeric effect leads to radical stabilization by resonance, which is the principal reason for the existence of stable radicals.[15] As shown in Scheme 7.1, benzylic conjugation is an example of the mesomeric effect. The effect is more significant if there are several π orbitals energetically close to the orbital containing the radical species, which is highly expected from lignin structure. The stability of a radical also arises from steric hindrance. For example, when there is a great deal of crowding around a radical center its reactivity substantially decreases and in turn the half-life of the radical greatly increases, becoming a persistent radical.[15]

7.2.2 Elementary Radical Reactions

As illustrated in Scheme 7.2 radical reactions are generally divided into three groups, which are the formation of radicals, the transformation of one radical into another, and the disappearance of radicals.[15] When it comes to chain

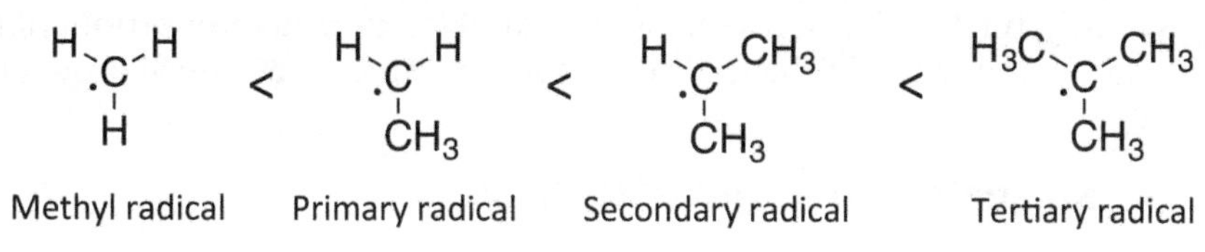

Figure 7.1 Radical stability by hyperconjugation.

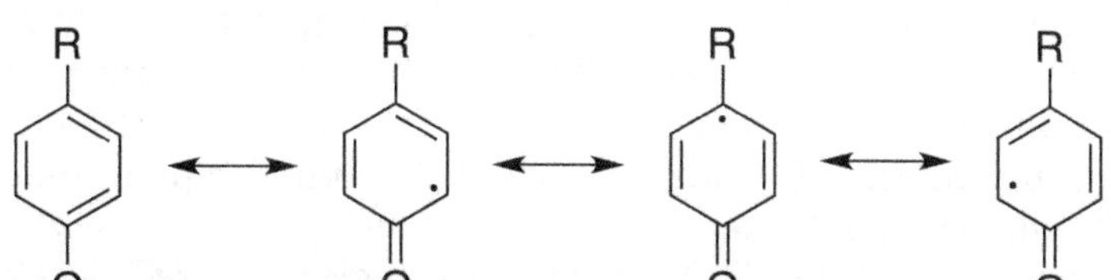

Scheme 7.1 Benzylic conjugation describing the mesomeric effect.

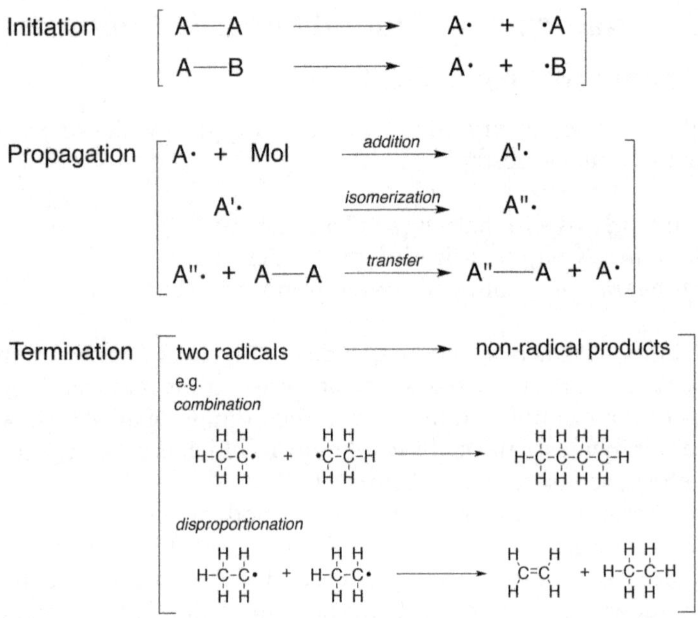

Scheme 7.2 The overall reaction mechanisms of a radical chain reaction.

mechanisms, it involves three steps: initiation, propagation and termination. The initiation phase describes the step that initially creates a radical species. For example, high temperature increases the vibrational energy of a molecule, which can lose this internal energy by collision and dispersion in the medium or by homolytic fission of one or several bonds.[15] In the propagation step, a generated free radical reacts with stable molecules to form new radicals. This step is critical because this phase is where the targeted products are formed. The first radical gives rise to one or more reactions. Propagation steps often involve addition, isomerization and hydrogen abstraction reactions. Lastly, the termination step involves dimerization, disproportionation, oxidation or reduction of the radicals to form a stable, non-radical product.

7.2.3 Magnetic Properties of Radicals

The question often arises as to whether a reaction in a complex process involves free radicals or not. The way of answering this question is dependent on whether the radical species are persistent, stable or short-lived.[15] For persistent and stable radicals, EPR spectroscopy is the most rapid and conclusive method of analysis. For short-lived radicals, *in situ* EPR detection or using a radical trapping adduct such as 5,5-dimethyl-pyrroline *N*-oxide (DMPO) could be considered.

A free radical is a molecule that contains a single, unpaired electron from the broken bond. Based on quantum mechanics, the electron spin is oriented randomly in space in the absence of an external magnetic field, which means there is no energy difference to measure. If a magnetic field is applied, however, the free electron acquires Zeeman energy with two orientations, which are parallel with the magnetic field (lower energy state) and anti-parallel with the magnetic field (higher energy state). The energies of the spin states diverge linearly as the magnetic field increases and the energy difference depends linearly on the magnetic field.[15,16] When irradiated by microwave, the majority of the electrons with an orientation of lower energy are driven to the higher energy state (from the parallel state to the anti-parallel state). This phenomenon results in the EPR absorption. EPR is very sensitive, being capable of detecting free radicals at concentrations of the order of 10^{-8} M.[15] An EPR spectrum can provide structural information about the radical species.

7.3 Free Radicals in Fast Pyrolysis of Biomass

Cellulose, hemicellulose and lignin have different thermal decomposition behavior due to their distinct inherent properties mainly attributed to structural differences. Thus, composition of biomass significantly affects reaction mechanisms during pyrolysis. Studying individual components is important to gain a comprehensive understanding of the biomass pyrolysis facilitating to establish a universal model of biomass pyrolysis. This chapter will address the free radical-related pyrolysis mechanisms of cellulose, lignin and the whole biomass.

7.3.1 Free Radicals in Lignin Pyrolysis

7.3.1.1 Study on Pyrolysis of Lignin Model Compounds

The structural complexity and inherent recalcitrance of lignin has limited the in-depth understanding of pyrolysis mechanisms. To address these challenges, pyrolysis of lignin model compounds has been widely studied. The lignin model compounds have a simple and representative structure. Typically, a model dimer linked by either a β-O-4 or α-O-4 bond (aryl-ether linkage) has been employed for understanding the mechanisms of lignin pyrolysis due to their abundance in lignin structure. Aromatic ether model compounds also have been studied.[17]

The pyrolysis of benzyl phenyl ether (BPE; α-O-4 model dimer) involves homolytic cleavage of the C–O linkage, the weakest bond in the molecule with a bond dissociation energy of 56.4 kcal mol^{-1}.[18] As shown in Scheme 7.3, homolytic cleavage of C–O forms the benzyl radical and the phenoxy radical participating in various reactions such as hydrogen abstraction reaction and termination reaction. For example, the benzyl and phenoxy radicals abstract hydrogen to form toluene and phenol, respectively. The benzyl radical is more stable than the phenoxy radical towards hydrogen abstraction because the bond dissociation energy of phenol is higher than that of toluene. As the phenoxy radical has significant unpaired-electron density delocalized into the aromatic ring, various ring-substituted products are formed from pyrolysis.[19] Benzyl radicals prefer termination reactions to form 2-benzylphenol or 4-benzylphenol. The formation of benzene, benzaldehyde and stilbene can be explained by β-scission of BPE radicals. Furthermore, chromatographic analyzes using gas chromatography and gel permeation chromatography showed the formation of large molecular weight compounds, which are formed by radical–radical coupling reactions followed by termination reactions.

Figure 7.2 shows the EPR spectrum of the condensed phase products of pyrolysis of the BPE at 500 °C. As shown, free radicals were observed in the condensed phase of BPE pyrolysis, indicating radical-involving reactions occurred. Although the analysis of condensed phase products does not provide information on highly reactive or short-lived and reactive free radicals, the observation of free radicals in the condensed phase of pyrolysis products supports the hypothesis that the pyrolysis of aryl-ether bond involves radical-generating homolytic cleavage reactions. The g-value obtained from the EPR spectrum of BPE was found to be 2.0035. The g-value determined in EPR analysis is commonly used to find the location where the radical species exist and to identify the radical structure. Considering the structure and reaction pathway of the model compound, the EPR signal is mainly associated with carbon-centered benzyl radicals and oxygen-centered phenoxy radicals.[19] It has been reported that the g-value of phenoxy radicals ranges from 2.0030 to 2.0040[20] and that of benzyl radicals is around 2.0030.[21] Although the g-value of an EPR spectrum can be used to identify radical species, the value can vary

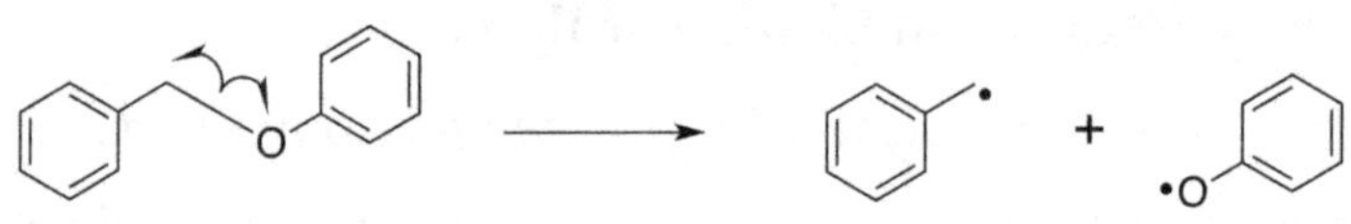

Scheme 7.3 The homolytic C–O cleavage of BPE forming benzyl radical and phenoxy radical.

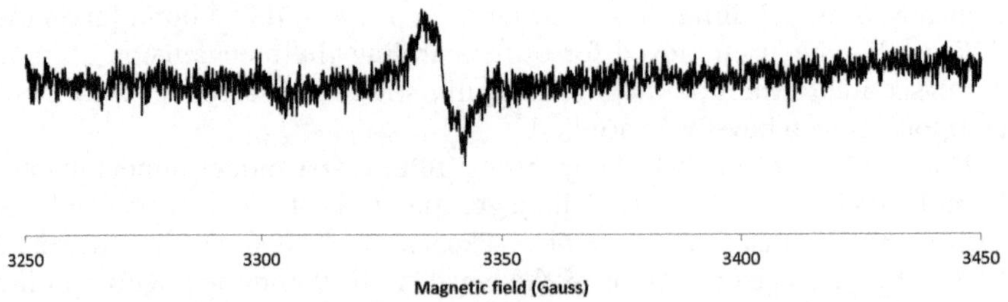

Figure 7.2 The EPR spectra of condensed-phase products from the pyrolysis of benzylphenylether (α-O-4). Reprinted from Journal of Analytical and Applied Pyrolysis, **110**, K. H. Kim, X. Bai, R. C. Brown, Pyrolysis mechanisms of methoxy substituted α-O-4 lignin dimeric model compounds and detection of free radicals using electron paramagnetic resonance analysis, 254–263, Copyright (2016) with permission from Elsevier.[19]

due to the reaction environment and only represents an integrated signal of all present species.

In addition to the α-O-4 dimer, β-O-4 model compounds have been extensively studied to understand pyrolysis mechanisms of lignin. The pyrolysis of 2-phenethyl phenyl ether (PPE), a model of the dominant β-O-4 linkage in lignin, has shown that it involves homolytic cleavage reactions.[22,23] Despite extensive studies, it is still unclear whether a homolytic unimolecular pathway is predominant over a concerted reaction mechanism. The presence of free radicals was empirically confirmed by photoionization time-of-flight mass spectrometry (PIMS).[23,24] However, computational analysis such as quantum mechanical calculation revealed that the concerted retro-ene reactions are dominant at low temperatures (<1000 °C).[23] Whereas the mechanism mentioned above is still under debate because of the lack of supporting evidence, advanced analytical techniques such as *in situ* EPR analysis of pyrolysis reactions have been developed and employed to enhance the understanding of the lignin decomposition.

Figure 7.3 shows the EPR spectra obtained from the *in situ* pyrolysis reaction in EPR using two β-O-4 dimeric model compounds; guaiacylglycerol-β-guaiacyl ether (GGE) and veratrylglycerol-β-guaiacyl ether (VGE) (Figure 7.4). As shown, both dimeric model compounds produced free radicals upon heating and the signal intensity clearly increases with increasing reaction temperature. Interestingly, the *in situ* EPR analysis shows that GGE creates more radical species and at a much lower temperature as compared to VGE. GGE, a phenolic form of dimer, showed a rapid increase in radical intensity that starts around 140 °C, whereas VGE, a blocked phenolic hydroxyl group in dimeric structure, starts around 320 °C. This means that the phenolic hydroxyl group plays an important role in thermal depolymerization of lignin. Once the radical species are produced as primary pyrolysis intermediates, they participate in a reaction network forming various products. Considering the high reactivity of free radicals, controlling the rate of radical formation is important to reduce undesirable secondary reactions. From the *in situ* radical observation of two model dimers, it is found that chemoselective blocking of phenolic hydroxyl group can suppress the radical formation during pyrolysis, which probably leads to a decrease in the formation of quinone methides. Finally, this strategy could lower the yield of undesirable solid products (char) from the lignin pyrolysis, because quinone methide is highly responsible for repolymerization reactions to large molecular weight compounds.[25,26] However, one has to consider that the observed radical reactions are highly dependent on the reaction environment. Different system pressures, for example, yield very different product selectivities. It was shown that ambient pressure during pyrolysis leads to more H-abstracting reactions as radical initiation, which then lead to recombination or rearrangement reactions of single phenols and aromatics.[17]

Another analytical method that enables *in situ* detection of radicals during pyrolysis of model compounds is molecular beam pyrolysis coupled with imaging photoelectron photoion coincidence spectroscopy (py-iPEPICO).[27,28]

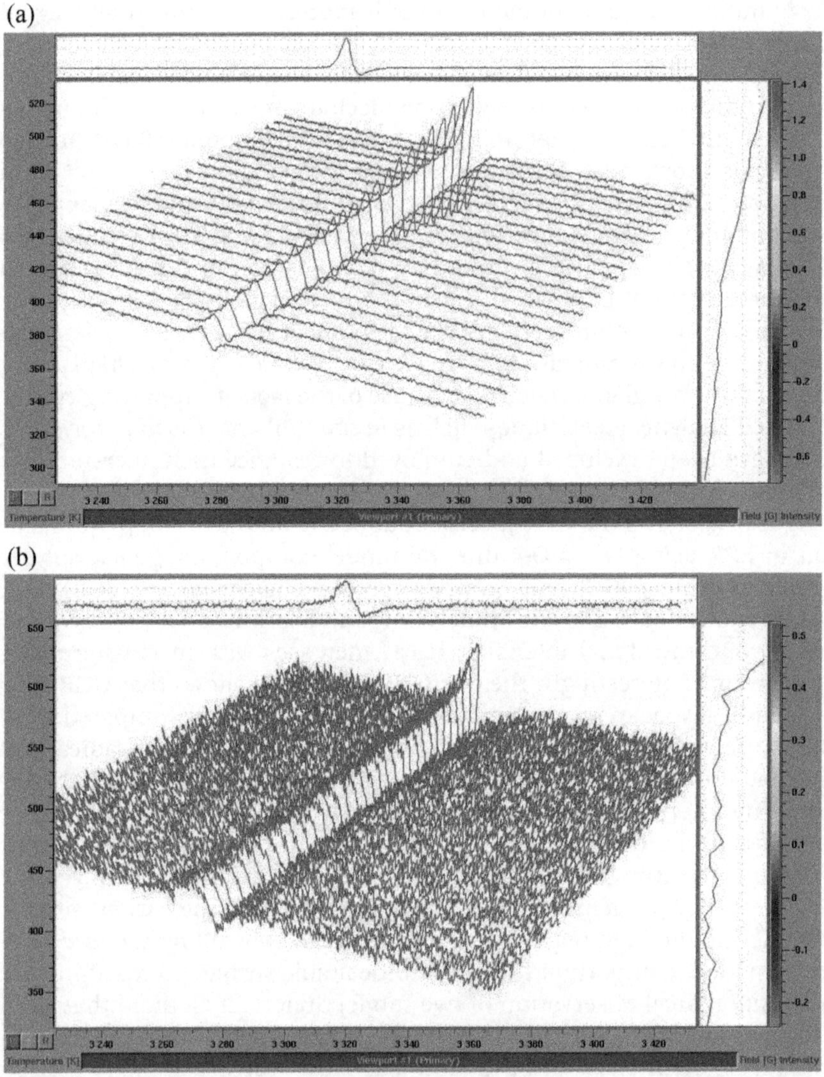

Figure 7.3 Signal intensity as a function of pyrolysis temperature. *In situ* EPR detection of reaction obtained from β-O-4 dimeric model compounds; (a) GGE and (b) VGE. (source: author of this chapter)

This technique is able to determine radicals and intermediates of pyrolysis isomer-specifically. Custodis *et al.*[17] recently studied pyrolysis mechanisms of aromatic ether model compounds using the combination of a micro-pyrolyzer coupled with iPEPICO. This iPEPICO technique allows for the measurement of photoion mass-selected threshold photoelectron spectra (ms-TPES),[29] which enable the isomer-selective identification of reactive intermediates and stable molecules. Since the analyzer is placed directly

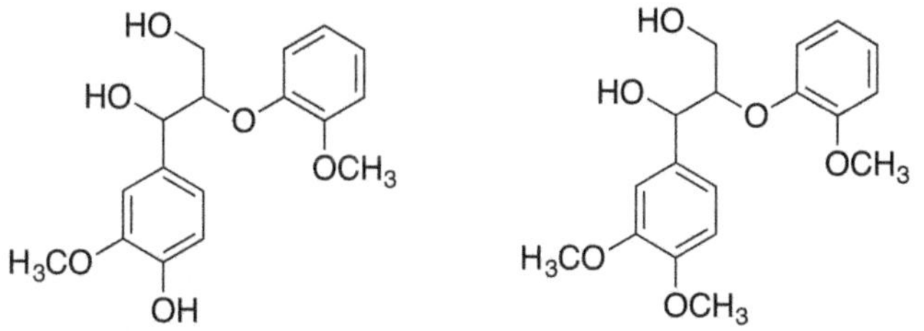

Figure 7.4 β-O-4 dimeric model compounds; Guaiacylglycerol-β-guaiacyl ether (GGE, (left)) and veratrylglycerol-β-guaiacyl ether (VGE, (right)).

behind the pyrolysis reactor connected by a skimmer, a sudden pressure drop of over eight orders of magnitude (to 10^{-6} mbar) quenches any further reactions in the detector. Diphenylether decomposes unimolecularly into phenyl and phenoxy radicals that further decarbonlyate into cyclopentadienyl radicals. Cyclopentadienyl is rarely detected during lignin pyrolysis, but readily recombines to form naphthalene.[30] At ambient pressure, however, radical formation is dominated by H-loss and recombination reactions. Substituted model compounds did not show this affinity for recombination. Guaiacol, for example, undergoes rearrangement to the corresponding benzaldehyde at ambient pressure and decarbonylation reactions at high vacuum. Comparison between the pyrolysis setup revealed the high dependence of radical formation on the system pressure and radical concentration in contrast to the unimolecular decomposition.[17,31,32] Therefore, pyrolysis studies concerning real lignin (biomass) are crucial for confirmation of the mechanism found with model compounds.

7.3.1.2 Study on Pyrolysis of Lignin and Lignin-Derived Bio-Oil

Lignin is synthesized through oxidative radical–radical coupling of monolignols, the most common of which are *p*-coumaryl, coniferyl, and sinapyl alcohols. The stable radical species have been observed in lignin structure. Figure 7.5 shows the EPR spectrum of organosolv lignin isolated from corn stover. The observation of radicals from unreacted lignin has triggered various discussions regarding the nature of the radicals in the pyrolysis products of lignin.

Thermal decomposition of lignin occurs by several competing, bond cleavage reactions at different temperatures depending on the bond dissociation energy.[33] As discussed above, pyrolysis behavior of α- and β-aryl-ether bonds (α-O-4 and β-O-4) has been frequently studied due to their prominent role in the lignin chemistry. However, research on free radicals involved in the pyrolysis of "real" lignin is important to understand some factors influencing pyrolysis in a beneficial way regarding the yield and selectivity of the products.[11] Kibet *et al.* reported the existence of gas-phase free radicals from lignin pyrolysis at 450 °C using an isothermal flow reactor in conjunction with a coldfinger-EPR system.[12] They analyzed radical species trapped at low

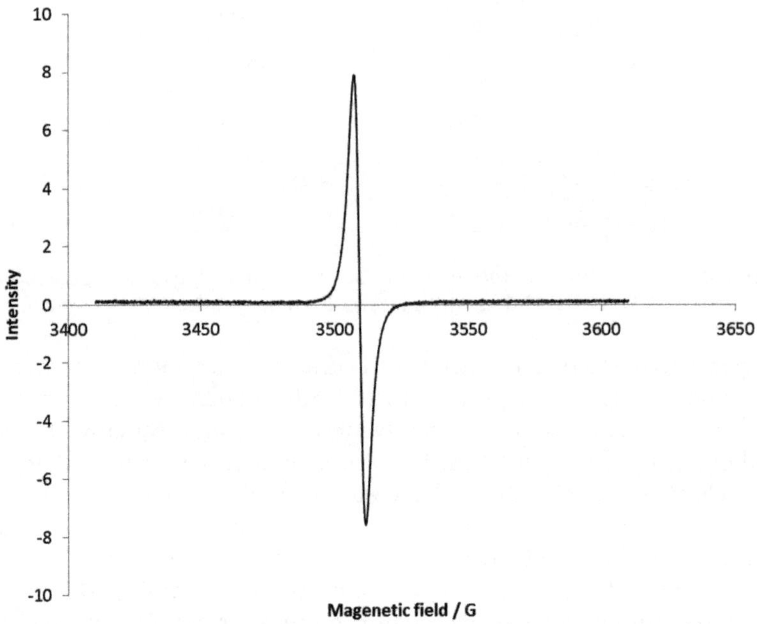

Figure 7.5 EPR spectrum of organosolv lignin isolated from corn stover. (source: author of this chapter)

temperature and found that methoxyl, phenoxy and substituted phenoxy radicals were present. More recently, an *in situ* EPR study of radical formation in the nonvolatile fraction during the lignin pyrolysis was reported.[11] In this work, hardwood lignin was found to form more radicals and volatile products than softwood lignin at temperatures ranging from 350 to 550 °C, which is explained by the different types of linkages in lignin structure and their respective bond energies. Typically, hardwood contains more β-O-4 linkages in lignin structure than softwood (60% *vs.* 40%),[34] this could lead to a higher chance of radical-related reactions during pyrolysis. Additionally, radicals formed from guaiacyl units (G-units) have a free 5-position on the benzene ring, whereas this position is occupied by a methoxyl group in syringyl units (S-units). Since softwood lignin is mainly composed of G-units, there is a higher probability of radical–radical coupling reactions forming new linkages such as 5-5 linkage, which also could lead to a reduced amount of free radicals in the pyrolysis products from softwood.[11]

Free radicals produced during the initial stage of lignin pyrolysis seem to be transformed into a stable form and they are detected in the condensed phase (bio-oil). The EPR spectrum of lignin bio-oil from the pyrolysis of organosolv lignin shows a strong singlet peak (Figure 7.6). The signal is very broad, featureless and does not show hyperfine splitting. This indicates that multiple radicals are present or strong matrix interactions occur between the products.[13,35] It is also attributed to the heterogeneity of the samples or if the radical species are not exclusively centered on a single atom.[36] Quantitative analysis of the

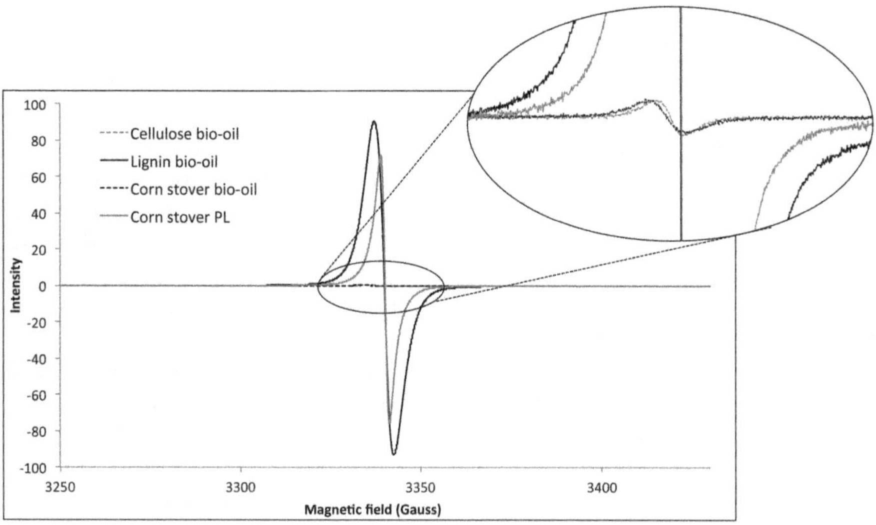

Figure 7.6 EPR spectra of bio-oil obtained from cellulose bio-oil, lignin bio-oil, corn stover bio-oil and pyrolytic lignin extracted from corn stover bio-oil. Reproduced from ref. 13 with permission from John Wiley and Sons, Copyright © 2015 WILEY-VCH Verlag GmbH & Co. KGaA, Weinheim.

condensed phase reveals that a significant amount of free radicals are present, which are very stable considering the half-life of free radicals. Lignin bio-oil had a concentration of 7.5×10^{20} spin per g bio-oil.[13] Comparing with a crystalline 2,2-dipehnyl-1-picrylhydrazyl (DPPH, $C_{18}H_{12}N_5O_6$), widely used as a representative stable radical standard with a radical concentration of 1.5×10^{21}, the amount of free radicals observed in lignin bio-oil is considerably high. The g-value of EPR signal of lignin bio-oil was reported to 2.0030.[13] In comparison, the g-value for the free radicals of asphaltene, an aromatic structure found in crude oil, ranges from 2.0028 to 2.0034[37] and free radicals observed from coal pyrolysis showed the g-value ranging from 2.0026 to 2.0035.[38] However, the g-value readily varies with the environment and interactions between products. Also, the broad and unstructured signal in EPR spectra indicates the presence of several different radicals that have slightly different g-values. Due to these reasons, it is very difficult to identify the radical structures and location. In lignin bio-oil, the free radicals are thought to be a combination of benzyl radicals, phenoxy radicals, and semiquinones based on the g-value and from the structural point of view.[35,39]

7.3.2 Study on Pyrolysis of Cellulose

Cellulose polymer is bound together *via* hydrogen bonding networks to form semicrystalline microfibrils.[40] The composition of cellulose, the β-1,4 linked linear glucose polymer (glucan), is relatively simple compared to other biomass components, which has facilitated extensive research on the mechanisms of cellulose pyrolysis. Clearly, levoglucosan is the primary direct product from cellulose pyrolysis, which is subsequently decomposed to lower molecular weight products or repolymerized into large molecular weight products, which can be carbonized to form char.[41] Despite dedicated works on cellulose pyrolysis, there are two different thermal decomposition mechanisms proposed, the homolytic reaction and the heterolytic (ionic) reaction. Some researchers reported that a concerted ionic mechanism is more favored over homolytic cleavage because it has a lower energy barrier.[42,43] Alternatively, active anhydroglucose was found to be obtained from a cellulose homolytic reaction at high temperatures (>200 °C).[41] The anhydroglucose can be transformed into levoglucosan through several elementary reactions.

Regardless of proposed pyrolysis mechanisms of cellulose, the presence of stable radicals in the condensed phase of cellulose pyrolysis was experimentally confirmed. As shown in Figure 7.6, cellulose bio-oil exhibits EPR signal although its peak height is significantly lower compared to lignin bio-oil. The observation of free radicals in the condensed phase indicates that cellulose pyrolysis is involved in homolytic cleavage reactions. The formation of furans and glycolaldehyde, the main products from cellulose pyrolysis along with levoglucosan, can be initiated by the homolytic cleavage of glycosidic bonds.[44]

The amount of free radicals in cellulose bio-oil was significantly lower (by two orders of magnitude) as compared to lignin bio-oil.[13] Currently, it is not possible to identify the accurate structure of radical species found in cellulose bio-oil. However, it is assumed that the stable radicals could be located in the oligosaccharide moiety present in cellulose bio-oil. Their stability is mainly attributed to the steric hindrance.[13]

7.3.3 Study on Pyrolysis of Biomass and Bio-Oil Analysis

Numerous studies on the pyrolysis of individual biomass components have demonstrated that pyrolysis of both cellulose and lignin involves radical-generating homolytic cleavage reactions. We discussed earlier that the frequency of homolysis in the vapor phase and the amount of radicals in the condensed phase are significantly different between cellulose and lignin. Table 7.1 shows the properties and amount of radicals in the condensed phase from pyrolysis of different biomasses and biomass components. Pyrolysis of whole biomass is expected to involve radical-generating reactions and stable radicals are eventually trapped in the condensed phases.

Recently, bio-oils produced from fast pyrolysis of corn stover[13] and loblolly pine[14] were analyzed to identify and quantify free radicals. The EPR analysis clearly reveals that the radicals are preferentially located in the lignin fraction in bio-oil, commonly called "pyrolytic lignin", regardless of the biomass origin, reactors and experimental conditions. As described above, the EPR signal intensity of pyrolytic lignin fraction extracted from corn stover bio-oil is comparable with that of bio-oil produced from corn stover derived organosolv lignin (see Figure 7.6).

Table 7.1 Properties and amount of radicals in the condensed phase from pyrolysis.

Source	Pyrolysis temperature (°C)	g-value	concentration (spin/g)	Ref.
Benzyl phenyl ether (α-O-4 dimer)	500	2.0035	1.2×10^{16}	19
Guaiacylglycerol-β-guaiacyl ether (β-O-4 dimer)	~300 (*in situ*)	2.0039	n/d	Author of this chapter
Organosolv Lignin (corn stover)	500	2.0030	7.5×10^{20}	13
Cellulose	500	2.0031	2.0×10^{18}	13
Corn stover	500	2.0028	5.5×10^{18}	13
Walnut shell	540	n/d	3.2×10^{16}	46
Corn cob	540	n/d	1.3×10^{16}	46
Pine	500	2.0026	n/d	14
Hydrolytic lignin	450	2.0071	n/d	12

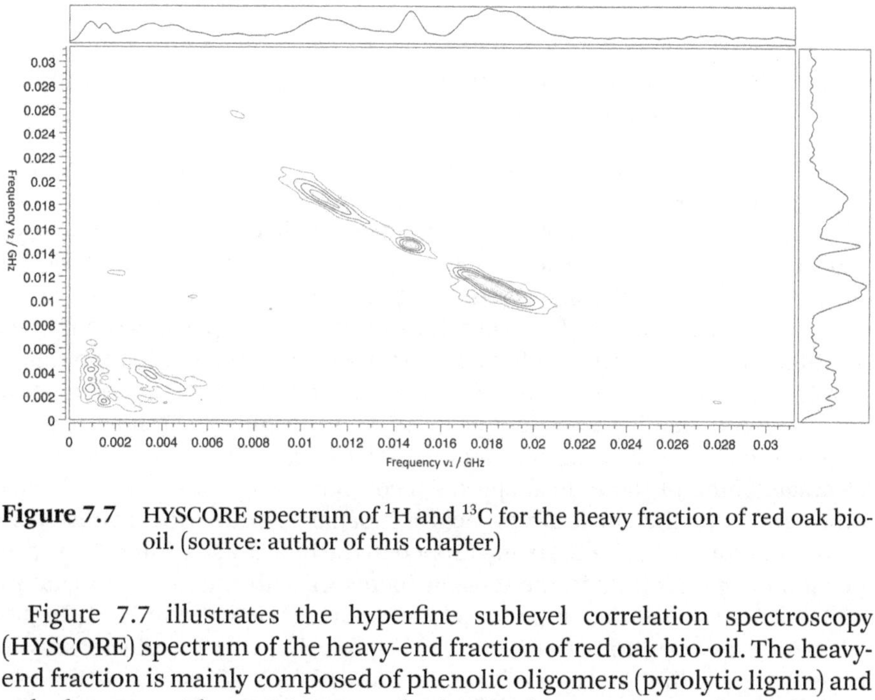

Figure 7.7 HYSCORE spectrum of ^{1}H and ^{13}C for the heavy fraction of red oak bio-oil. (source: author of this chapter)

Figure 7.7 illustrates the hyperfine sublevel correlation spectroscopy (HYSCORE) spectrum of the heavy-end fraction of red oak bio-oil. The heavy-end fraction is mainly composed of phenolic oligomers (pyrolytic lignin) and anhydrosugars. The HYSCORE method, one of the two-dimensional electron spin-echo envelope modulation spectroscopies, allows analysis of the broad hyperfine structure. As discussed above, the EPR signals of radical species observed in the bio-oil show intense and broad spectra without well-resolved hyperfine structure, which makes chemical characterization of radicals very difficult. However, the HYSCORE technique enhances the frequency resolution of the spectrum and the hyperfine structure under the broad EPR signals can be observed.[45] Figure 7.7 shows the typical spectrum obtained for radicals in the phenolic oligomers in bio-oil where the most of radicals are located. The spectrum possesses a group of strong lines grouped around the intense diagonal peak corresponding to the proton Zeeman frequency at ν_1 = 15 MHz and ν_2 = 15 MHz. For radicals in pyrolytic lignin, the broad ^{1}H hyperfine spectrum perpendicular to the diagonal $\nu_1 = \nu_2$ axis was clearly detected from 9 to 21 MHz through the matrix line, indicating that various radicals with ^{1}H hyperfine coupling constants smaller than about 10 MHz are present in phenolic oligomer. In addition to the ^{1}H hyperfine structure, ^{13}C signal of radicals clearly appeared from the HYSCORE spectrum. The series of correlation peaks around ν_1, ν_2 = 1–6 MHz are identified to the ^{13}C hyperfine spectrum showing the bending and straight spectral ridges. The straight ridge on the $\nu_1 = -\nu_2$ axis with its center at ν_1 = 3.5 MHz and ν_2 = 3.5 MHz arises from a variety of radical species with distinct hyperfine splittings. The deviation of the signal from the $\nu_1 = -\nu_2$ axis is attributed to the anisotropic hyperfine interaction with ^{13}C nuclei.[45] The main radical species in bio-oil are considered as two-dimensionally π-conjugated molecules although an in-depth investigation is required to fully characterize the radicals.

7.3.4 Detection of Free Radicals in Biochar and Their Impact on Soil Application

Previously, we discussed that free radicals are produced during pyrolysis and they are converted into stable forms, which are detectable in the condensed phase. In addition, those stable radical species are likely transformed to solid biochar during pyrolysis reactions. In this section, we discuss an overview of biochar production focusing on the reactions involving free radicals.

Biochar, a co-product of fast pyrolysis of biomass, typically accounts for 15–20 wt% of the overall pyrolysis products. Biochar obtained from biomass pyrolysis can benefit society as a carbon sequestration agent and soil amendment.[47] Upon rapid heating under an inert atmosphere, thermal degradation of biomass, including free radical processes, is initiated by homolytic cleavage. Some of the radical species recombine in various ways with each other and with the substrate to yield a biochar residue with a substantial amount of free radical fragments trapped in the biochar structure.[48] The proportion of free radicals in the biochar increases with the final temperature and reaches a maximum at pyrolysis temperatures of 500–600 °C. At higher temperatures, the number of free radicals in biochar structure decreases due to defect-annealing processes.[39] The amount of free radicals, as well as the biochar yield from pyrolysis reactions, highly depends upon heating rate, residence time, final temperature and feedstock composition. The g-value of the EPR signal obtained from biochar is similar to that of bio-oils, ranging from 2.0036 to 2.0053.[47] As previously discussed, the reported g-value suggests that a mixture of oxygen-centered radicals, such as semiquinones and phenoxy radicals, and carbon-centered radicals, such as benzylic radicals (aromatic radical) are present in biochar structure. Figure 7.8 shows proposed structures of biochar.[49]

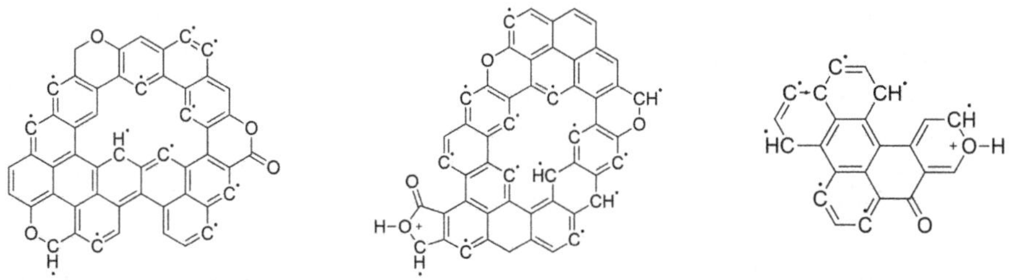

Figure 7.8 Proposed structure of biochar. Reproduced with permission from ref. 49, Copyright (2007) American Chemical Society.

Table 7.2 Properties and amount of radicals in the biochar from pyrolysis biochars.

Source	Pyrolysis temperature (°C)	g-value	concentration (spin/g)	Ref.
Corn straw	500	2.0029	3.0×10^{19}	50
Wheat straw	500	2.0029	2.9×10^{19}	50
Pine needle	550	2.0028	1.4×10^{19}	50
Corn stalk rice straw Wheat straw	200–500	2.0053 (200 °C)–2.0036 (500 °C)	n/d	47
Pectin	500	2.0023	1.9×10^{19}	51
Tobacco	500	2.0030	n/d	52

It is important to understand the effect of the stable radicals in biochar when the biochar is applied to agricultural purposes. Surprisingly, free radicals in biochar were recently found to inhibit the germination of crop seedlings and retard root and shoot growth.[47] This recent study clearly indicates that the long-lasting free radicals in biochar structure can induce reactive oxygen species in water, and they could lead to a negative effect on plant growth. The correlation between a significant free radical signal and toxic impacts was confirmed by plasma membrane tests in plant roots. In-depth research on the impact of free radicals in biochar, especially when applied to soil, must be conducted for its safe application and to derive value from biochar. Table 7.2 shows the properties and amount of radicals in the biochar from pyrolysis biochars.

7.4 Potential Role of Free Radicals in Condensed-Phase Polymerization

Recent studies on EPR analyses of bio-oil revealed that a significant amount of free radicals are present in bio-oil, especially in the lignin fraction. Also, those free radicals detected in the condensed phase are stable against aging.[13,14] The thermal stability of free radicals in the condensed phase can be determined by aging tests. Sealed bio-oil sample is simply stored at room temperature for certain period of time. Typically, the study of bio-oil thermal stability during storage is conducted under accelerated aging conditions at 80 °C for 24 hr. Interestingly, the amount of free radicals in bio-oils does not change upon aging, although the bio-oil can undergo significant repolymerization reactions.[13,14] If radicals underwent radical–radical coupling reactions to form large molecular weight products, the amount of free radicals in the bio-oil was expected to decrease by radical termination reactions. No correlation between the high degree of polymerization and concentration of free radicals in bio-oil was found, implying that the free radicals do not participate in repolymerization reactions in bio-oil.[13] In addition, the thermal behavior of radicals during aging was not affected even with the addition of radical scavengers to bio-oil.[14] These results clearly show that the radical species in the condensed phase are extremely stable towards repolymerization reactions. Instead, acid-catalyzed repolymerization in reactive functionalities such as carbonyl and vinyl groups is responsible for these secondary reactions. Figure 7.9 describes the concentration of free radicals in bio-oils from lignin, cellulose and red oak as a function of aging time at room temperature. Clearly, no significant changes are found from bio-oils after a 10 week aging test, which is consistent with the previous findings above.

As discussed, radicals trapped in the organized aromatic structure are stable due to the resonance effect. Highly delocalized radicals make the radical species stable in the aromatic matrix, which is the principal reason for the high stability of radicals. The reactive or short-lived radicals generated in the initial stage of pyrolysis participate in vapor phase reactions in a short span of time. However, once they are trapped into the highly organized aromatic matrix, the radicals are highly stable.

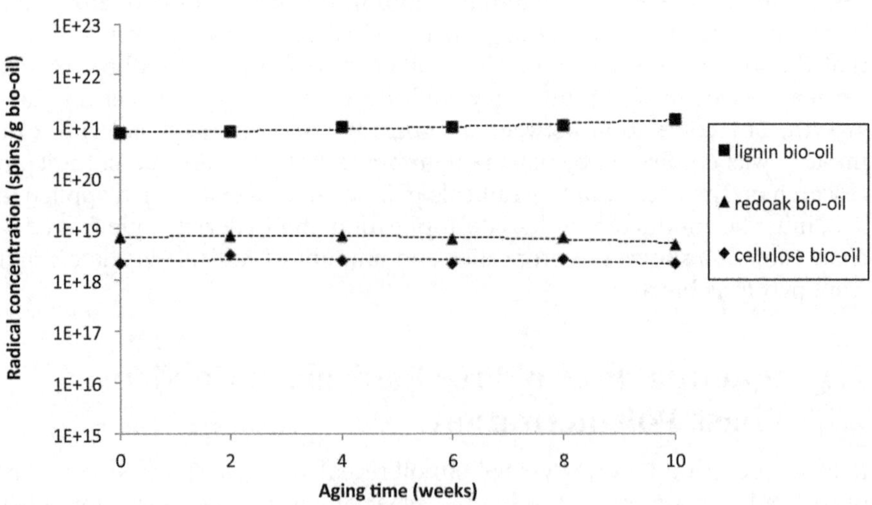

Figure 7.9 Concentration of free radicals in lignin bio-oil, cellulose bio-oil and red oak bio-oil as a function of aging time up to 10 weeks at room temperature. (source: author of this chapter)

7.5 Free Radicals in Catalytic Pyrolysis

Catalytic fast pyrolysis (CFP) is fast pyrolysis in the presence of a catalyst, which is most commonly a zeolite. It is thought that CFP is a two-step process, with the biomass first pyrolyzing to vapors that diffuse and adsorb onto the catalyst. Very likely, radicals participating in reactions over the catalyst surface are mostly, if not exclusively, generated during the pyrolysis step.

The micropores of zeolites together with the active acid sites are believed to be efficient and active in the conversion of phenolic compounds derived from the pyrolysis of the lignin component of biomass. *In situ* EPR studies on the catalyst-lignin mixture show a large increase in radical concentrations in the presence of zeolite. The stabilizing effect of zeolites on lignin radicals is already significant at a catalyst loading of 25 wt%. At relatively low temperatures (350 °C), the effect of zeolites is small, while at higher temperatures radicals in the solid reside of pyrolysis are prominent, but do not correlate to liquid yield.[53] Silicalite, which has the porous structure of ZSM-5 but has no acid sites, yields the highest radical concentration. This implies that the microporous structure stabilizes the phenoxy radicals generated during lignin pyrolysis.

Simple silica, which does not have a porous structure, has no significant effect on the radical concentration during pyrolysis. The 4- to 5-fold increase in radical concentration when zeolites are present shows that zeolites enable radical formation and lower the apparent activation energy for radical formation. Whether this effect originates from enhancing radical reactions or inhibiting radical recombination (terminating) reactions cannot be distinguished.[53]

Many ether bonds present in lignin generally lead to oxygen-based radicals, which may be stabilized and less converted over zeolites. Generally, lignin from hardwoods containing more syringol subunits may produce more stable radicals that can diffuse into zeolite micropores.[54] Experiments have shown a slight increase in liquid yield when comparing hardwood and softwood lignins extracted using the same separation method.[55] This higher yield correlates with higher conversion into benzene, toluene, and xylenes (BTX) compared to the softwoods over H-USY zeolite. The larger the degree of deoxygenation of products, the more water produced.[56,57] This means that main deoxygenation reactions remove the oxygen from phenolic compounds in the form of water, which does not contain radicals. This explains the absence of correlation between yield/BTX selectivity and radical concentration in CFP.

Yields from CFP of lignin from comparable kinds of biomass can vary by 20 wt% depending upon the techniques used to extract it. Lignin from solvent extraction using dioxane, which mainly starts with neutral radicals at ambient pressure,[55] produces many phenolic compounds and a low degree of deoxygenation, while for organosolv lignin extracted with ethanol, which showed the highest intrinsic radical concentration, deoxygenation is more complete. Catalytic pyrolysis of Kraft lignin, which is produced under very harsh temperatures and acid concentrations and contains sulfur compounds that might deactivate the zeolite catalyst, does not deoxygenate, only producing phenolic compounds.

Zeolites appear to stabilize radicals during CFP as evidenced by the large radical concentration found in the solid residue from pyrolysis. Pore size, acid site concentrations and acid site strength of zeolites can potentially impact radical activity during CFP. However, to date no correlations have been observed between catalytic properties and radicals during CFP of lignin.

7.6 Future Research and Opportunities

Despite extensive studies, the fundamental chemistry of biomass pyrolysis is still lacking and challenges remain due to the structural complexity of biomass and the multiphase nature of the process. Advanced analytical techniques such as *in situ* EPR analysis, PIMS and py-IPEPICO have revealed the presence of free radicals during pyrolysis. It appears that homolytic cleavage of ether linkages in lignin are primarily responsible for free radical generation during biomass pyrolysis. Those radical species, as primary reaction intermediates, participate in various subsequent reactions forming multiple compounds.

Considering their short half-life and high reactivity, it is important to capture free radicals in the very early stages of pyrolysis to suppress radical–radical coupling reactions that could lead to the formation of large molecular weight products. Because free radicals observed post-condensation (both in liquid and solid) are highly stable in the aromatic matrix and do not participate in secondary reactions, pretreatment of biomass prior to pyrolysis reaction or co-pyrolysis of biomass with radical scavengers or radical trapping agent should be considered. This requires improved understanding of reaction mechanisms of biomass pyrolysis. Significant research is required to identify the best way for understanding the complex pyrolysis chemistry and develop a more viable process for industrial application.

7.7 Chapter Summary

- Biomass pyrolysis involves homolytic cleavage reactions, and aryl-ether linkages in lignin structure are highly responsible for radical generation
- Free radicals participate in various reaction pathways forming multiple compounds in the very early stage of pyrolysis, then transformed into stable form when condensed
- In the condensed phase, free radicals are manly located in lignin fraction where the highly organized aromatic matrix contributes radical stability by resonance effect
- In lignin pyrolysis, the presence of catalyst, extraction process and biomass origin of lignin could affect the concentration and properties of radical

References

1. A. V. Bridgwater, D. Meier and D. Radlein, *Org. Geochem.*, 1999, **30**, 1479–1493.
2. M. S. Mettler, D. G. Vlachos and P. J. Dauenhauer, *Energy Environ. Sci.*, 2012, **5**, 7797–7809.
3. J. E. White, W. J. Catallo and B. L. Legendre, *J. Anal. Appl. Pyrolysis*, 2011, **91**, 1–33.
4. H. P. Yang, R. Yan, H. P. Chen, D. H. Lee and C. G. Zheng, *Fuel*, 2007, **86**, 1781–1788.
5. A. Demirbas, *Energy Convers. Manage.*, 2000, **41**, 633–646.
6. P. R. Patwardhan, R. C. Brown and B. H. Shanks, *ChemSusChem*, 2011, **4**, 636–643.
7. Y. S. Choi, R. Singh, J. Zhang, G. Balasubramanian, M. R. Sturgeon, R. Katahira, G. Chupka, G. T. Beckham and B. H. Shanks, *Green Chem.*, 2016, **18**, 1762–1773.
8. W. Boerjan, J. Ralph and M. Baucher, *Annu. Rev. Plant Biol.*, 2003, **54**, 519–546.
9. P. R. Patwardhan, R. C. Brown and B. H. Shanks, *ChemSusChem*, 2011, **4**, 1629–1636.
10. X. L. Bai, K. H. Kim, R. C. Brown, E. Dalluge, C. Hutchinson, Y. J. Lee and D. Dalluge, *Fuel*, 2014, **128**, 170–179.
11. C. Bahrle, V. Custodis, G. Jeschke, J. A. van Bokhoven and F. Vogel, *ChemSusChem*, 2014, 7, 2022–2029.
12. J. Kibet, L. Khachatryan and B. Dellinger, *Environ. Sci. Technol.*, 2012, **46**, 12994–13001.
13. K. H. Kim, X. L. Bai, S. Cady, P. Gable and R. C. Brown, *ChemSusChem*, 2015, **8**, 894–900.
14. J. Meng, T. I. Smirnova, X. Song, A. Moore, X. Ren, S. Kelley, S. Park and D. Tilotta, *RSC Adv.*, 2014, **4**, 29840–29846.
15. J. Fossey, D. Lefort and J. Sorba, *Free Radicals in Organic Chemistry*, Wiley, Paris, 1995.
16. G. R. Eaton, S. S. Eaton, D. P. Barr and R. T. Weber, *Quantitative Epr*, Springer Science & Business Media, 2010.
17. V. B. F. Custodis, P. Hemberger, Z. Q. Ma and J. A. van Bokhoven, *J. Phys. Chem. B*, 2014, **118**, 8524–8531.
18. R. Parthasarathi, R. A. Romero, A. Redondo and S. Gnanakaran, *J. Phys. Chem. Lett.*, 2011, **2**, 2660–2666.
19. K. H. Kim, X. Bai and R. C. Brown, *J. Anal. Appl. Pyrolysis*, 2014, **110**, 254–263.
20. E. Vejerano, S. Lomnicki and B. Dellinger, *Environ. Sci. Technol.*, 2010, **45**, 589–594.
21. N. J. Turro, V. Ramamurthy and J. C. Scaiano, *Modern Molecular Photochemistry of Organic Molecules*, University Science Books, Sausalito, 2010.
22. P. F. Britt, A. Buchanan, M. J. Cooney and D. R. Martineau, *J. Org. Chem.*, 2000, **65**, 1376–1389.

23. M. W. Jarvis, J. W. Daily, H.-H. Carstensen, A. M. Dean, S. Sharma, D. C. Dayton, D. J. Robichaud and M. R. Nimlos, *J. Phys. Chem. A*, 2011, **115**, 428–438.
24. T. He, Y. Zhang, Y. Zhu, W. Wen, Y. Pan, J. Wu and J. Wu, *Energy Fuels*, 2016, **30**, 2204–2208.
25. X. Huang, T. I. Korányi, M. D. Boot and E. J. Hensen, *ChemSusChem*, 2014, **7**, 2276–2288.
26. T. Nakamura, H. Kawamoto and S. Saka, *J. Wood Chem. Technol.*, 2007, **27**, 121–133.
27. A. Bodi, P. Hemberger, T. Gerber and B. Sztaray, *Rev. Sci. Instrum.*, 2012, **83**, 083105.
28. P. Hemberger, G. da Silva, A. J. Trevitt, T. Gerber and A. Bodi, *Phys. Chem. Chem. Phys.*, 2015, **17**, 30076–30083.
29. A. Bodi, B. Sztaray, T. Baer, M. Johnson and T. Gerber, *Rev. Sci. Instrum.*, 2007, **78**, 084102.
30. A. V. Friderichsen, E. J. Shin, R. J. Evans, M. R. Nimlos, D. C. Dayton and G. B. Ellison, *Fuel*, 2001, **80**, 1747–1755.
31. A. M. Scheer, C. Mukarakate, D. J. Robichaud, M. R. Nimlos and G. B. Ellison, *J. Phys. Chem. A*, 2011, **115**, 13381–13389.
32. A. M. Scheer, C. Mukarakate, D. J. Robichaud, M. R. Nimlos, H.-H. Carstensen and G. B. Ellison, *J. Chem. Phys.*, 2012, **136**, 044309.
33. E. Jakab, O. Faix and F. Till, *J. Anal. Appl. Pyrolysis*, 1997, **40**, 171–186.
34. E. A. Capanema, M. Y. Balakshin and J. F. Kadla, *J. Agric. Food Chem.*, 2004, **52**, 1850–1860.
35. B. Dellinger, S. Lomnicki, L. Khachatryan, Z. Maskos, R. W. Hall, J. Adounkpe, C. McFerrin and H. Truong, *Proc. Combust. Inst.*, 2007, **31**, 521–528.
36. G. L. Squadrito, R. Cueto, B. Dellinger and W. A. Pryor, *Free Radical Biol. Med.*, 2001, **31**, 1132–1138.
37. C. L. B. Guedes, E. Di Mauro, V. Antunes and A. S. Mangrich, *Mar. Chem.*, 2003, **84**, 105–112.
38. L. Petrakis and D. W. Grandy, *Fuel*, 1981, **60**, 115–119.
39. J.-W. Feng, S. Zheng and G. E. Maciel, *Energy Fuel*, 2004, **18**, 1049–1065.
40. Y. Horikawa, T. Itoh and J. Sugiyama, *Cellulose*, 2006, **13**, 309–316.
41. X. Zhang, J. Li, W. Yang and W. Blasiak, *Energy Fuel*, 2011, **25**, 3739–3746.
42. H. B. Mayes and L. J. Broadbelt, *J. Phys. Chem. A*, 2012, **116**, 7098–7106.
43. T. Hosoya, Y. Nakao, H. Sato, H. Kawamoto and S. Sakaki, *J. Phys. Chem. A*, 2009, **74**, 6891–6894.
44. M. S. Mettler, S. H. Mushrif, A. D. Paulsen, A. D. Javadekar, D. G. Vlachos and P. J. Dauenhauer, *Energy Environ. Sci.*, 2012, **5**, 5414–5424.
45. T. Ikoma, O. Ito, S. Tero-Kubota and K. Akiyama, *Energy Fuel*, 1998, **12**, 1363–1368.
46. W. J. He, Q. Y. Liu, L. Shi, Z. Y. Liu, D. H. Ci, C. Lievens, X. F. Guo and M. X. Liu, *Bioresour. Technol.*, 2014, **156**, 372–375.
47. S. Liao, B. Pan, H. Li, D. Zhang and B. Xing, *Environ. Sci. Technol.*, 2014, **48**, 8581–8587.

48. J. Lehmann and S. Joseph, *Biochar for Environmental Management: Science, Technology and Implementation*, Routledge, 2015.
49. J. Bourke, M. Manley-Harris, C. Fushimi, K. Dowaki, T. Nunoura and M. J. Antal, *Ind. Eng. Chem. Res.*, 2007, **46**, 5954–5967.
50. G. D. Fang, J. Gao, C. Liu, D. D. Dionysiou, Y. Wang and D. M. Zhou, *Environ. Sci. Technol.*, 2014, **48**, 1902–1910.
51. J. W. Feng, S. K. Zheng and G. E. Maciel, *Energy Fuel*, 2004, **18**, 560–568.
52. S. Zheng, J.-W. Feng and G. E. Maciel, *Energy Fuels*, 2005, **19**, 2247–2253.
53. C. Bährle, V. Custodis, G. Jeschke, J. A. van Bokhoven and F. Vogel, *ChemSusChem*, 2016, **9**, 2397–2403.
54. E. Dorrestijn, L. J. Laarhoven, I. W. Arends and P. Mulder, *J. Anal. Appl. Pyrolysis*, 2000, **54**, 153–192.
55. V. B. Custodis, C. Bährle, F. Vogel and J. A. van Bokhoven, *J. Anal. Appl. Pyrolysis*, 2015, **115**, 214–223.
56. J. Zakzeski, P. C. Bruijnincx, A. L. Jongerius and B. M. Weckhuysen, *Chem. Rev.*, 2010, **110**, 3552–3599.
57. R. Rinaldi, R. Jastrzebski, M. T. Clough, J. Ralph, M. Kennema, P. C. Bruijnincx and B. M. Weckhuysen, *Angew. Chem., Int. Ed.*, 2016, **55**, 8164–8215.

CHAPTER 8

Bio-Oil Stabilization

ANJA OASMAA[a]

[a]VTT Technical Research Centre of Finland, Finland, P.O.Box 1000, FI-02044 VTT, Finland
*E-mail: anja.oasmaa@vtt.fi

8.1 Introduction

Fast pyrolysis is the rapid thermal degradation of carbonaceous material under inert atmosphere. The main product from the fast pyrolysis of biomass is a highly polar liquid, commonly known as bio-oil or pyrolysis liquid.[1–3] Biomass composition is the main variable affecting the yields and composition of bio-oils. Organic liquid yields are highest for wood biomass and lowest for herbaceous biomass.[11] This is due to the different amounts of lignin, cellulose, and hemicelluloses but also due to the higher content of ash and alkali metals in herbaceous biomass compared to wood. Alkali metals catalyse pyrolysis reactions resulting in higher yields of water, char, and non-condensable gases.[4–11]

Unlike petroleum, bio-oils are highly polar and reactive, making them unstable during handling and storage.[12–16] Extended times at elevated temperatures result in a severe increase in viscosity and phase separation affecting their pumping and atomization behaviour.[12,18] New phases can be formed during the storage of bio-oils because of the changes in bio-oil composition. Even a short time at high temperatures can lead to irreversible changes in bio-oil to form separated phases.

Various methods have been developed to stabilize bio-oils. They include dilution, neutralization, esterification, use of additives, and mild hydrotreating. The stability phenomenon and stabilization methods are described in this chapter.

Green Chemistry Series No. 50
Fast Pyrolysis of Biomass: Advances in Science and Technology
Edited by Robert C. Brown and Kaige Wang

Published by the Royal Society of Chemistry, www.rsc.org

8.2 Physico-Chemical Composition of Bio-Oil

The composition of bio-oil is a function of the biomass from which it is produced[11] but also of pyrolysis process conditions, product condensation procedures, and storage conditions. Bio-oils are complex mixtures of mainly polar compounds (Table 8.1).[11,15,16,19] Wood-derived bio-oils are typically composed of 70–80 wt% water-soluble (WS) cellulose- and hemicellulose-derived oxygenated compounds, mainly aldehydes, ketones, carboxylic acids, and carbohydrate-derived sugars. The organic part of the water-insoluble (WIS) fraction of fresh bio-oil is for the most part lignin-derived oligomers.

The chemical composition of bio-oil affects its physical properties (Table 8.2).[15,16] Water content affects several properties including density, viscosity, and heating value. The high molecular weight of bio-oil contributes to its high density and viscosity while the high content of acids is responsible for its low pH and high Total Acid Number (TAN).

8.3 Aging

High amounts of reactive oxygen-containing functionalities, a low molar H/C ratio, and the presence of an acidic environment and catalytically-active alkali metals, contribute to the instability of bio-oils.[7,14–16,20] This instability, typically characterized as "aging", arises from condensation and polymerisation of reactive organic components in the bio-oil to form larger molecules that are responsible for changes in chemical composition and physical properties of the bio-oil,[16] such as an increase in average molecular weight (M_w) and viscosity.[7,13–17,20–22,24,25] Aging effects begin immediately after pyrolysis vapours are quenched to liquid, proceed rapidly during the first months of storage, and gradually slow over longer time frames.[22] Because aging is an exothermic process, heat generation is also strongest for fresh oil, decreasing gradually over time.[26] Temperature has a significant effect on the aging rate. Aging reactions speed up with temperature (Figure 8.2a); thus storage at moderate temperatures is advised.[15,16,20] Heat generation in bio-oils correlates with changes in the chemical composition and physical properties of bio-oil.[26] Alkali metals catalyse the aging reaction and may yield products with increased water content and prone to phase separation.[11,17] Carboxylic acids do not react for aging below 90 °C[23] but they provide a suitable environment for acid catalysed reactions.[20] Formic acid is claimed to be more efficient at catalysing polymerization reactions than acetic acid.[24] Besides carboxylic acids, phenolic compounds also contribute to the acidity of bio-oil.[23,24] Most of these reactions are exothermic.[26] Several bio-oil aging reactions have been proposed:[15]

- Aldehydes and ketones are known to be highly reactive with each other and with phenolics, forming oligomers, resins and water
- Carbohydrates may react with phenolic compounds forming larger molecules, or react in acidic environment steps-wise into reactive furfurals

Table 8.1 Chemical composition and oxygen distribution of bio-oil from pine sawdust and forest residue[17] "Top" is the extractives separating spontaneously from bio-oil and "bottom" is the bio-oil after extractives have been separated.

	Pine		Forest Residue			
	Bottom 97%		Bottom 89%		Top 11%	
Fast Pyrolysis of Bio-Oils	Wet Wt%	Dry Wt%	Wet Wt%	Dry Wt%	Wet Wt%	Dry Wt%
Water	*23.9*	*0*	*24.4*	*0*	*19.5*	*0*
Acids	*4.3*	*5.6*	*3.3*	*4.4*	*6.4*	*7.9*
Formic acid		1.51		1.46		
Acetic acid		3.38		7.35		6.01
Propionic acid		0.20		0.18		0.18
Glycolic acid		0.55		0.33		
Alcohols	*0.23*	*0.93*	*0*	*0*	*0.16*	*0.20*
Methanol		0.63		0		0
Ethylene glycol		0.30		0		0.20
Aldehydes, ketones, furans, pyrans	*17.4*	*22.3*	*20.4*	*27.0*	*12.8*	*15.9*
Acetaldehyde, hydroxy-		8.93		8.66		8.94
Propionaldehyde, 3-hydroxy		0.75		1.17		1.02
Hydroxypropanone (acetol)		2.84		2.55		2.25
Butanone, 1-hydroxy-2-		0.23		0.22		0.20
Butandial or Propanal		0.29		0.73		0.64
Cyclopentene-1-one, 2-hydroxy-2-		0.84		0.21		0.19
Cyclopentene-3-one, 2-hydroxy-1-methyl-1-		0.53		0.50		0.41
Furanone, 2(5H)-		0.69		0.73		0.65
Furaldehyde, 2-		0.54		0.67		0.55

Furaldehyde, 5-(hydroxymethyl)-, 2-		1.14		0.44		0.38
Pyran-4-one, 3-hydroxy-5,6-dihydro-, (4H)-		0.72		1.38		1.18
Pyran-4-one, 2-Hydroxymethyl-5-hydroxy-2,3-dihydro-, (4H)-		0.39		0.14		0.12
Sugar type compounds	*34.4*	*45.3*	*28.8*	*38.1*	*21.9*	*27.1*
Anhydro-β-D-arabino-furanose, 1,5-		0.27		0.17		0.12
Anhydro-β-D-xylofuranose, 1,5-		0		0.33		0.31
Anhydro-β-D-glucopyranose (Levoglucosan)		4.01		3.48		3.31
Dianhydro-α-D-glucopyranose, 1,4:3,6-		0.17		0.15		0.14
Cellobiosan		1.3		NA		NA
Cellotriosan		0.1		NA		NA
LMM lignin	*13.4*	*17.7*	*12.0*	*15.8*	*15.5*	*19.2*
Catechols		0.06		0.09		0.08
Lignin derived phenols		0.09		0.22		0.19
Guaiacol		0.52		0.28		0.25
Guaiacol, 4-methyl-		0.49		0.15		0.14
Guaiacol, 4-propenyl-(trans) (isoeugenol)		0.40		0.12		0.13
Vanillin		0.50		0.23		0.21
Homovallinin (Phenylacetaldehyde, 4-hydroxy-3-methoxy-)		0.27		0.09		0.09
Acetoguaiacon (Phenylethanone, 4-hydroxy-3-methoxy-)		0.22		0.09		0.08
Coniferylaldehyde		0.26		0.05		0.04
Syringol		0		0.40		0.36
Syringol, 4-methyl-		0		0.29		0.26
Syringol, 4-allyl		0		0.28		0.06
Syringol, 4-(1-Propenyl)-, trans		0		0.28		0.27
Syringaldehyde		0		0.41		0.37
Sinapaldehyde (trans)		0		0.57		0.54
HMM lignin	*1.95*	*2.6*	*4.3*	*5.6*	*7.6*	*9.5*
*Extractives**	*4.35*	*5.7*	*2.8*	*3.7*	*16.4*	*20.4*
Solids	*0.011*	*0.014*	*0.040*	*0.053*	*2.90*	*3.60*

Table 8.2 Properties of fast pyrolysis bio-oils.[3]

Property	Unit	Typical Range	Applicable test methods
Water	wt%	20–30	ASTM E203
HHV	MJ kg^{-1}	14–19	DIN51900, ASTM D240
LHV	MJ kg^{-1}	13–18	DIN51900, ASTM D240, ASTM D5291 for H
pH	-	2–3	ASTM E70
TAN (CAN + PN)	mg KOH g^{-1}	150–200	Modified ASTM D664
Kinematic viscosity at 40 °C	$mm^2\ s^{-1}$	15–40	EN ISO 3104, ASTM D445
Density at 15 °C	kg dm^{-3}	1.11–1.30	EN ISO 12185, ASTM D4052
Pour point	°C	−9 to −36	EN ISO 3016, ASTM D97
Carbon	wt% on d.b.	50–60	ASTM D5291
Hydrogen	wt% on d.b.	7–8	ASTM D5291
Nitrogen	wt% on d.b.	<0.5	ASTM D5291
Sulfur	wt% on d.b.	<0.05	EN ISO 20846, ASTM D 5453
Chlorine	ppm	<75	Colorimetric titration. No validated method
Oxygen	wt% on d.b.	35–40	As difference
Solids	wt%	<1	ASTM D7579
MCR. CCR	wt%	17–23	ASTM D4530, ASTM D189
Ash	wt%	<0.3	EN ISO 6245
Na. K. Ca. Mg	wt% on d.b.	<0.06	EN ISO 16476
Flash point	°C	40–110	EN ISO 2719, ASTM D93B
Sustained combustibility	-	Does not sustain	EN ISO 9038

- The reactivity of lignin monomers and oligomers depends on their substitution. Condensation and polymerization of reactive fragments can take place

These aging reactions result in reductions in carbonyl compounds, aldehydes, ketones, and sugars, contributing to a decrease in the volatility of bio-oil and an increase in the heavy WIS fraction, which increases the viscosity of the bio-oil.[15,16] The molecular weight and viscosity are closely correlated (Figure 8.1a). Figure 8.1b illustrates how viscosity increases with temperature. The effect of storage temperature on the viscosity and on the carbonyl content is shown in Figure 8.2.[27]

Meng *et al.*[20] studied bio-oil aging reactions using bio-oil fractions (Figure 8.3). They presented both single-fraction and multi-fraction aging pathways. They found that aging of ether insolubles (EIS), representing the sugar fraction, decreased the amount of carbonyl carbons suggesting condensation reactions similar to aldol condensation (Figure 8.3a) and that the amount of C=C carbon increased, an indication of the sugar degradation and repolymerisation path. They also reported both the acid-catalysed (Figure 8.3c) and free radical reaction paths (Figure 8.3d) of pyrolytic lignin fraction.

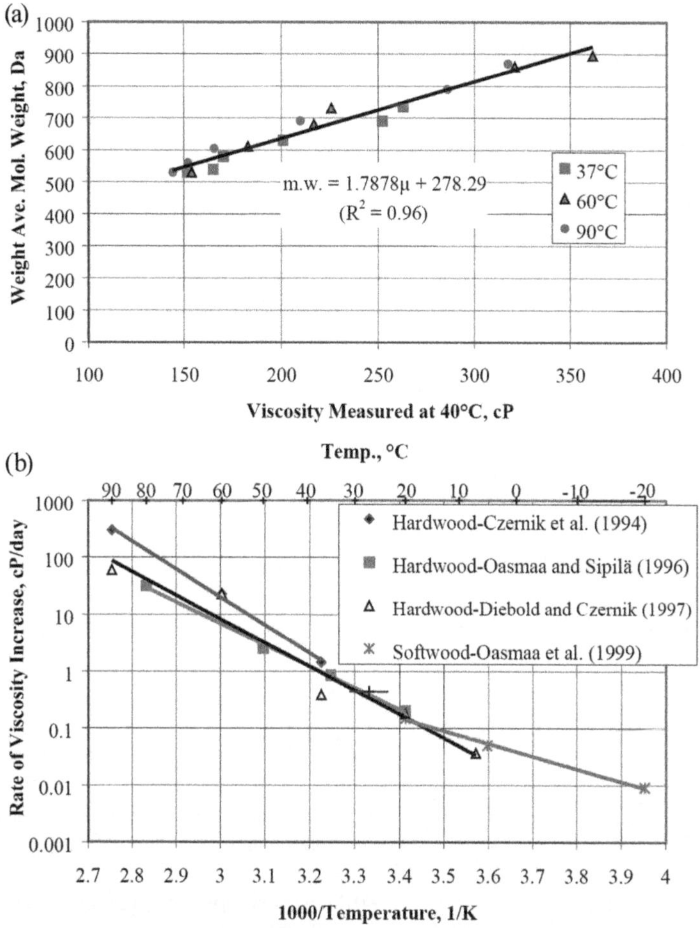

Figure 8.1 (a) Correlation of increase in molecular weight with increase in viscosity.[9] (b) Effect of temperature on viscosity increase. Reprinted with permission from ref. 15.

A microstructural characterization of bio-oils obtained from biomass was studied by Fratini *et al.*[28] using small-angle neutron scattering (SANS). Bio-oils are reported to be nanostructured fluids constituted of a complex continuous phase and nanoparticles mainly formed by the association of units of pyrolytic lignins. The aggregation of these units over time produces branched structures with a fractal dimension (Df) of between 1.4 and 1.5, which are responsible for bio-oil aging. The values obtained for the Df indicate quite an open structure, compatible with the arrangement reported on the left side of Figure 8.4. The results support the thermal ejection theory formulated by Piskorz *et al.*[1]: during the pyrolysis process, the well-packed lignin molecules contained in the feedstock undergo partial cracking and expel lignin oligomers that reassemble to form pyrolytic lignins. Recently,

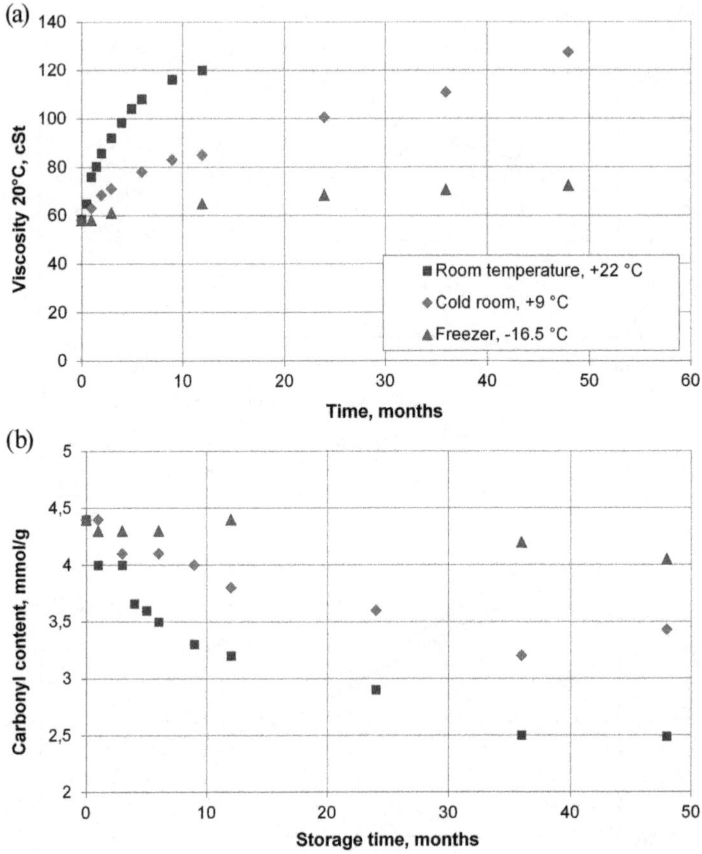

Figure 8.2 Change in (a) viscosity and (b) carbonyl content (wet basis) over time at various storage temperatures for the same pine pyrolysis liquid (water content 24 wt%). Reprinted with permission from ref. 27. Copyright (2011) American Chemical Society.

Bai *et al.* proposed another theory.[79] They present that at least a major fraction of lignin depolymerizes to monomers that recombine in the vapour phase producing oligomers that could look like the packed structure shown in the Figure 8.4. Meng *et al.*[20,29,76] reported that acid-catalysed condensation reactions may be initiated from reactive radical species in bio-oil (Figure 8.3b). The aggregation of pyrolytic lignins is a fundamental process in the aging of pyrolysis oils. One key point in increasing bio-oil stability is the slowing of polymerisation among lignin fragments.

8.4 Stability Indicators

According to Diebold[15] the aging rate of softwood bio-oil is about the same as for hardwood bio-oils at 20 °C. Czernik[7,14] stated that the molecular weight of bio-oils correlates very well with viscosity during aging at 37–90 °C. On the

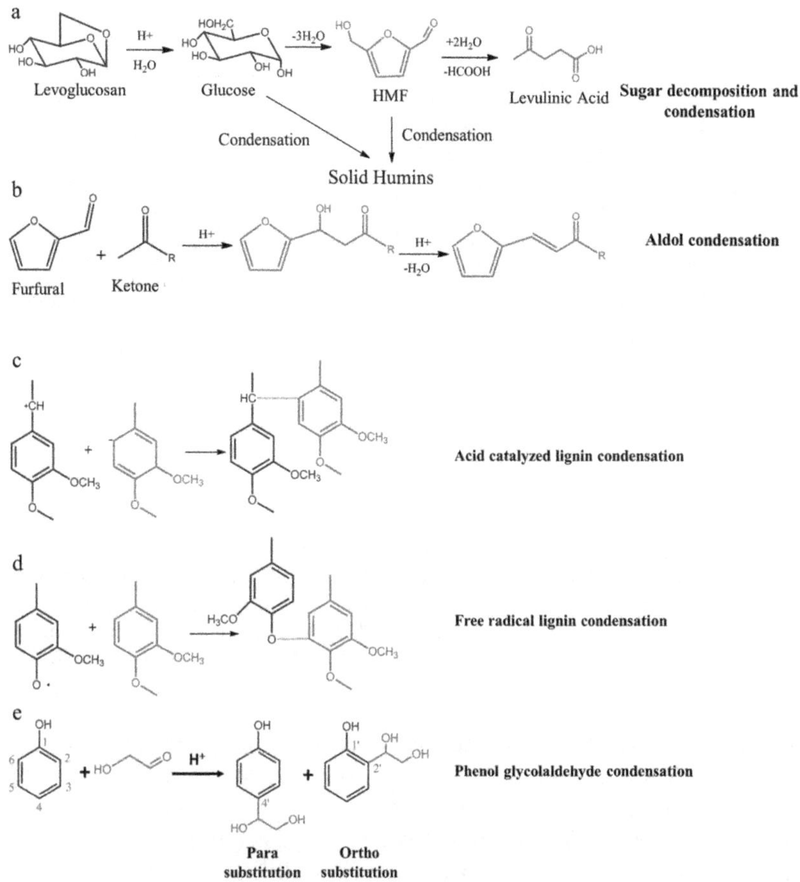

Figure 8.3 Tentative condensation of sugar and lignin fractions during bio-oil aging: (a) sugar decomposition and condensation forming humins; (b) aldol condensation of furfural and ketone; (c) acid-catalyzed lignin condensation; (d) radical initiated lignin condensation; and (e) phenol glycolaldehyde coupling. Aromatic carbon showed as C_1', C_2', and C_4' after glycolaldehyde substitution. Reprinted with permission from ref. 20. Copyright (2014) American Chemical Society.

basis of the good correlation for bio-oil aging data, relatively similar chemical reactions appear to occur over this temperature range. This is the basis for conducting accelerated aging research at elevated temperatures and then applying the results to predict storage stability of bio-oils at lower temperatures. The advantage of accelerated aging tests is the short time required to demonstrate the aging properties of a particular bio-oil.

There is no standardized method for measuring the stability of bio-oil. The main changes, increase in viscosity, WIS, and molecular weight, and reduction in carbonyl compounds can be used as stability indicators.[15–17,27] The stability test based on viscosity changes at 80 °C for 24 h is often preferred for its simplicity.[15,16] The viscosity increase in the test at 80 °C for 24 hours

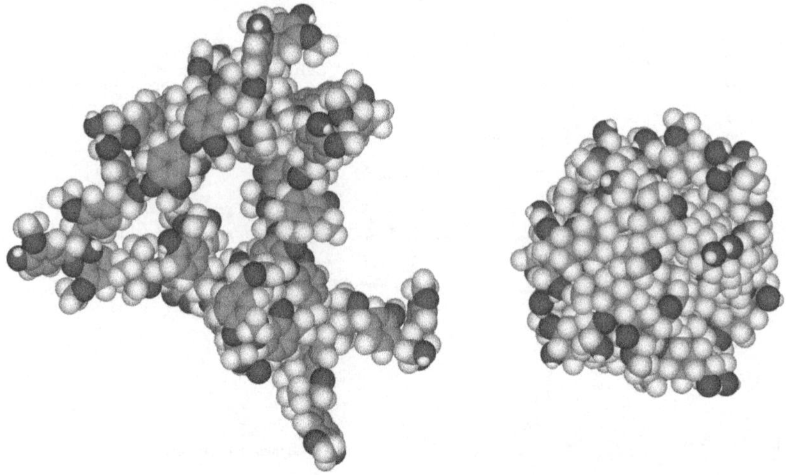

Figure 8.4 Three-dimensional structure of a globular lignin molecule (on the right) and a hypothetical pyrolytic lignin molecule characterized by the same molecular weight (M_w = 6500 g mol^{-1}) of the globular molecule, but formed by the association of 9 G-lignin tetramers (Carbon atoms: cyan; Oxygen atoms: red; Hydrogen atoms: white). Reprinted with permission from ref. 28. Copyright (2006) American Chemical Society.

correlates approximately to one year's storage at room temperature.[20,21] Several round robin experiments have been carried out to validate the viscosity increase-based stability test (24 h 80 °C).[30,31]

There is a good correlation between various stability indicators.[15,22,27] Recently, it was demonstrated that bio-oil stability indicators can also be used to estimate the heat generation potential of bio-oil in storage. In particular, the change in the concentration of carbonyl compounds exhibits a clearly linear correlation with heat generation.[26]

8.5 Phase Separation

The phase stability of bio-oil is closely related to the chemical composition and solubility distributed between its various constituents.[17,32–37] It has been suggested that, as a rule of thumb, the ratio of water, WS, and WIS should be around 25:50:25 to avoid phase separation.[38] There are two main types of phase separation that are related to the type of feedstock used to produce the bio-oil. The first type of phase separation is the formation of a waxy hydrophobic upper phase in bio-oils produced from biomass rich in extractives, like forest residues[17,19,36] or bark.[32–35] The top phase is waxy and contains less water and more solids (char) than the bottom phase.[19,32–37] The second type of phase separation is observed in bio-oils derived from wet or high-ash content lignocellulosic biomass, and is easily recognized by the formation of an upper layer with a high water content (typically 40–50 wt%) and a heavy tar-like phase at the bottom with a much lower water content (typically less than

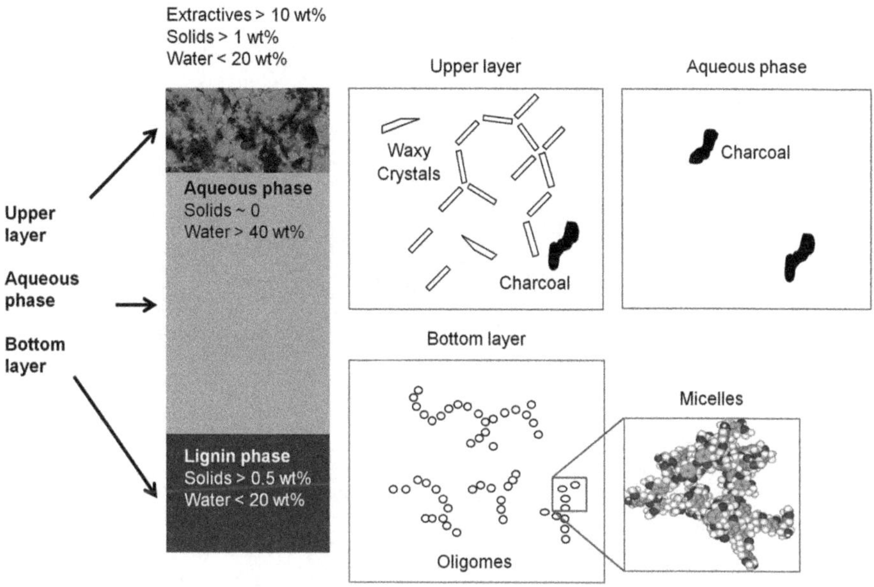

Figure 8.5 Phase separation of forest residue bio-oil. Reprinted with permission from ref. 17,28,35. Copyright (2015, 2006, 2006) American Chemical Society.

25 wt%).[19] This tar-like phase is rich in lignin-derived oligomers. Phase separation of bio-oil through the second mechanism can happen immediately after pyrolysis vapours have condensed to bio-oil if excess water is present in the biomass feedstock, or during storage because of changes in the chemical composition of the oil. The water content in fresh bio-oil can be too high because of high starting moisture of the feedstock (>10 wt%) or high produced water arising from high ash content (>1 wt%) in the feedstock, which catalyses dehydration reactions. Figure 8.5 presents all of these phases. Figure 8.6 shows the effect of high ash content on phase separation.[17]

Other influences on bio-oil water content include low volatile content (less than about 75 wt%) in the feedstock and high pyrolysis processing temperatures (greater than about 550 °C), which can increase the fragmentation of primary products. Fragmentation increases the yield of gases and reduces the content of organic compounds in the bio-oil, which causes the organic phase to become more hydrophobic because of the deoxygenation reactions. Before and after the phase separation takes place, the change in chemical composition increases the average molecular weight, viscosity, WIS fraction, and water content and decreases the amount of carbonyl compounds.

8.6 Stabilization

Instability of bio-oil is connected with the high amount of reactive oxygen functionality, low molar H/C ratio, free radicals, and presence of aging catalysts. There are several ways to improve stability of bio-oils such as removing

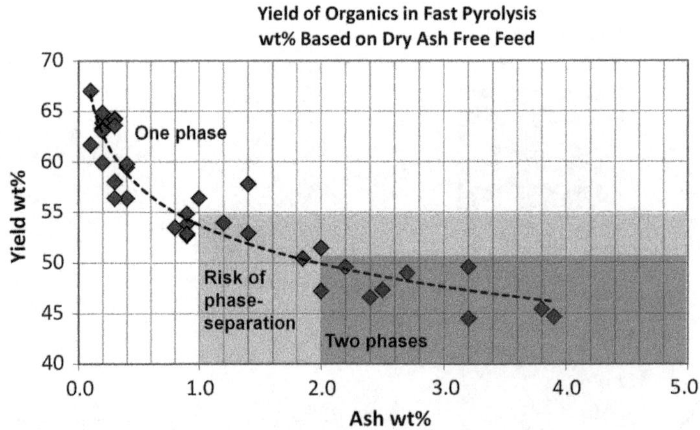

Figure 8.6 Effect of feedstock ash on phase separation tendency of bio-oil. Reprinted with permission from ref. 17. Copyright (2015) American Chemical Society.

the compounds that catalyse aging reactions or modifying the chemical composition of the bio-oil.[39] All of these techniques possess technical and/or economic challenges.

8.6.1 Removal of Alkali Metals that Catalyse Aging Reactions

Alkali metals are known to catalyse aging reactions. Alkali metals can be removed by washing the biomass feedstock, by hot vapour filtration of vapours, or by filtration and ion-exchange of fresh bio-oil.

8.6.1.1 Removal of Alkali Metals from Feedstock[40–43]

Biomass feedstock can be washed with acidic water to remove alkali metals from feedstock. After leaching, high yields of single phase bio-oils can be obtained even from high-ash-containing straw and hay. The efficiency of the washing liquor decreases in the following order: nitric acid > acetic acid > water.[41] Higher temperatures (50 °C) are more effective for removal of the inorganics. The challenge here is the large amount of washing liquor needed.

8.6.1.2 Removal of Alkali Metals by Hot Vapour Filtration[9,10,15]

Previous studies, especially at the National Renewable Energy Laboratory, showed that alkali metals play a major role in the long-term stability of bio-oils. Pyrolysis vapours have been filtered through ceramic candle filters at temperatures of around 450 °C. Improved quality bio-oil with better stability was obtained with an organic loss of about 10 wt%. Hot-vapour filtration to remove the char fines works well on a small-scale because of the various available technical solutions and easy maintenance. In continuous large-scale operations, the challenges include preventing filter fouling over long operating times, finding a durable non-reactive filter element, and minimising the loss of organics.

8.6.1.3 Removal of Alkali Metals by Post-Processing

Alkali metals are attached to char particles, which elutriate with pyrolysis vapours and aerosols from the pyrolyzer. Once the vapours are condensed to liquid the alkali metals can leach into the bio-oil. Bio-oils are not readily filtered because of their high viscosity and sticky consistency. Very limited published results[44–47] on filtering are available. Bio-oils contain sticky material which typically block the filters and forms a thick sludge in filtration and centrifugation, which leads to a high loss of organics. Parallel 900 micron rotary filters in oil cycles have been claimed[45] to catch the largest particles, if a process disturbance occurs. On-line centrifugation[44] has also been accomplished, however with lower separation efficiencies. The filtration efficiency is better for fresh bio-oil than aged bio-oil. The addition of alcohol, such as 2–5% isopropylalcohol, improves filtration efficiency (mass loss of 2–3% and separation efficiency of 65–90%). Microfiltration of bio-oil (tubular ceramic membranes with nominal pore sizes of 0.5 and 0.8 μm, 38 to 45 °C, three different trans-membrane pressures) has been claimed[45,46] to be efficient for the removal of the solid particles. Membrane cleaning was achieved with methanol, sodium hydroxide, and acetic acid. However, this new filtration method is expensive, is likely to have a low throughput, and is at an early stage of development.

Ion exchange is a common process used to change the ionic form of the material. The ion-exchange of bio-oil has been developed *e.g.* at UOP and NREL;[48,49] these studies reported significantly lower metal levels in bio-oil.

8.6.2 Torrefaction

The goal of biomass torrefaction is to degrade hemicelluloses for the purpose of lowering yields of acids and furfural during pyrolysis.[29,50,51] The physicochemical characteristics of biomass change significantly, especially at 250 °C. The dehydration of hydroxyls, dissociation of *O*-acetyl branches, and breakage of ether linkages are the main reactions. Meng *et al.*[29] suggested that aggregation of pyrolytic lignin to form large droplets may be the dominant factor for the rapid increase in viscosity observed during aging. It was also discovered that bio-oil viscosity is more strongly correlated to the WIS content than the molecular weight of the bio-oil. In addition, both accelerated and natural aging methods revealed that bio-oils made from torrefied wood had more stable chemical compositions and phase behavior than the raw bio-oil. These results suggest that pretreating the pyrolysis feedstock for bio-oil production could potentially slow down the rate of aging. However, this advantage is off-set by the lower organic liquid yields and energy recovery from the pretreated biomass.

8.6.3 Effect of Solvent Addition

Addition of solvents, like alcohols, is known to improve the homogeneity, and stability (Figure 8.7) by decreasing the viscosity increase during aging of bio-oil.[15,52,53] Sundqvist *et al.*[26] showed that addition of alcohol also had a limited but clear effect on heat generation (Figure 8.8).

8.6.4 Esterification, Acetalization

In esterification, alcohol is added to bio-oil in the presence of a catalyst at moderate temperatures (50–170 °C) to convert carboxylic acids into esters.[54,57] In acetalization, carbonyl groups are converted into acetal groups. There has been some work on simultaneous esterification and acetalization of aldehydes.[55] Also esterification under supercritical CO_2 conditions has been claimed to be beneficial over conventional esterification.[56] In these processes excess alcohol (methanol, ethanol, *n*-butanol) is needed, and the ratio of bio-oil to alcohol is less than 1 : 1. Catalysts used include: 732- and NKC-9-type ion exchange resins, solid acid catalyst Nafion SAC13, and Amberlyst-70.

Simultaneous esterification and acetalization has the potential to sufficiently improve bio-oil quality to replace light fuel oil and simplify upgrading. Decreases in viscosity, TAN, carbonyl content, and aging rate can be achieved while improving volatility and heating value, acetal formation, and increasing miscibility with diesel fuels.[47] However, there remain several challenges, including: 1. limited conversion to esters and acetals; 2. loss of light compounds during removal of water; 3. decrease in pH due to acid catalysts;

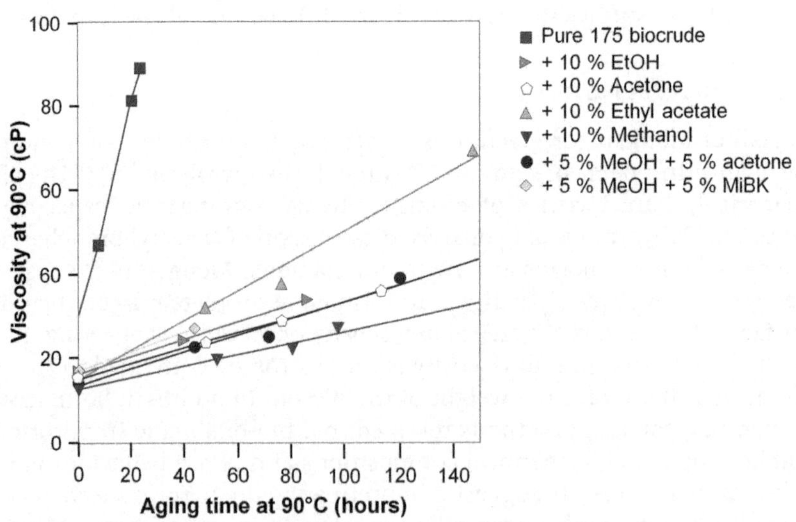

Figure 8.7 Effect of solvent addition (wt%) on bio-oil stability. Reprinted with permission from ref. 52. Copyright (1997) American Chemical Society.

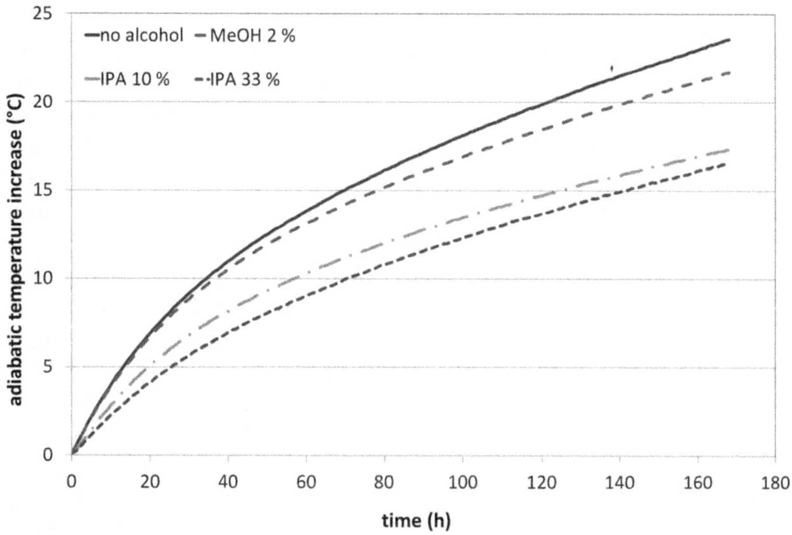

Figure 8.8 Effect of alcohol addition on adiabatic temperature increase (°C) of bio-oil during aging treatment at 50 °C (dilution eliminated). Reprinted with permission from ref. 26. Copyright (2016) American Chemical Society.

4. high alcohol requirements; 5. challenges in recovery of catalysts; and 6. separation of organic and water phase due to changes in composition.

8.6.5 Stabilizing Bio-Oil with Antioxidants

Antioxidants or hydrogen donors can decrease bio-oil aging. Meng *et al.*[20,29] studied the impact of radicals on bio-oil aging. They reported that besides acid-catalysed aging reactions, condensation reactions might be initiated by reactive radical species in bio-oil. Radicals were identified by EPR (electron paramagnetic resonance spectroscopy) and they were preferentially located in the bio-oil lignin fraction, especially in the higher molecular weight lignin. They reported that the bio-oil radical concentration did not decrease significantly in accelerated aging (at 80 °C). In comparison to the use of radical scavengers (1 wt% eugenol, isoeugenol, guaiacol, ascorbic acid, 2,6-ditert-butylphenol, butylated hydroxytoluene), diluting the radical concentration with methanol showed a more profound effect on controlling the lignin condensation. Kim *et al.*[58] reported similar findings: that the radicals formed during pyrolysis are situated almost entirely in the lignin fraction, and are relatively stable carbon-centred radicals. Resonance-stabilized radicals are formed *via* homolytic cleavage of the interunit ether bonds of lignin. The primary cause of instability is believed to be heat-induced homolytic cleavage of lignin fragments into radical intermediates during pyrolysis. The feedstock characteristics, evolving physical and chemical properties of biomass particles, and residence times of both solid and gas phases in reactors hold the key to the desired performance of the pyrolysis process. Liu *et al.*[59] reported that lignin-based radicals are specifically derived from the cleavage of weak (ether) and strong (carbon–carbon) linkages. They also showed that the hydrogen-deficient radicals can be efficiently stabilised by providing a source of hydrogen atoms, such as dihydrophenanthrene (DHP). If such added H-atom or electron donors are not present, the radicals can abstract hydrogen from the lignin and carbohydrate structures present in the mixture, leading to the formation of new radical species.

Udomsap *et al.*[60] used antioxidants propyl gallate (PG), *tert*-butyl hydroquinone (TBHQ), and butylated hydroxyanisole (BHA) at a concentration of 1000 ppm in bio-oil for stabilization. They reported lower aging rates (viscosity increase 17–20%) with antioxidants compared to crude bio-oil (45% increase in viscosity) during accelerated aging test. The effectiveness of antioxidants increased in the following order: BHA, TBHQ, and PG. Ma *et al.*[61] prepared antioxidants and concluded that dichloromethane extraction of pyroligneous acid (DEP) showed superior characteristics to BHA and butylated hydroxytoluene (BHT) for scavenging free radicals. Dobele *et al.*[62] showed that the WIS fraction of bio-oil can be regarded as a promising polymer antioxidant.

Pretreatment of biomass with calcium formate has been claimed to significantly affect the physical and chemical properties of bio-oil, including dramatically lowering the oxygen content and increasing its stability. This formate-assisted fast pyrolysis (FAsP) has been demonstrated on lignin.[63] By pretreating the feedstock with calcium formate, a significantly deoxygenated bio-oil is produced. This effect is likely attributable to the activity of calcium formate as a hydrogen donor under pyrolysis conditions. The added calcium compound is claimed to be recycled after the char is burned off.

8.6.6 Transfer Hydrogenation

Transfer hydrogenation is a catalytic addition of hydrogen from a non-gaseous hydrogen source, referred to as reducing agent or hydrogen donor. Catalytic transfer hydrogenation is a mild and safe operation and has been widely used in industry and in organic synthesis. Metals commonly used for heterogeneous and homogeneous catalytic hydrogen transfer include palladium, ruthenium, rhodium, iridium and nickel. Reducing agents include formic acid to form formate salts, alcohols to form ketones (isopropanol to acetone), hydrazine (N_2H_2 from N_2H_4) to form diimide, or cycloalkanes to form aromatic structures.[64]

A study of catalytic transfer hydrogenation for stabilizing bio-oil that evaluated many combinations of donor and catalysts failed to discover useful reactions.[60] Tested catalysts were 10% Pd-Degussa, 5% Ru-Alfa Aeser, 10% Pd-BASF1, 5% Ru-Engelhard, 10% Pd-BASF2, 5% Rh-Johnson Matthey, 50% Ni-Sud Chemie, 5% Rh-Degussa, 5% Pt-Strem, 5% Pd + 5% Ru-Engelhard, 5% Pt-Engelhard, and 5% Pd + 5% Rh-Engelhard.

8.6.7 Catalytic Hydrotreatment

Oxygen is associated with almost every compound in bio-oil, which makes its removal challenging. Two commonly used catalytic thermal processes to reject oxygen from organic molecules or introduce hydrogen into bio-oil are catalytic hydrotreatment and catalytic cracking.[65–72] The former uses hydrogen to remove oxygen in the form of water while the latter accomplishes the removal of oxygen in the form of water and carbon oxides.

Mild hydrotreatment can be considered a stabilization method. In order to avoid coking of bio-oils during catalytic hydrodeoxygenation (HDO), low-temperature stabilization (below 300 °C) has been proposed.[67] The low-temperature hydrotreatment reduces carbonyl groups associated with polymerisation reactions, and saturates vinyl groups to aliphatic groups, enabling longer term upgrading of pyrolysis oil. Olarte *et al.*[72] have demonstrated HDO (with over 1400 h operation) in a two-stage 400 mL fixed bed continuous reactor. The acidity and oxygen content of bio-oil was significantly decreased by HDO (Figure 8.9). Also, coking tendency measured as micro carbon residue (MCR) decreased. Under severe process conditions both the oxygen content and the acidity of the bio-oil was significantly reduced.[78]

Mild HDO has been reported to take place at room temperature (10% Pd/C, 21 °C, 1 bar).[68] The treatment converted carbonyl groups to alcohols and saturated vinyl groups to aliphatic groups. Viscosity of samples was significantly reduced which was affected by the feedstock. Feedstock also affected hydrogen consumption, which was higher with agro-biomass than with wood. The stability of the hydrotreated bio-oils was clearly improved.

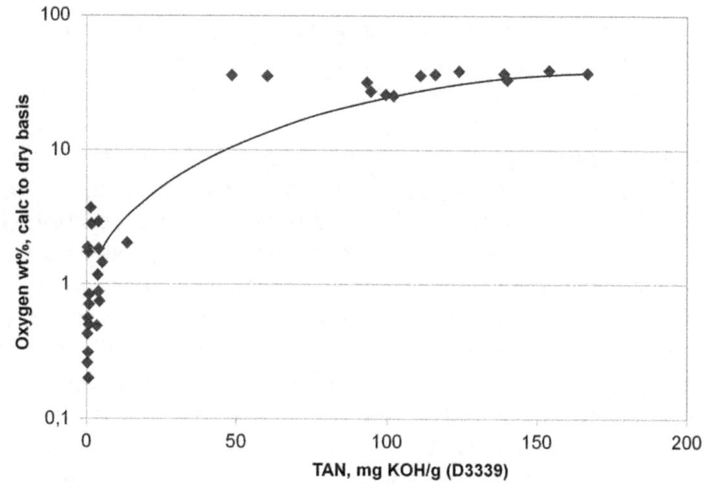

Figure 8.9 Relation of TAN (mg of KOH g^{-1} of oil according to ASTM D3339) to the oxygen content (wt%) in bio-oil and hydrotreated bio-oil. Reprinted with permission from ref. 23. Copyright (2010) American Chemical Society.

8.6.8 Miscellaneous

The tendency of bio-oil to age can be exploited to stabilize it. A one week pre-thermal treatment at 40 °C reduces the bio-oil reactivity, improving its subsequent stability.[25] High-pressure homogenization (HPH) has been claimed to improve the physicochemical properties and storage stability of bio-oil.[75] The viscosity, and mass average molecular weight of the bio-oil decreased significantly. The viscosity of bio-oil homogenized at 100 MPa increased by only 13.9% after storage at 40 °C for 60 days, whereas that of unhomogenized bio-oil increased 56% after the same storage period. High pressure thermal treatment (HPTT)[77] is a process where bio-oil undergoes a phase split yielding a gas phase, an aqueous phase and an organic phase, resulting in bio-oil with lower oxygen and water content and higher energy content.

Bio-oil fractionation by step-wise condensation is one way to control the thermal stability of bio-oils.[73] The most reactive fractions can be separately stabilized and used in various applications.

Electrocatalytic hydrogenation (ECH) is claimed to offer a new approach for bio-oil stabilization.[74] After electrocatalytic hydrogenation of the WS fraction of bio-oil (H_2, room temperature, a ruthenium supported on activated carbon cloth) aldehydes and ketones were reduced to the corresponding alcohols or diols. Carbon recovery into the liquid product was more than 80%. Besides stabilization of bio-oil for subsequent fuel production, hydrogen and valuable diols were produced during ECH.

8.7 Conclusion

Stability of bio-oils is a complex phenomenon, especially as numerous compounds with different reactivities are involved. During aging, acid-catalysed condensation and polymerisation reactions take place which increase the average molecular weight and viscosity of bio-oil. The decrease in carbonyl compounds during aging is one of the main chemical changes. Condensation reactions initiated by reactive lignin-derived radical species also appear to be responsible for aging. Condensation and polymerization reactions increase the ratio of WIS to WS compounds leading to phase-separation of bio-oil.

Aging can be slowed by: dilution of bio-oil with solvents; removing of alkali metals by washing, hot filtration, or ion-exchange; removing hemicelluloses by torrefaction; or adding antioxidants to stabilize lignin radicals. In some cases, controlled aging can be used to stabilize bio-oil. Enhanced stabilization can be achieved by esterifications and acetalization of carboxylic acids and carbonyls, respectively. Stage-fractionation of bio-oil and separate stabilization of these fractions is one alternative approach. In low-temperature catalytic hydrodeoxygenation, reactive carbonyl compounds are stabilized by hydrogen.

All these stabilization methods have advantages and disadvantages. The appropriate stabilization method will depend on the bio-oil application. When hydrocarbon fuels are the desired product, more expensive hydrotreatment (HDO) routes to stabilization might be justified. When targeted as feedstock for an oil refinery, thermal stability is a key factor. An increase in viscosity would be an indicative factor on the polymerisation tendency of the bio-oil, and also the oxygen content, distillability, and MCR. Stability of bio-oils is improved by decreasing the amount of highly reactive species, especially carbonyl compounds and furfurals, adding anti-oxidants to decrease reactivity of lignin-derived compounds; removing components like acids and alkali metals that have catalytic activity, and increasing the relative amount of solubilisation agents like alcohols.

More research is needed to understand the aging of bio-oil and to be able to control it. Also, the correlation of aging with heat generation of bio-oil should be further studied. The heat generation of bio-oils might have serious consequences. When storing larger quantities of bio-oil in insulated containers, the heat transfer between the tank and the surrounding area decreases and the bio-oil temperature increases. Hence, the heat generation capacity of fresh bio-oils should be taken into account when designing industrial storage and utilization systems. Especially low temperature (20–80 °C) catalytic hydrotreatment should be more studied to clarify the reactions taking place in various fast pyrolysis bio-oils.

References

1. D. S. Scott and J. Piskorz, *J. Chem. Eng.*, 1982, **60**, 666–674.
2. D. Meier, B. van de Beld, A. V. Bridgwater, D. C. Elliott, A. Oasmaa and F. Preto, State-of-the-art of fast pyrolysis in IEA bioenergy member countries, *Renewable Sustainable Energy Rev.*, 2013, **20**, 619–641E.
3. A. Oasmaa, B. Van De Beld, P. Saari, D. C. Elliott and Y. Solantausta, *Energy Fuels*, 2015, **29**(4), 2471–2484.
4. D. S. Scott, L. Paterson, J. Piskorz and D. Radlein, *J. Anal. Appl. Pyrolysis*, 2001, **57**, 169–176.
5. D. C. Elliott, *Biomass Bioenergy*, 1994, **7**, 179–185.
6. F. A. Agblevor, S. Bessler and R. J. Evans, Inorganic Compounds in Biomass Feedstocks: Their Role in Char Formation and Effect on the Quality of Fast Pyrolysis Oils, in *Proceedings of Biomass Pyrolysis Oil Properties and Combustion Meeting, September 26–28, 1994, Estes Park, CO*, ed. T. A. Milne, Golden, CO: National Renewable, NREL-CP-430-7215, 1994.
7. S. Czernik, Storage of Biomass Pyrolysis Oils, *Proceedings of Specialist Workshop on Biomass Pyrolysis Oil Properties and Combustion, Estes Park, CO, Sept. 26–28, 1994, NREL Paper No. CP-430-7215, pp. 67–76*, Energy Laboratory, 1994, pp. 77–89.
8. F. A. Agblevor, S. Bessler and R. J. Evans, *Influence of Inorganic Compounds on Char Formation and Quality of Fast Pyrolysis Oils, Abstracts of the ACS 209th National Meeting*, Anaheim, CA, April 2–5, 1995.
9. J. P. Diebold, J. W. Scahill, S. Czernik, S. D. Phillips and C. J. Feik, *Progress in the Production of Hot-Gas Filtered Biocrude Oil at NREL*, CPL Press, Newbury, UK, 1996.

10. R. M. Baldwin and C. J. Feik, *Energy Fuels*, 2013, **27**, 3224–3238.
11. A. Oasmaa, Y. Solantausta, V. Arpiainen, E. Kuoppala and K. Sipilä, *Energy Fuels*, 2010, **24**, 1380–1388.
12. A. Oasmaa, A. Källi, C. Lindfors, D. C. Elliott, D. Springer, C. Peacocke and D. Chiaramonti, *Energy Fuels*, 2012, **26**(6), 3864–3873.
13. B. Scholze, Long-term stability, catalytic upgrading, and application of pyrolysis oils–improving the properties of a potential substitute for fossil fuels, [Doctoral dissertation], University of Hamburg, Hamburg, Germany, 2002.
14. S. Czernik, D. Johnson and S. Black, *Biomass Bioenergy*, 1994, **7**(1–6), 187–192.
15. J. Diebold, *A Review of the Chemical and Physical Mechanisms of the Storage Stability of Fast Pyrolysis Bio-Oils*, 2002, http://www.nrel.gov/docs/fy00osti/27613.pdf.
16. A. Oasmaa and C. Peacocke, *A Guide to Physical Property Characterisation of Biomass-Derived Fast Pyrolysis Liquids. A Guide*, Espoo: VTT, 79 p. + app. 46 p., VTT Publications, 731 ISBN 978-951-38-7384-4, 2010, http://www.vtt.fi/inf/pdf/publications/2010/P731.pdf.
17. A. Oasmaa, T. Sundqvist, E. Kuoppala, M. G.-P. Manuel, Y. Solantausta, C. Lindfors and V. Paasikallio, *Energy Fuels*, 2015, **29**(7), 4373–4381.
18. J. Lehto, A. Oasmaa, Y. Solantausta, M. Kytö and D. Chiaramonti, *Fuel Oil Quality and Combustion of Fast Pyrolysis Bio-Oils*, 2013, VTT Technology: 87, Espoo, VTT, 79 p. ISBN 978-951-38-7929-7 (Soft back ed.), 978-951-38-7930-3 (PDF), http://www.vtt.fi/inf/pdf/technology/2013/T87.pdf.
19. A. Oasmaa, E. Kuoppala and Y. Solantausta, *Energy Fuels*, 2003, **17**(2), 433–443.
20. J. Meng, A. Moore, D. Tilotta, S. Kelley and S. Park, *ACS Sustainable Chem. Eng.*, 2014, **2**(8), 2011–2018.
21. A. Oasmaa, *Fuel Oil Quality Properties of Wood-Based Pyrolysis Liquids*, Academic dissertation, Department of Chemistry, University of Jyväskylä, Jyväskylä, 2003, 32 p. + app. 251 p., Research Report Series, Report 99, Doctoral thesis.
22. A. Oasmaa and E. Kuoppala, *Energy Fuels*, 2003, **17**(4), 1075–1084.
23. A. Oasmaa, D. C. Elliott and J. Korhonen, *Energy Fuels*, 2010, **24**, 6548–6554.
24. X. Hu, Y. Wang, D. Mourant, R. Gunawan, C. Lievens, W. Chaiwat, M. Gholizadeh, L. Wu, X. Li and C.-Z. Li, *AIChE J.*, 2013, **59**(3), 888–900.
25. D. Meier, G. Jesussek and S. Radtke, Chemical Stability of Wood Fast Pyrolysis Liquids, in *Pyrolysis and Gasification of Biomass and Waste, Strasbourg, France, Sept. 30–Oct. 1, 2002*, ed. A. V. Bridgwater, CPL Press, Newbury, U.K., 2003.
26. T. Sundqvist, Y. Solantausta, A. Oasmaa, L. Kokko and V. Paasikallio, *Energy Fuels*, 2016, **30**(1), 465–472.
27. A. Oasmaa, J. Korhonen and E. Kuoppala, *Energy Fuels*, 2011, **25**(7), 3307–3313.

28. E. Fratini, M. Bonini, A. Oasmaa, Y. Solantausta, J. Teixeira and P. Baglioni, *Langmuir*, 2006, **22**(1), 306–312.
29. J. Meng, A. Moore, D. C. Tilotta, S. S. Kelley, S. Adhikari and S. Park, *Energy Fuels*, 2015, **29**, 5117–5126.
30. D. C. Elliott, A. Oasmaa, F. Preto, D. Meier and A. V. Bridgwater, *Energy Fuels*, 2012, **26**(6), 3769–3776.
31. D. C. Elliott, A. Oasmaa, D. Meier, F. Preto and A. V. Bridgewater, *Energy Fuels*, 2012, **26**(12), 7362–7366.
32. T. Ba, A. Chaala, M. Garcia-Perez, D. Rodrigue and C. Roy, *Energy Fuels*, 2004, **18**, 704–712.
33. T. Ba, A. Chaala, M. Garcia-Perez and C. Roy, *Energy Fuels*, 2004, **18**(1), 188–201.
34. A. Chaala, T. Ba, M. Garcia-Perez and C. Roy, *Energy Fuels*, 2004, **18**(5), 1535–1542.
35. M. Garcia-Perez, A. Chaala, H. Pakdel, D. Kretschmer, D. Rodrigue and C. Roy, *Energy Fuels*, 2006, **20**, 364–375.
36. A. Oasmaa, E. Kuoppala, Y. Solantausta and S. Gust, *Energy Fuels*, 2003, **17**(1), 1–12.
37. A. Oasmaa, I. Fonts, M. R. Pelaez-Samaniego, M. E. Garcia-Perez and M. Garcia-Perez, *Energy Fuels*, 2016, **30**(8), 6179–6200.
38. A. Oasmaa and D. Meier, *Analysis and Characterisation, Conference & Meeting Reports. Analysis and Characterisation*, Subject Group Workshop, Montpellier, France, 23–24 April 1999, PyNe Newsletter 1999, Issue 8.
39. A. V. Bridgwater, *Biomass Bioenergy*, 2012, **38**, 68–94.
40. D. S. Scott, J. Piskorz, D. Radlein and P. Majerski, Process for the production of anhydrosugars from lignin and cellulose containing biomass by pyrolysis, US5395455 A, 1995.
41. S. R. G. Oudenhoven, R. J. M. Westerhof, N. Aldenkamp, D. W. F. Brilman and S. R. A. Kersten, *J. Anal. Appl. Pyrolysis*, 2013, **103**, 112–118.
42. S. D. Stefanidis, E. Heracleous, D. Th. Patiaka, K. G. Kalogiannis, C. M. Michailof and A. A. Lappas, *Biomass Bioenergy*, 2015, **83**, 105–115.
43. S. Renkonen, Removal of alkali metals for the enhancement of fast pyrolysis process, Master's thesis for the degree of Master of Science in Technology, Aalto University, Finland, September, 2016, Pages 98+9.
44. A. Källi, Pyrolyysiöljyn kiintoaineen poisto linkouksella, Master's thesis, Aalto-University, School of engineering, Department of Energy Technology, 2012.
45. J. Innanen, Solids removal in pyrolysis, Master's thesis, Lappeenranta University of Technology, Finland, 2013, p. 73.
46. E. Butler, G. Devlin, D. Meier and K. McDonnell, *Renewable Sustainable Energy Rev.*, 2011, **15**(8), 4171–4186.
47. S. Xiu and A. Shahbazi, *Renewable Sustainable Energy Rev.*, 2012, **16**, 4406–4414.
48. S. J. Frey, R. Gattupalli and T. A. Brandvold, 2012, US 2012/0317871 A1.

49. Nrel, http://www.energy.gov/sites/prod/files/2015/04/f21/thermochemical_conversion_abdullah_231401.pdf, accessed 27.01.2015.
50. B. Ru, S. Wang, G. Dai and L. Zhang, *Energy Fuels*, 2015, **29**, 5865–5874.
51. A. A. Boateng and C. A. Mullen, *J. Anal. Appl. Pyrolysis*, 2013, **100**, 95–102.
52. J. P. Diebold and S. Czernik, *Energy Fuels*, 1997, **11**(5), 1081–1091.
53. A. Oasmaa, E. Kuoppala, J.-F. Selin, S. Gust and Y. Solantausta, *Energy Fuels*, 2004, **18**(5), 1578–1583.
54. R. N. Hilten, B. P. Bibens, J. R. Kastner and K. C. Das, *Energy Fuels*, 2010, **24**(1), 673–682.
55. X. Li, R. Gunawan, C. Lievens, Y. Wang, D. Mourant, S. Wang, H. Wu, M. Garcia-Perez and C.-Z. Li, *Fuel*, 2011, **90**(7), 2530–2537.
56. H.-Y. Cui, J.-H. Wang, S.-Q. Wei, S.-P. Zhuo, Z.-H. Li, L.-H. Wang and W.-M. Yi, *J. Fuel Chem. Technol.*, 2010, **38**(6), 673–678.
57. T. Sundqvist, A. Oasmaa and A. Koskinen, *Energy Fuels*, 2015, **29**(4), 2527–2534.
58. K. H. Kim, X. Bai, S. Cady, P. Gable and R. C. Brown, *ChemSusChem*, 2015, **8**, 894–900.
59. M. Liu, J. Yang, Z. Liu, W. He, Q. Liu, Y. Li and Y. Yang, *Energy Fuels*, 2015, **29**, 5773–5780.
60. P. Udomsap, Y. H. Yeinn, J. T. H. Hui, B. Yoosuk, S. B. Yusuf, and S. Sukkasi, Towards Stabilization of Bio-oil by Addition of Antioxidants and Solvents, and Emulsification with Conventional Hydrocarbon Fuels, *Utility exhibition on power and energy systems: Issues & Prospects for Asia (ICUE)*, 28-30 Sept 2011, pp. 1–5.
61. C. Ma, K. Song, J. Yu, L. Yang, C. Zhao, W. Wang, G. Zu and Y. Zu, *J. Anal. Appl. Pyrolysis*, 2013, **104**, 38–47.
62. G. Dobele, T. Dizhbite, I. Urbanovich, A. Andersone, J. Ponomarenko and G. Telysheva, *J. Anal. Appl. Pyrolysis*, 2009, **85**, 81–86.
63. P. A. Case, M. C. Wheeler and W. J. DeSisto, *Bioresour. Technol.*, 2014, **173**, 177–184.
64. D. C. Elliott, S.-J. Lee and T. R. Hart, *Stabilization of Fast Pyrolysis Oil: Post Processing*, PNNL-21549, March 2012.
65. E. Laurent, C. Pierret, P. Grange and B. Delmon, Control of the Deoxygenation of Pyrolytic Oils by Hydrotreatment, *Biomass for Energy, Industry and Environment, 6th EC Conference*, Athens, 1991.
66. R. Maggi, A review on catalytic hydrotreating processes for the upgrading of liquids produced by flash pyrolysis, *Stud. Surf. Sci. Catal.*, 1997, 99–112, ISSN: 0167-2991.
67. D. C. Elliott, *Energy Fuels*, 2007, **21**, 1792–1815.
68. M. R. Rover, P. H. Hall, P. A. Johnston, R. G. Smith and R. C. Brown, *Fuel*, 2015, **153**(1), 224–230.
69. R. H. Venderbosch, A. R. Ardiyanti, J. Wildschut, A. Oasmaa and H. J. Heeres, *J. Chem. Technol. Biotechnol.*, 2010, **85**, 674–686.
70. A. Oasmaa, E. Kuoppala, A. Ardiyanti, R. H. Venderbosch and H. J. Heeres, *Energy Fuels*, 2010, **24**(9), 5264–5272.

71. R. H. Venderbosch, A. R. Ardiyanti, J. Wildschut, A. Oasmaa and H. J. Heeres, *J. Chem. Technol. Biotechnol.*, 2010, **85**(5), 674–686.
72. M. V. Olarte, A. H. Zacher, A. B. Padmaperuma, S. D. Burton, H. M. Job, T. L. Lemmon, M. S. Swita, L. J. Rotness, G. N. Neuenschwander, J. G. Frye and D. C. Elliott, *Top. Catal.*, 2016, **59**, 55–64.
73. M. R. Haverly, K. V. Okoren and R. C. Brown, *Energy Fuels*, 2016, **30**, 9419–9426.
74. Z. Li, S. Kelkar, L. Raycraft, M. Garedew, J. E. Jackson, D. J. Miller and C. M. Saffron, *Green Chem.*, 2014, **16**, 844.
75. R. He, X. P. Ye, F. Harte and B. English, *Fuel Process. Technol.*, 2009, **90**, 415–421.
76. M. Jiajia, T. I. Smirnova, X. Song, A. Moore, X. Ren, S. Kelley, S. Park and D. Tilott, *RSC Adv.*, 2014, **4**, 29840–29846.
77. F. de Miguel Mercader, M. J. Groeneveld, S. R. A. Kersten and J. A. Hogendoorn, *Fuel*, 2010, **89**(10), 2829–2837.
78. R. H. Venderbosch and H. J. Heeres, in *Biofuel's Engineering Process Technology*, ed. M. A. dos Santos Bernardes, ISBN 978-953-307-480-1, Published: August 1, 2011 under CC BY-NC-SA 3.0 license, © The Author(s), ch. 17, Pyrolysis Oil Stabilisation by Catalytic Hydrotreatment.
79. X. Bai, K. H. Kim, R. C. Brown, E. Dalluge, C. Hutchinson, Y. J. Lee and D. Dalluge, Formation of phenolic oligomers during fast pyrolysis of lignin, *Fuel*, 2014, **128**, 170–179.

CHAPTER 9

Extraction of Value-Added Chemicals from Bio-Oil Products

D. MEIER[a]

[a]thermophil international, Langbergring 60, 21033 Hamburg, Germany
*E-mail: info@thermophil.com

9.1 Introduction

The chemical nature of bio-oils (liquid products from fast pyrolysis of lignocellulosic biomass) has been described comprehensively in the literature.[1–5] The thermal degradation of the main biomass constituents holocellulose and lignin leads to the formation of hundreds of organic monomeric molecules. In addition, oligomeric molecules and water are formed. In a simplified model, four fractions can be distinguished: (1) Water (*ca.* 25 wt%) is the main single component. It stems from both unavoidable reaction water and adjustable moisture content in the feedstock; (2) monomeric components (*ca.* 40 wt%) as measured by gas chromatography (GC) and high performance liquid chromatography (HPLC), representing the largest fraction; (3) Another important fraction (*ca.* 15 wt%) is represented by water-insoluble, oligomeric lignin-derived components known as "pyrolytic lignin"; (4) Unknown components (*ca.* 20 wt%) may comprise oligomeric sugars and numerous other small molecules which are not determined in standard analytical protocols.

Green Chemistry Series No. 50
Fast Pyrolysis of Biomass: Advances in Science and Technology
Edited by Robert C. Brown and Kaige Wang

Published by the Royal Society of Chemistry, www.rsc.org

Although the main focus for the utilization of bio-oils is on the production of heating and transportation fuels, there is an increasing interest in the production of chemicals as this is expected from future biorefineries. In addition, this would positively contribute to the overall process economics. Despite the complicated chemical composition of bio-oils, numerous attempts have been made to enrich certain functional groups through fractionation or to extract single molecules (*e.g.* as acetic acid). In the following chapter the strategies for obtaining value-added chemicals from bio-oils are presented, both at laboratory scale and at technial scale.

9.2 Equilibrium-Based Separation

9.2.1 Distillation

9.2.1.1 Ordinary Distillation

Ordinary distillation is perhaps the oldest separation method mankind has used to enrich and separate components of interest, such as water, ethanol, essential oils, tars, *etc.* The basic concept is based on evaporation and condensation of volatile components which exhibit different vapor pressures at a given temperature. Vapor-liquid equilibrium is commonly encountered in distillation.[6,7] Nowadays, ordinary distillation as a common separating technology is probably the most mature and abundant industrial process and the first treatment of petroleum in refineries.[8] Every petroleum refinery introduces crude oil into a distillation tower, which separates petroleum into fractional cuts based on boiling temperature.[9] This separation allows for downstream upgrading and refining to occur at conditions appropriate for each fractional cut.[10]

Due to their chemical composition, bio-oils show a very wide range of boiling temperatures. In addition to water and volatile organic components, biomass pyrolysis oils contain substantial amounts of non-volatile materials such as sugars and oligomeric phenolics.[11] In addition, the slow heating of the oils during distillation results in the polymerization of some of the reactive components. Consequently, the oils start boiling below 100 °C but the distillation stops at 250–280 °C, leaving 35–50% of the starting material as residue.[12] Because traditional atmospheric column distillation of bio-oil was more or less regarded as impossible, research into innovative and novel bio-oil distillation techniques is an ongoing topic.

9.2.1.2 Fractional Distillation

Fractional distillation at ambient and reduced pressure was employed with corn stover bio-oil and yielded three fractions with different boiling point ranges. Characterization of the distilled organic fraction showed that it had a much better quality than the crude bio-oil, containing little water and fewer oxygenated compounds, and having a higher heating value.[13] Aromatic and oxygenated compounds were distributed in the light and middle fraction, whereas phenolic compounds were concentrated in the heavy fraction. Overall distillate recovery amounted to around 84% for atmospheric distillation and around 73% for vacuum distillation.[14] Successful fractional distillation at ambient pressure was reported when bio-oils from tail gas reactive pyrolysis (TGRP) were used. These oils are said to provide higher quality oils with increased thermal stability. All bio-oils produced from TGRP contained significantly less acid (0.2–4% total), trace amounts of aldehydes, and significantly higher concentrations of hydrocarbons and phenolics (5–20% each). Distillates were categorized according to boiling temperature and chemical components: benzene–toluene–xylenes (BTX), styrene, indene, pyridine, phenols, naphthalenes, acetic acid, and fluorene/anthracene. The relatively narrow product distributions allowed for the isolation of pure naphthalene by recrystallization from a naphthalene-rich fraction.[15] A zero-waste process based on atmospheric distillation was proposed by Zhang through combining distillation with co-pyrolysis of the resulting distillation residue.[16]

9.2.1.3 *Molecular Distillation (Short Path Distillation)*

Traditional distillation technology separates components by differences in their boiling points. However, molecular distillation (or short path distillation) is quite different and precisely relies on the various mean free paths of different substances.

Molecular distillation technology has been widely used in the chemical, pharmaceutical, and foodstuff industries, as well as in scientific research to concentrate and purify organic chemicals. It is a feasible process for the separation of thermally unstable materials, taking into account that it only takes a few seconds to complete the separation process. Due to the mild separation conditions (low temperature and short heating time) molecular distillation is quite suitable for the separation of thermally sensitive compounds. In fact, this distillation technology has been widely used in the fine chemical, pharmaceutical, and food-processing industries, such as in the extraction of tocotrienol from palm fatty acid distillates, the recovery of tocopherol from rapeseed oil, and the purification of octacosanol and monoglycerides.[17]

The process is distinguished by the following three features: (1) short residence time in the zone of the molecular evaporator exposed to heat; (2) low operating temperature due to vacuum in the space of distillation; (3) a characteristic mechanism of mass transfer in the gap between the evaporating and condensing surfaces. The liquid to be distilled flows over the cylinder as a thin-wiped-film with a thickness of 0.05–2 mm according to the peripheral liquid load and the viscosity, while being evenly distributed around its perimeter and stirred by the action of the wiper (see Figure 9.1). A residence time of 10^{-1}–10^{1} s is achieved by formation of a thin liquid film through wiping or rotation of the evaporating surface. Removal of the vapor barrier above the surface of evaporation, at a pressure reduced to 10^{-1}–10^{1} Pa, causes the operating temperature to be reduced by up to 250 °C compared with normal

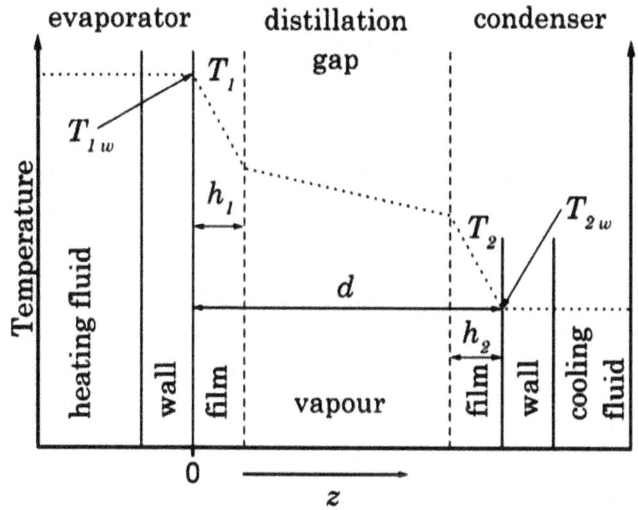

Figure 9.1 Temperature profile in a molecular evaporator. Reprinted from Chemical Engineering Journal, **85**(2–3), J. Lutišan, J. Cvengroš, M. Micov, Heat and mass transfer in the evaporating film of a molecular evaporator, 225–234, Copyright (2002) with permission from Elsevier.[19]

pressure. The short distance of 10–50 mm between the evaporating and condensing surfaces enables a large portion of the molecules in the vapor state to reach the cooling surface and condense without being disturbed. In the literature, a frequently-given condition is that the width of the gap should be less than, or comparable with, the mean free path of the molecules evaporated. The passage of the molecules through the distillation space should then be collision free.[18,19] Molecules will move more rapidly when the liquid mixture is heated. Surface molecules will overcome intermolecular forces and escape as gas molecules when they obtain sufficient energy. With an increased amount of gas molecules above the liquid surface, some molecules will return to the surface. Under certain conditions, the molecular motion will achieve dynamic equilibrium, which is manifested as equilibrium on a macroscopic scale.[13]

Molecular distillation technology was introduced into the field of bio-oil separation by Wang *et al.*[20–22] The separation characteristics of bio-oil at 70, 100, and 130 °C were investigated. A bio-oil distillation yield of about 85% was achieved without coking at 130 °C. In another experiment a bio-oil fraction rich in carboxylic acids was produced at 50 °C and the water in the residual bio-oil fraction was completely removed at the same time.[23] Distillation evaluation models, as well as bio-oil distilled fractions, were investigated in a two-step bio-oil molecular distillation process.[21] Here, the distance between the cooling and heating surfaces is less than the mean free path for a light molecule, but greater than that for a heavy molecule. Therefore, the light molecules escaping from the heating surface can easily reach the cooling surface and be condensed. The dynamic balance is thereby broken, and the light molecules are continuously released from the liquid phase. On the contrary, the heavy molecules are not released and return to the liquid phase. In this way, the light and heavy molecules are effectively separated.

Studies on the distribution of acidic compounds in bio-oil fractions has been carried out at 50 °C distillation temperature. The carboxylic acid content in the refined bio-oil was used to estimate the separation efficiency. Guo *et al.*[24] chose five major acids in bio-oil and studied their separation characteristics. The amount of acetic acid, the most abundant acid in bio-oil, was reduced to 1.9 wt% and 0.96 wt% in the middle fraction (MF) and high fraction (HF), respectively. The results indicated that acidic compounds could be effectively separated from the crude bio-oil by means of molecular distillation technology.

In addition, the distribution characteristics of several chemicals in three fractions (low, LF; MF; HF) was measured. Six chemicals were selected as being representative of ketones, aldehydes, phenols, and sugars, respectively. 1-Hydroxy-2-propanone, the most abundant ketone in bio-oil, could not be detected in the MF or HF after separation, indicating that it was extremely enriched in the LF. The content of furfural in the HF was just a little higher than that in the MF, but much lower than that in bio-oil. The distributions of these two compounds reflected the enrichment of small ketone and aldehyde molecules in the LF. Phenol appeared to be difficult to separate as there were similar distributions in bio-oil, MF, and HF. In contrast, compounds of higher molecular weight tended to be enriched in the MF and HF. For example, 1,2-benzenediol and 3-methyl-1,2-benzenediol were more abundant in the MF and HF than in the bio-oil before separation. In particular, the relative content of 1,2-benzenediol amounted to 11.73 wt% in HF, about five times higher than that in bio-oil (2.45 wt%).[13]

Based on the operation experiences gained with the molecular distillation apparatus, a larger-scale joint reduced pressure and molecular distillation set-up was established at the State Key Laboratory of Clean Energy Utilization, Zhejiang University. The flow diagram of this joint distillation system is illustrated in Figure 9.2. The processing capacities of the reduced pressure distillation and molecular distillation units were both 8–10 kg h^{-1}, and they could be run at temperatures up to 300 °C and pressures down to 50 Pa. The reduced pressure distillation unit could be operated separately to remove the water from bio-oil as well as to obtain bio-oil fractions. When these two units were assigned to run together, the pre-treated bio-oil from the first reduced pressure distillation unit could be pumped directly into the molecular distillation unit.

Recently, tail gas reactive pyrolysis (TGRP)[9] of spirulina algal biomass and cyanobacteria was performed resulting in more stable oils which could be distilled with a short path distillation unit in an attempt to concentrate and separate valuable commodity chemicals.[15,25,26]

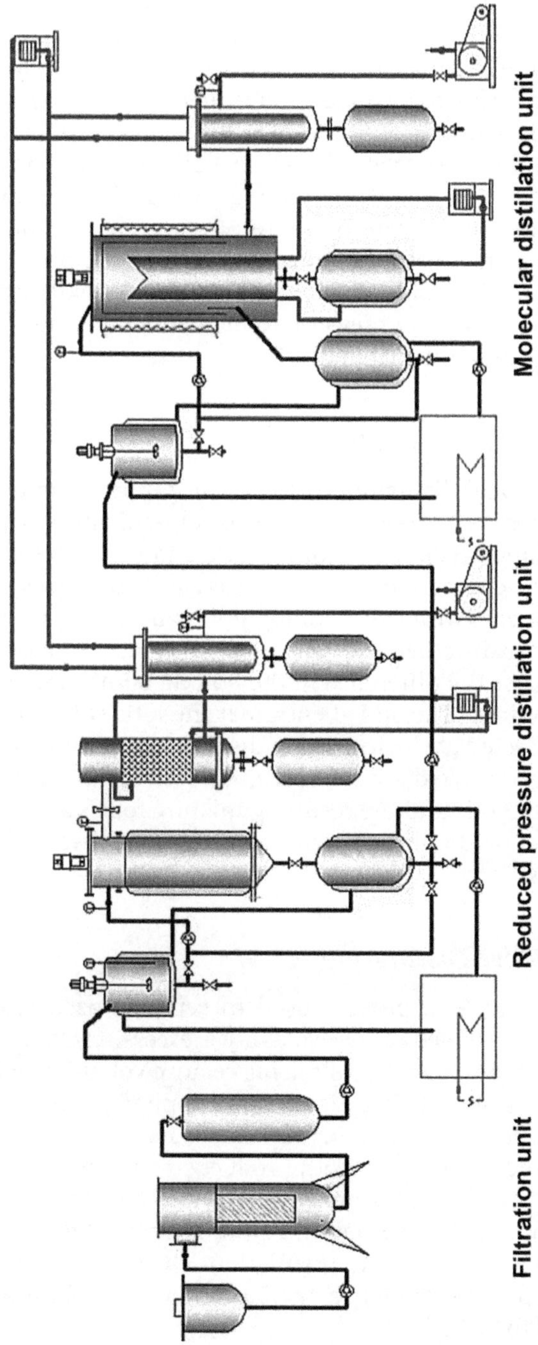

Figure 9.2 Schematic diagram of the joint distillation system.[13]

9.2.1.4 Vacuum Distillation

In contrast to atmospheric pressure distillation, vacuum distillation has a strong advantage, in that it operates at much lower temperatures.[27] Hence, it is an appropriate method for heat-sensitive feeds whereby the pressure above the liquid mixture to be distilled is reduced to less than its vapor pressure (usually less than atmospheric pressure) causing evaporation of the most volatile liquid(s) (those with the lowest boiling points). This distillation method works on the principle that boiling occurs when the vapor pressure of a liquid exceeds the ambient pressure. Vacuum distillation is used with or without heating the mixture and is a key part of the petroleum refining process. The distilled organic fraction from vacuum distillation of bio-oils contains almost no water and fewer oxygenated compounds, although the method uses large amounts of energy.[28]

9.2.1.5 Azeotropic Distillation

Although conventional distillation is the most important industrial separation process, there are situations where solvent-enhanced separation, *e.g.* azeotropic distillation, is a more appropriate choice. In azeotropic distillation a third volatile component (separating agent, entrainer) is added to separate two components with close boiling points by means of the formation of an azeotropic mixture between one of the original components and the entrainer to increase the difference in the boiling points and facilitates separation by distillation.[9] The added entrainer must be sufficiently volatile so that it can be collected at the top of the distillation column. A combination of reactive and azeotropic distillation is also possible. For example, alcohols can be added to bio-oils and the resulting mixture forms an azeotrope while the alcohol reacts predominantly with acids and aldehydes to form stable esters and acetals.[29,30]

9.2.1.6 Extractive Distillation

Extractive distillation is a process used to separate azeotrope-containing systems or systems in which relative volatility is excessively low. An additive, or separating agent, that is capable of raising relative volatility and eliminating azeotropes in the system being distilled is supplied to the column, where it mixes with the feed components and exerts its effect. The agent is subsequently recovered from one or both product streams by a separate process and recycled for reuse. The extractive agent and less volatile components flow to the bottom of the distillation column, where the extracted component is recovered by a subsequent distillation. The non-extracted species are distilled to the top of the extractive distillation tower. So far this method has not been applied to bio-oils.

Often bio-oils separate into an aqueous and tarry phase. When aqueous solutions are involved the separation is typically a liquid–liquid extraction, because distillation-based techniques require the evaporation of large amounts of water in most cases. Recently, the use of ionic liquids (IL) as solvents in extractive distillation and extraction processes was considered.[31] It was concluded that co-extraction of water is a crucial factor and should be minimized. Another important factor is the recovery of ILs. Although this technique has not yet been applied to bio-oils it offers interesting perspectives for future research.

9.2.1.7 Steam Distillation

Steam distillation is a conventional commercially-utilized process for isolating volatile organic compounds that are sensitive to high heat. Different from the earlier separation methods, steam distillation is normally used for direct separation of desirable components from solid biomass feedstock, not liquid mixture. In this method, steam is introduced by heating water, and passed through oil-containing plant material. With the addition of steam, the oil–water mixture boils at a lower temperature (<100 °C at atmospheric pressure) allowing heat-sensitive compounds to be separated with less decomposition. Steam distillation is suitable for extracting light components whose vapor pressures are relatively high (≥1.33 kPa at 100 °C). For components whose vapor pressures at 100 °C are between 0.67 kPa and 1.33 kPa, superheated steam is used for the distillation.[32]

Saturated steam in contact with pyrolysis oil provides heat to the oil, reducing the viscosity and facilitating the vaporization of a portion of the oil into the vapor phase which will be subsequently transferred with steam. In such a case, each volatile compound contained in the oil exerts a partial pressure, which is equal to its vapor pressure. Vasalos *et al.*[33] performed steam distillation of the acidic fraction of a pyrolytic oil sample after neutralization with 8% NaOH and extraction with CH_2Cl_2. The authors demonstrated the potential of steam distillation as a recovery method of light phenolic compounds from a biomass-derived pyrolytic oil.[34]

The instability of pyrolysis oils can mainly be ascribed to the presence of acids, aldehydes, ketones and phenols. All these substance classes are prone to condensation reactions. However, most of them are known to be steam-distillable. Therefore, steam distillation tests were conducted by Meier *et al.* to remove the "dangerous" components from the bio-oil.[35] It was assumed that the temperature for steam distillation (100 °C) is not too high to promote recondensation reactions of the oil which should be less probable due to the constant removal of condensable components. About 20 g of oil was steam-distilled for 120 min, until *ca.* 1 L of distillate was produced. This fraction was extracted in a separatory funnel with dichloromethane (DMC), and the solvent as well as the residual water evaporated to obtain two fractions: (1) extract of distillate and (2) residue of evaporation. The remaining water in the steamed flask was also extracted with dichloromethane and the two phases treated as described above. A tar fraction remained in the flask. All fractions were dried over sodium sulfate and gravimetrically determined.

It was observed that the extract of the distillate comprises mainly monophenols, the evaporation residue of the distillate contains ketones and the evaporation residue of the water phase comprises mainly levoglucosan. The water extract showed very low GC-signals. Accordingly it seems possible to use steam distillation as a procedure for levoglucosan isolation.

Steam distillation of a birch wood vacuum pyrolysis oil was described by Murwanashyaka *et al.*[34] They studied the recovery of phenols at various steam : pyrolysis oil ratios. A 14.9 wt% volatile pyrolysis oil fraction (based on the total feed oil) was recovered at a steam : oil ratio of 27. The distillate was analyzed by GC-MS after acetylation and showed 21.3 wt% of phenolic compounds on the pyrolysis oil basis. The distillate was further distilled under a total pressure of 0.7 kPa. About 16 sub-fractions were recovered. The steam-distilled fractions were found to be chemically and thermally stable when subjected to further solvent fractionation and purification processes. The 2.6-dimethoxyphenol (syringol)-rich fraction was separated and purified. Syringol with a purity of 92.3% could be obtained.

9.2.2 Liquid–Liquid Extraction

9.2.2.1 Introduction

Liquid–liquid extraction (LLE), also known as solvent extraction, is a typical ternary system in the chemical engineering field, which separates chemicals from one solution to another based on the different solubility of the solute chemical in two solvents. Three components are involved in the LLE process: solute, diluent, and extractant. Solute, which is dissolved in the diluent, is extracted from the diluent and dissolved into another solvent, the extractant.

The requirements for extractants are as follows:

- the extractant should have specific selectivity for the solute so that most of the desirable solute can be dissolved in the extractant from the diluent and *vice versa* for undesirable compounds
- the solute should be easily separated from extractant after LLE
- a high partition ration is another property allowing effective separation of a desirable compound from the diluent
- the extractant should have high solubility for the target solute and low solubility for the diluent
- high interfacial tension should be present for rapid coalescence for extractant droplets after agitation
- low toxicity for the extractant and diluents is needed for industrial use[36]

Multiple factors need to be considered when applying the LLE process to bio-oils. Determining how best to choose an extractant is usually the key to the successful design of LLE separation. With the development of new types of extraction processes, the use of LLE in biorefineries field will have a promising future.[7,36]

The solvents for bio-oil extractions include water, ethyl acetate, paraffins, ethers, ketones, and alkaline solutions. In recent years, some special solvents, such as supercritical CO_2, have also been used. By selecting appropriate solvents for extraction of the desired products, good separation of bio-oil can be achieved.[13]

9.2.2.2 Polarity-Based Separations

9.2.2.2.1 Extraction with Organic Solvents and Liquid CO_2. Extraction of whole bio-oil with solvents of different polarity is rarely described in literature. Selective extraction of crude bio-oil obtained from hydrothermal liquefaction of willow wood using nine solvents with different polarities were investigated in detail.[37] The bio-oil extraction yield of the nine solvents were from high to low: tetrahydrofuran (THF) > toluene > ethyl acetate > acetone > ether > methylene chloride > methanol > petroleum ether > *n*-hexane.

Based on their extraction yield, an efficient solvent combination of *n*-hexane, ethyl acetate, and tetrahydrofuran was used to separate the bio-oil through multistep extraction into three parts: light oil (26.1 wt%), mid-weight oil (54.2 wt%), and heavy oil (19.7 wt%). These fractions were characterized by gas chromatography-mass spectrometry, Fourier transform infrared spectroscopy, ^{1}H-nuclear magnetic resonance spectroscopy, and thermogravimetric analysis. The results showed that most of the valuable chemicals were contained in the light oil such as phenols and cyclopentenones; the mid-weight oil consisted of aromatic oligomers derived from the decomposition of lignin, which could be a promising candidate for partial substitute for petroleum- asphalt binder; the heavy oil was found to be rich in alkanes. As well as single solvent extractions a multistep sequential solvent extraction approach was also applied using tetrahydrofuran, followed by ethyl acetate, and finally by *n*-hexane.

The oil yield differed greatly with the polarity of the solvents. The highest yield (45 wt%) of oil extraction was obtained by using THF as the extraction solvent. Toluene also had a high extraction yield on bio-oil. Ethyl acetate, acetone, dichloromethane, ether, and methanol extracted nearly 30% bio-oil from the bio-crude, while *n*-hexane had an extraction yield of only 8.27 wt%. The yield of extracted bio-oil did not vary linearly with the polarity of the solvent, but fluctuated up and down. Therefore, solvent extraction not only can be considered as an efficient way to separate highly valued chemicals with oligomers, but also as a promising method to separate different chemical families from bio-oil.

According to Table 9.1, ketones and phenolic compounds were easier to extract by solvents with high polarity, such as THF, EAC, acetone, and methanol. Ketones and phenolics are polar compounds, and their extraction efficiency increased with increased extraction solvent polarity, which is in accordance with the principle of dissolution in a similar material structure.

A solvent extraction scheme for the grouping of bio-oils into different chemical families was reported by Garcia-Perez *et al.*[38] based on vacuum pyrolysis

Table 9.1 Extraction of main chemical families with solvents of different polarity. Reprinted with permission from X. Yang, L. Hang, K. Chen, X. Zhu, S. Zhang and J. Chen, *BioResourses*, 2014, **9**, 5219–5233.[37,a]

	Relative extraction percentage *via* various solvents (%)								
Chemical family	PE	HEX	TL	Ether	DCM	THF	EAC	DMK	MeOH
Ketones	50.2	40.2	50.4	59.5	85.0	100.0	89.4	94.5	94.1
Phenolics	68.2	45.8	51.9	65.9	70.9	91.4	88.9	90.1	100.0
Alkoxyphenolics	80.9	67.5	72.4	59.0	83.3	100.0	90.1	99.2	89.8
Indanones	59.8	43.3	29.4	34.5	50.6	74.7	100.0	92.6	90.3
Indoles	100.0	76.9	50.3	34.2	34.2	56.2	60.1	61.9	55.7
Esters	100.0	50.0	31.0	35.1	48.0	36.2	18.3	45.4	45.7
Fatty acids	100.0	30.4	24.7	35.4	54.6	23.0	19.2	26.8	27.6
Alkanes	100.0	55.1	70.5	71.3	60.1	74.2	44.7	30.9	6.2

[a]PE: petroleum ether (boiling point 60–90 °C); HEX: *n*-hexane; TL: toluene; DCM: dichloromethane; THF: tetrahydrofuran; EAC: ethyl acetate; DMK: dimethyl ketone (acetone); MeOH: methanol. Note: Peak areas of the main chemical families were compared after normalization.

oils obtained from softwood bark and hardwood. Seven grams of bio-oil were first extracted using 200 mL of toluene. This operation was performed in order to separate some of the wood-extractive derived compounds that interfere with heavy compounds precipitating. Waxy materials remaining in suspension in the toluene-soluble fraction were separated using a Whatman N1 42 filter (this type of filter was used in all filtrations). The toluene-insoluble fraction was solubilized in 200 mL of MeOH and filtrated in order to remove the char, non-polar waxy materials and other very heavy oligomeric compounds. The solvents were eliminated from the filtrates in a rotary evaporator and the dry residue was weighed. This residue was solubilized again in MeOH (5 mL of MeOH per 1 g of residue). 10 g of the MeOH–oil mixture was added to 300 mL ice-cooled distilled water dropwise under agitation. The water-insoluble fraction was removed by filtration. The solid residue was washed with distilled water for 1 h then was further extracted with dichloromethane until the filtrate was colorless. The solid remaining in the filter was dried at 105 °C overnight. The water-soluble fraction was further extracted with 300 mL of diethylether in a separation funnel. Diethyl-ether-soluble and CH_2Cl_2-soluble fractions were evaporated in a rotary evaporator at 40 °C. Low molar mass compounds were also removed during the solvent removal. These losses were reported as volatile compounds.

Table 9.2 presents the composition of hardwood rich in fibers (HWRF) and softwood bark residue (SWBR) bio-oils by LLE following the sequence shown in Figure 9.3. The upper layer from SWBR is mainly composed of compounds soluble in toluene (79.0 mass%). This fraction has been attributed to the "wood extractives-derived compounds". The water-insoluble fraction was separated into two sub-fractions (Fractions 5 and 6, Figure 9.3) depending on their solubility in CH_2Cl_2. The CH_2Cl_2-soluble sub-fraction (Fraction 6, Figure 9.3) is an important portion of SWBR bottom layer with 12.5 mass% and HWRF bottom layer with 13.7 mass%. This fraction is rich in low molar

Table 9.2 Fractionation of bio-oils using different solvents (wt% on bio-oils as received basis). Reprinted from Biomass and Bioenergy, **31**(4), M. Garcia-Perez, A. Chaala, H. Pakdel, D. Kretschmer, C. Roy, Characterization of bio-oils in chemical families, 222–242, Copyright (2007) with permission from Elsevier.[38,a]

Fraction	SWBR (Run: H-67) Upper layer	SWBR (Run: H-67) Bottom layer	HWRF (Run: G 823–830) Bottom layer
Water	3.50	14.60	13.00
Toluene solubles	78.95	1.29	7.76
MeOH insolubles	0.99	2.09	0.36
Water solubles	6.04	41.26	46.85
Diethylether soluble	0.74	7.94	4.78
Diethylether insoluble	5.30	33.32	42.07
Water insolubles	8.65	25.51	20.86
CH_2Cl_2 soluble	6.74	12.54	13.70
CH_2Cl_2 insoluble	2.18	12.97	7.16
Volatile loss (by difference)	1.87	15.25	11.17

[a]SWBR: softwood bark residue. HWRF: hardwood rich in fibers.

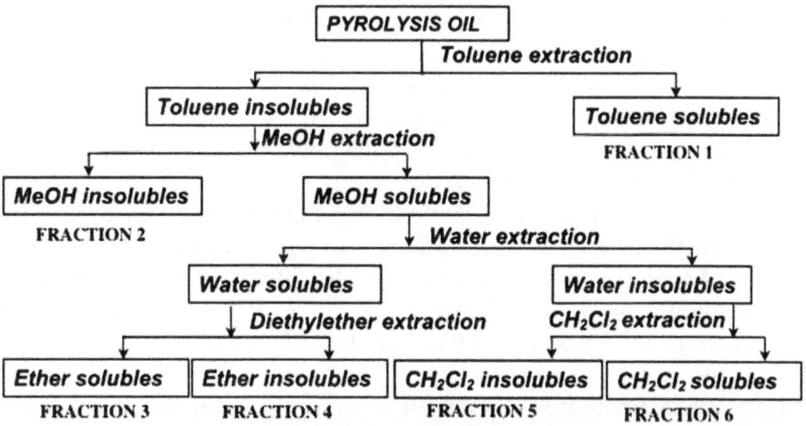

Figure 9.3 Fractionation scheme for bio-oil grouping into chemical families. Reprinted from Biomass and Bioenergy, **31**(4), M. Garcia-Perez, A. Chaala, H. Pakdel, D. Kretschmer, C. Roy, Characterisation of bio-oils in chemical families, 222–242, Copyright (2007) with permission from Elsevier.[38]

mass "pyrolytic lignin".[11,39,40] The CH_2Cl_2-insoluble sub-fraction (Fraction 5, Figure 9.3) has been referred to as the "high molar mass pyrolytic lignin" by Sipilä *et al.*[39] The fractions obtained by solvent extraction were clearly mixtures. Bio-oils were described as mixtures of eight macro-families.

Whole bio-oil extraction with two organic solvents (*n*-hexane and acetone) and liquid CO_2 was described by Feng and Meier in a comparative study also involving supercritical CO_2.[41] The latter method will be described later.

The pyrolysis liquid used in the experiments was provided by a commercial BBQ-charcoal producer (proFagus, Bodenfelde, Germany). It is a by-product of charcoal production through slow pyrolysis and it is named "SPL" (Slow Pyrolysis Liquid). SPL had a water content of around 0.7 wt% in contrast to fast pyrolysis liquids with a typical water content in the range of 25–28 wt%. Extraction of chemicals was carried out in a standard Soxhlet apparatus first with hexane and subsequently with acetone. Extraction ended when solvents appeared clean (after *ca.* 20 h). After extraction, solvents were removed with a rotary evaporator.

As the liquid samples could not be directly used in the Soxhlet apparatus, silica gel (0.2–0.5 mm) was used as carrier material to adsorb the bio-oil. The porous property and the polar nature of SiO_2 enables the absorbance of a high amount of pyrolysis liquid. Because intermolecular forces between pyrolysis liquid and SiO_2 could influence the selectivity of the extraction, two loadings of SiO_2 : oil (100 : 80 and 100 : 40) were chosen for comparison. SPL and SiO_2 were mixed in a glass beaker with a glass rod. After complete adsorption on SiO_2, the sample was transferred into an extraction thimble, topped with cotton wool and placed into the Soxhlet extractor.

Hexane has similar solvent properties to CO_2. Although *n*-hexane is a non-polar solvent, many polar substances in pyrolysis liquids are slightly soluble in it and could be finally extracted. The yield of hexane extracts was determined as 41.4 wt% for 40% loading and 56.2 wt% for 80% loading. Results showed that higher oil loading (80%) gives higher yields of hexane extracts. The intermolecular forces between pyrolysis liquids and SiO_2 could be totally reduced with higher loading.

Because pyrolysis liquids are almost completely miscible with acetone, it was used for the extraction of the residue remaining after hexane extraction. The yield of acetone extracts was 53.7 wt% for 40% loading and 40.1 wt% for 80% loading, resulting in a total recovery for solvent extraction of 94.8 wt% (40% loading) and 96.3 wt% (80% loading). This shows that components which could not be extracted with hexane were effectively extracted with acetone. Approximately 5 wt% remained on the SiO_2 due to strong intermolecular forces.[41]

Extraction with liquid CO_2 was carried out in a pressure cylinder equipped with a Soxhlet glass extractor. As with the solvent extraction, samples with two different oil loadings (40% and 80%) were used in the experiments. Around 300 g dry ice was introduced into the stainless steel extractor and the lid quickly closed. Afterwards, the apparatus was immersed 6 cm into a 40 °C water bath. The condenser finger was maintained at 0 °C. The equilibrium pressure was 4 MPa. In order to enable complete extraction, extraction time was set to 20 h. At the end of the extraction, CO_2 was carefully released with the help of needle valve in order to avoid any loss of solid sample from the thimble and of volatile components.

9.2.2.2.2 Extraction with Water and Organic Solvents. Many of the solvent-based extraction methods first start with a separation into water-soluble and water-insoluble fractions followed by subsequent extraction with organic solvents. Already in the early development phase of the fast pyrolysis technology, a simple water extraction technique was found to be adequate for the fractionation of the homogeneous, single-phase bio-oil into a water-rich phase and a water-poor phase.[11] Later Sipilä *et al.*[39] presented a scheme where the water-soluble fraction was treated in a subsequent step with diethylether. The yellow-colored water-soluble fractions comprised a considerable part of the pyrolysis oils. The proportion of water-solubles ranged from about 60 to 80 wt% of the oils. All the compounds in pyrolysis oil eluted by GC were also found in the water-soluble fraction. The amount of diethylether solubles determined as evaporation residues ranged from 4 to 8 wt% of the pyrolysis oils. The residues were bright yellow non-viscous liquids. The diethylether solubles contained low molecular mass degradation products of lignin and polysaccharides. Only a small part of the diethylether insoluble fractions was eluted by GC. The main compounds were levoglucosan and cellobiosan. The water-insoluble fractions contained high molecular compounds that were not eluted on GC.

Figure 9.4 shows a further development of the scheme as presented earlier by Sipilä *et al.*[39] It is extended by the introduction of a further separation of the water-insoluble fraction into dichloromethane- (DCM) solubles (low molecular mass lignin) and DCM-insolubles (high molecular mass lignin).[42–44]

Bennett *et al.*[45] investigated water extraction parameters of bio-oil from softwood in order to obtain high concentrations of levoglucosan for subsequent hydrolysis into glucose and fermentation to ethanol. Optimal selection of water-to-oil ratio, temperature and contact time yielded an aqueous phase containing a levoglucosan concentration of up to 87 g L^{-1}, a yield of 7.8 wt% of the bio-oil.

Vitasari *et al.*[46] studied water extraction parameters (water-to-oil ratio and stirring rate) on the yields of various bio-oil components such as glycolaldehyde, acetic acid, acetol, furfural, furanone, levoglucosan, syringol, and guaiacol. They concluded that stirring rate determines the time to reach equilibrium, but does not influence the equilibrium composition. The extent of water addition corresponds to water dilution. The distribution coefficient and extraction yield of a compound are determined by its polarity and solubility, water-to-oil ratio, and the nature of pyrolysis oil. They found that water extraction is not a selective method, but very useful to recover 80–90% polar compounds and to reduce the complexity of pyrolysis oil prior to further isolation steps.

Based on their findings a profitable process design of an integrated biobased acetic acid, glycolaldehyde, and acetol production in a pyrolysis oil-based biorefinery was developed and discussed.[47] The process design is aimed at recovering all acetic acid and glycolaldehyde with a purity >99% and comprises three extractions, three distillations, and a five-effect flash evaporation step[47] (see Figure 9.5).

The first step of the designed process is water extraction of pyrolysis oil. In the water extraction, the nature of the bio-oil determines the amount

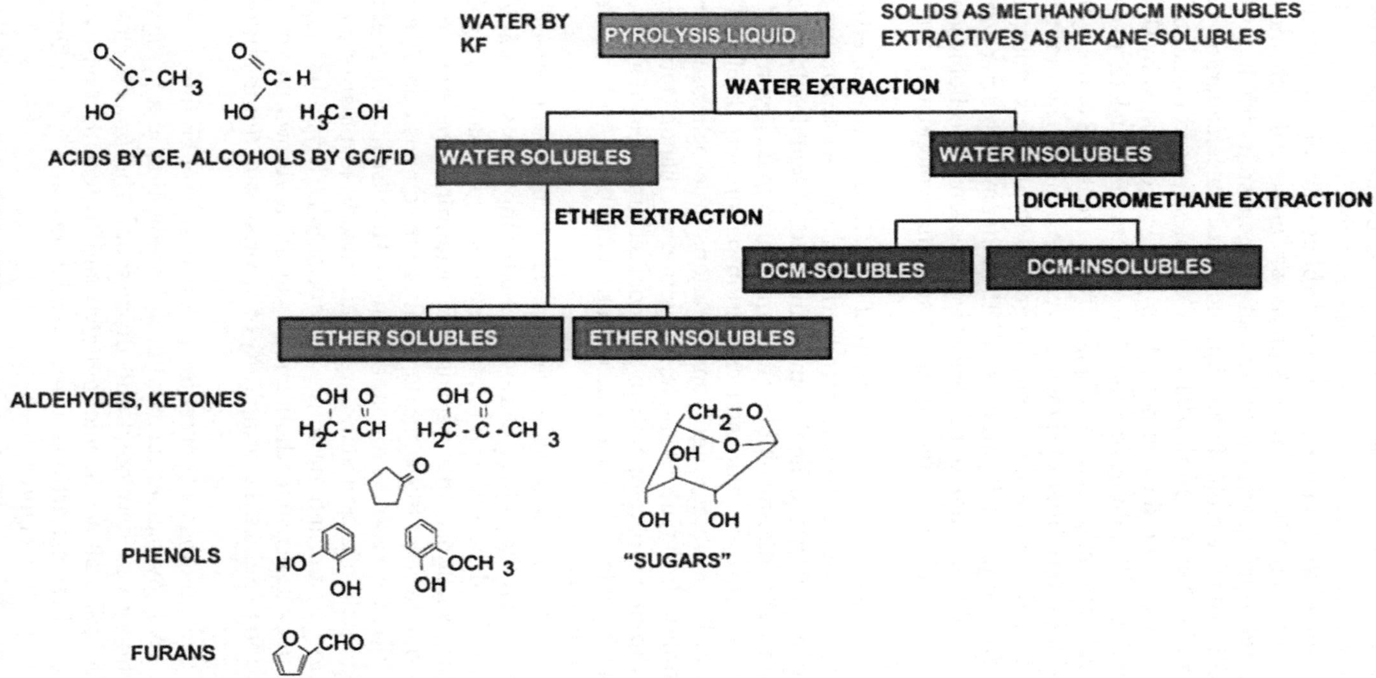

Figure 9.4 Solvent fractionation scheme for fast-pyrolysis liquids. Reprinted with permission from ref. 42, Copyright (2010) American Chemical Society.

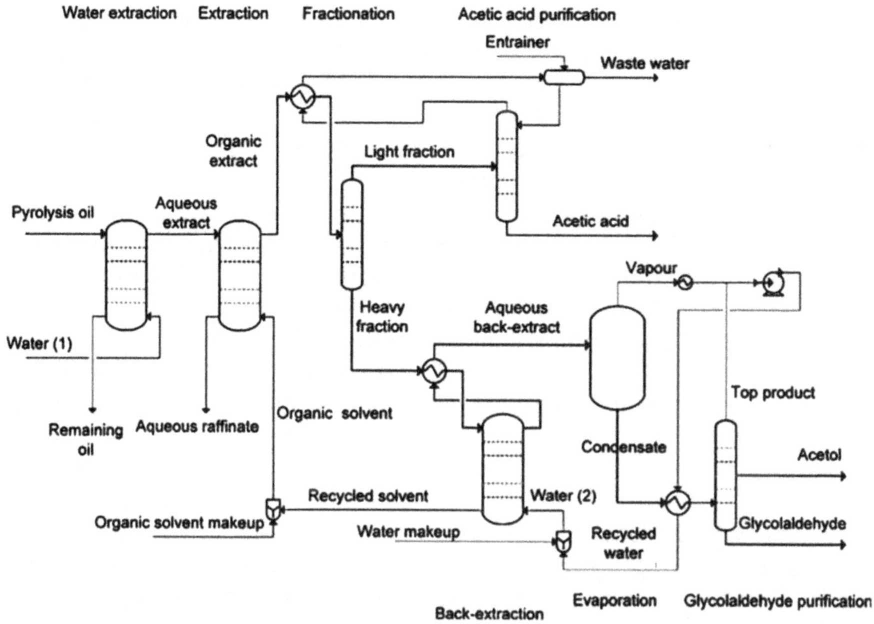

Figure 9.5 Integrated design process scheme for the separation of acetic acid, acetol, and glycolaldehyde. Reprinted from Chemical Engineering Research and Design, **95**, C. R. Vitasari, G. W. Meindersma, A. B. de Haan, Conceptual process design of an integrated bio-based acetic acid glycoaldehyde and acetol production in a pyrolysis oil-based biorefinery, 133–143, Copyright (2015) with permission from Elsevier.[47]

of necessary water. Forest-residue-derived pyrolysis oil required 50% more water compared to pine-derived pyrolysis oil to extract >99% acetic acid, glycolaldehyde, and acetol. Forest-residue-derived pyrolysis oil has an optimum single stage extraction yield at a water-to-oil ratio in the range of 0.65–0.7, while the optimum yield of pine-derived pyrolysis oil is achieved in a single stage extraction at the lowest feasible water-to-oil ratio (0.5).[46]

In a second step, the aqueous stream is extracted with 2-ethyl-1-hexanol. A solvent-to-feed ratio of 6 is needed to extract all acetic acid and glycolaldehyde from pyrolysis oil derived aqueous streams. This calculation result was based on the assumption that the distribution coefficients are composition independent.[48] Moreover, about two thirds of the acetol is co-extracted. Hence, the aqueous raffinate of the extraction column contains a considerable amount of residual acetol and levoglucosan. This stream may be subsequently fermented to produce itaconic acid[49] or citric acid.[50] Levoglucosan may be also hydrolyzed to glucose followed by fermentation to produce ethanol.[45] However, as the effect of acetol on the fermentation is still unknown, the treatment of this aqueous raffinate stream was excluded in the process design and considered as waste.[47]

Due to low extraction selectivity, the organic extract from the extraction column contains water, acetic acid, glycolaldehyde, and acetol in 2-ethyl-1-hexanol. Hence, after the extraction, the aqueous stream is split into light and heavy fractions based on their boiling points. In the fractionation column, only water and acetic acid are evaporated, while keeping the higher boiling point compounds in the bottom product.

Acetic acid is separated from water by heterogeneous azeotropic distillation with isobutyl acetate as the entrainer.[51] Pre-concentration is unnecessary prior to distillation, although the distillation feed contains around 70% water.[52] Both reboiler duty and entrainer make-up flow rate were varied to achieve acetic acid purity >99% and to minimize the entrainer loss with the top water stream.[51] This water stream is saturated with 0.7 wt% isobutyl acetate. Since the effect of isobutyl acetate on the extraction and distillation is yet unknown, the water stream is considered as waste.

The heavy fraction of the fraction column is subjected to further treatment (Figure 9.5). Back-extraction with water was chosen over vacuum distillation to separate glycolaldehyde and acetol from 2-ethyl-1-hexanol. This selection was based on the batch vacuum distillation experiments. The experimental results showed that some part of 1-octanol co-evaporated with glycolaldehyde and acetol, providing an organic distillate containing 1-octanol. Since the boiling point of 2-ethyl-1-hexanol is lower than that of 1-octanol (195 °C), it is expected that 2-ethyl-1-hexanol will be co-evaporated as well.

Therefore, the bottom stream of the fractionation column is back-extracted with water to recover glycolaldehyde and acetol. The aqueous extract is very dilute, containing more than 95% water. Thus, pre-concentration by water evaporation is needed prior to purification.

The five-effect flash evaporation is conducted at a reduced pressure. The last separation step is glycolaldehyde purification, which is a ternary distillation with a side acetol product stream and glycolaldehyde bottom product. This product split is in accordance with their vapor pressure data, in which acetol has generally been reported to have a higher vapor pressure than glycolaldehyde at the same temperature.[47]

It is well known that water is the most prominent single component in bio-oil. The typical water content ranges from 20–30 wt%. Mixing with water is possible up to about 45 wt%, then phase separation occurs and a tar-like product separates as a bottom phase and an aqueous phase as a top phase. Scott[11] and Radlein *et al.*[53] were the first to name the tarry phase as "pyrolytic lignin."

Pyrolytic lignin is a frequently used term for the high molecular weight water-insoluble parts of bio-oil. Scholze and Meier[40] reported on the isolation of a powder-like solid material from bio-oil for the first time by adding small amounts of oil to an excess of water (weight ratio >1 : 200). This approach is suitable for quantitative precipitation of hydrophobic lignin-derived fragments, which results after drying as a brown powdery material.

The precipitation of pyrolytic lignin is performed as follows: 60 mL pyrolysis oil is added dropwise to 1 liter of ice-cooled water while stirring at 6000 rpm with an Ultra-Turrax homogenizer (oil to water ratio must be a least

1:100). After slowly adding pyrolysis oil to water the precipitation of pyrolytic lignin begins. Subsequently, the pyrolytic lignin is filtered over a Buchner funnel, re-suspended in water and slowly stirred for about 4 h in order to remove further water-soluble material. Finally, the suspension is filtered again and the wet pyrolytic lignin is dried carefully under vacuum at room temperature.

Pyrolytic lignin precipitation is material- and time-consuming and in the majority of cases simply done to obtain the content of PL. Therefore, simpler methods are needed to get satisfactory reproducibility and to make the process more flexible towards the amount of available sample material. Hence, for PL quantification the following modified precipitation methods can also be applied:

(A) A laboratory mixer (22 000 rpm) filled with distilled water (500 mL) is used to precipitate PL through dropwise feeding 1 mL of bio-oil. The obtained oil-in-water suspension is filtered under slight vacuum (500 mbar) using Blauband filter paper. Sticky residues in the mixer are dissolved in ethanol, which is removed under vacuum (78 °C; 200 mbar) with a rotary evaporator.

(B) A rapid application can also be carried out by a careful introduction of a 100 µl bio-oil sample in 30 mL water using a micro Ultra-Turrax homogenizer (20 000 rpm) and a 60 mL plastic syringe. The syringe is first coupled to a disposable syringe filter (1 µm) connected *via* a Luer lock connection and then mounted on a vacuum manifold normally used for solid phase extraction (SPE). After filtering the suspension, the PL residue on the filter is rinsed with another 30 mL of water, dried under vacuum and weighed to get the PL-content by weight difference.

9.2.2.2.3 Extraction with Supercritical CO_2. Supercritical fluids, especially CO_2, are widely used in the extraction of natural matter, such as essential oils, herbal drugs, antioxidants and other materials.[54] The extraction and fractionation of bio-oils with supercritical fluids is gaining increasing importance.[27,55–61] Compared with expensive organic solvents, CO_2 proves to be a good supercritical medium for the extraction of thermally sensitive components as encountered in bio-oil with relative low pressure (7.38 MPa) and temperature (31.1 °C). CO_2 in its supercritical state ($scCO_2$) behaves like a hydrocarbon solvent with very low polarizability.[62]

As already mentioned above, pyrolysis liquids are a complex mixture of several hundred compounds and many of them are in low concentrations. The separation of single components is technically feasible but economically unattractive. Most of the compounds contained in pyrolysis liquids are polar in nature and many of them have a relatively low solubility in $scCO_2$ or liquid CO_2. As a result, $scCO_2$ extraction comprises the extraction of both easily accessible solutes and solutes with low solubility. Table 9.3 shows the solubility of some selected chemical substances measured by Francis in liquid CO_2.[63]

Table 9.3 Mutual solubilities of some chemicals with liquid carbon dioxide. Reprinted with permission from ref. 63, Copyright (1954) American Chemical Society.[a]

Compounds	*x*	*y*
Formic acid	M	M
Acetic acid	M	M
Acetone	M	M
2-Butanone	M	M
Methanol	M	M
Isopropylalcohol	M	M
Ethyleneglycol	7	0.2
Ethyl formate	M	M
Ethyl acetate	M	M
Acetaldehyde	M	M
Benzene	M	M
Phenol	3	3
o-Cresol	30	2
m-Cresol	20	4
p-Cresol	30	2
Eugenol	38	10
Water	6	0.104

[a]M = miscible; *x*-value: solubility of liquid CO_2 in substance in wt%; *y*-value: solubility of substance in liquid CO_2 in wt%.

scCO_2 extraction was adopted for a feasibility study of the separation of phenol-rich oil from bio-oil by Patel *et al.*[60] In their study, the scCO_2 extraction was performed with bio-oils derived from sugarcane bagasse and cashew nut shells with a pressure range of 120–300 bar, temperature range of 30–60 °C and scCO_2 mass flow rates of 0.7–1.2 kg h^{-1}. A phenolic-rich oil was extracted, the concentration of extracted phenolic compounds of which was about 72%, with a total oil yield of 15%. Cardanol-rich oils were also extracted from the bio-oils derived from cashew nut shells. A higher concentration of cardanol (86%) was observed in the extract, along with approximately 5% phenols (5%). The oils obtained through the supercritical fluid extraction route is free from all acidic and alcoholic components. This is due to the fact that CO_2 is non-polar, thus the solubility of highly polar substances like acids and alcohols in scCO_2 is negligible.

Naik *et al.* studied the scCO_2 fractionation of bio-oil produced from mixed biomass of wheat and wood sawdust[58] and wheat-hemlock.[57] Their scCO_2 extracts were less acidic and moisture free with a higher calorific value (around 40 kJ kg^{-1}). Extraction exhibited some selectivity: low pressure favored the extraction of furanoids, pyranoids and oxygenated benzenoids while high pressure stimulated the extraction of fatty acids and high molecular mass alcohols.

Wang *et al.*[59] employed scCO_2 extraction for the separation of simulated and real bio-oils. Weakly polar components could be selectively enriched in the scCO_2 phase by changing the temperature, pressure and the use of different adsorbents, which can affect the intermolecular forces. After scCO_2

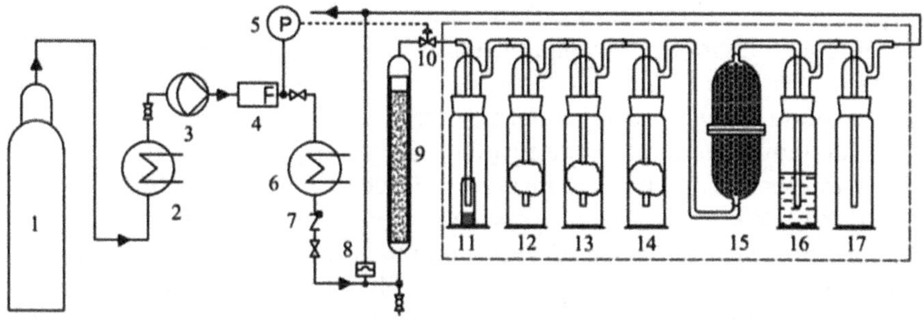

Figure 9.6 Scheme of supercritical CO_2 extraction unit. 1 CO_2 cylinder. 2 Cooler. 3 Membrane pump with cooled head. 4 CO_2 mass flow meter. 5 Pressure gauge. 6 Pre-heater. 7 Check valve. 8 Rupture disc. 9 Extractor. 10 Back pressure regulator. 11 Collector. 12–14 Cotton wool gas filters. 15 Activated carbon filter. 16 Water wash bottle. 17 Gas wash bottle. Reprinted from Biomass and Bioenergy, **65**, Y. Feng, D. Meier, Comparison of supercritical CO_2, liquid CO_2, and solvent extraction of chemicals from a commercial slow pyrolysis liquid of beech wood, 346–354, Copyright (2016) with permission from Elsevier.[41]

extraction, the quality of bio-oil could be improved by the reduction of water content to one fifth, a two-fold increase of the calorific value, and a decrease in acidity together with improved stability and appearance.

Feng & Meier[64] investigated the competing effects of volatility and solubility of organic components of $scCO_2$ extraction for both slow pyrolysis and fast pyrolysis liquids with different water contents. Pyrolysis liquids were adsorbed on silica gel and extraction was carried out in a commercial $scCO_2$ extraction system (HDT Sigmar Mothes, Berlin, Germany) depicted in Figure 9.6. The extractor volume is 640 mL, the length 750 mm, and internal diameter 33 mm. For sample preparation, 40 g bio-oil was adsorbed on 100 g SiO_2 (40% loading) and 80 g bio-oil was adsorbed on 100 g SiO_2 (80% loading). Pyrolysis oil and SiO_2 were mixed in a glass beaker with a glass rod. After complete adsorption the sample was transferred into a basket made of stainless steel mesh and placed into the extractor. CO_2 from the cylinder was cooled before pumping to maintain liquid conditions. After pressurization by the membrane pump, it flowed first into the pre-heater to reach supercritical conditions and afterwards passed the extractor.

The extraction pressure was set at 20 MPa and controlled automatically by a back pressure regulator. The extraction temperature was set to 50 °C together with the pre-heater and all the pipes before the needle valve. In order to avoid the formation of dry ice in the needle valve (the Joule-Thomson effect), the temperature of the needle valve was heated to 200 °C. Total extraction time was 6 h. CO_2 flow rate was set to 500 g h^{-1} and controlled by the membrane pump. The extracts were collected in glass vials at room pressure and temperature and sampled every 1 h. Therefore, the mass of extracts and the mass of CO_2 consumed per 1 h could be observed to evaluate the extraction process.

The extracts collected show a transparent reddish color with pungent odors due to the high amount of volatile compounds. They are clear, oil-like liquids with lower viscosity compared to the original pyrolysis liquid. No chars or solids are found in the extracts rendering them more stable than pyrolysis liquid. Pyrolytic lignin remains in the residue. No further phase separation occurs in the extracts as determined by visual inspection. The color of the extracts becomes darker after several days when stored at room temperature, indicating possible reactions of chemical compounds with light and air. This process is slowed down when the samples are stored in dark and cold conditions.[41]

It was noted that higher CO_2 flow rates result in higher amounts of available fresh CO_2 which can dissolve more compounds. Furthermore, it was observed that a flow rate of 600 g min^{-1} CO_2 gave a yield up to 51.1% under the same conditions.[64] In the subsequent extraction period, the extraction efficiency decreases because the easily accessible solute fraction becomes exhausted. Within the first three hours of extraction, 19.4% extracts were obtained for 40% loading and 35.0% for 80% loading. But during the next three hours of extraction, only 4.0% and 6.0% extracts were obtained for 40% and 80% loading, respectively. The amount of extracts obtained from the 80% loading was higher than that of 40% loading.

Ketones were found to be the most abundant group in all the extracts, followed by aromatics, furans, acids, esters, acetates and pyrans. Based on GC-identified compounds, the chemical groups of slow pyrolysis liquid, $scCO_2$ and liquid CO_2 extracts are similar but different from hexane and acetone extracts. It could be shown that that the most abundant group (ketones) is effectively enriched in both scCO2 and liquid CO_2 extracts. Comparing $scCO_2$ and liquid CO_2 extraction, guaiacols content is higher in scCO2 extracts, indicating that scCO2 has an improved solubility of guaiacols compared to liquid CO_2. Non-aromatic compounds such as ketones, acids, furans, esters, pyrans and acetates could be effectively enriched in $scCO_2$ and liquid CO_2 extracts. Aromatic compounds are almost completely extracted with hexane.[41]

Recently, Cheng *et al.*[55] presented a three-step supercritical CO_2 extraction for selective fractionation of fast pyrolysis bio-oil. With appropriate optimization of extraction parameters, lipids, hemicellulose, lignin, and condensed aromatics were enriched in three different fractions. $scCO_2$ extraction of bio-oil was carried out with an accelerated supercritical fluid extraction system (MV-10 ASFE, Waters, USA). One gram of bio-oil was homogeneously mixed with 8 g diatomaceous earth in a 25 mL stainless steel extraction cell. The extracts were collected in methanol to minimize the loss of volatile compounds. The whole SFE procedure was carried out successively under three different conditions, in three consecutive steps. The first step lasted 5 min under 10 MPa using pure $scCO_2$ as the extraction solvent; the second step lasted 25 min under 20 MPa using 90% scCO2 and 10% methanol as the extraction solvent; the third step lasted 40 min under 30 MPa using 75% $scCO_2$ and 25% methanol as the extraction solvent. The extraction temperature was kept at 50 °C and the total solvent flow rate was 1.0 mL min^{-1}. The yields for the three extracts were 13.4 wt%, 24.8 wt%, and 32.9 wt%, which accounted for 71.1 wt% of the raw bio-oil. If the water content (26.9 wt%) of raw bio-oil was counted, the total yield of $scCO_2$ extraction should be 98 wt%. It was concluded that with appropriate optimization of extraction parameters, pyrolysis bio-oil can be roughly extracted into compound-type-specific fractions. Results indicated that CO_2 SFE has the potential for bio-oil preprocessing to meet demands for fuel production and fine chemical production.[55]

9.2.2.2.4 Extraction with Switchable Hydrophilicity Solvents (SHS). Removal of solvents from products by distillation is a common industrial practice that, unfortunately, requires the use of a volatile solvent. Evaporation of volatile solvents contributes to smog formation, inhalation toxicity, flammability and high insurance costs. Jessop *et al.* sought a new non-distillative route for the separation of solvents from products, so that volatile solvents would no longer have to be used. They recently published the first switchable hydrophilicity solvent (SHS), and proposed that its unusual phase behavior with water could form the basis of a non-distillative method for solvent removal.[65,66] An SHS is a solvent that has very little miscibility with water in the absence of CO_2 but complete miscibility with water in the presence of CO_2 (Figure 9.7).

The change in miscibility is caused by a chemical reaction of the CO_2 and water with the SHS, giving a water-soluble bicarbonate salt of the protonated SHS. The SHS must therefore be sufficiently basic to be protonated by carbonic acid. CO_2 is preferred as the trigger for the switching process because it is non-toxic, benign, inexpensive and easily removed.

Jessop *et al.*[66] reported the first SHS, a liquid amidine, *N*,*N*,*N*′-tributylpentanamidine. They showed that it could be used as a substitute for volatile solvents and could be removed from organic products (such as soybean oil) without distillation, by extraction of the solvent with carbonated water. However, because *N*,*N*,*N*′-tributylpentanamidine is commercially unavailable, very difficult to synthesize on any reasonable scale, potentially hydrolytically unstable and sufficiently hydrophobic to bioaccumulate, it is impractical as a solvent. Hence Jessop *et al.* identified several simple tertiary amines that can serve as SHSs and are either commercially available or easily prepared.

Switchable solvents can be switched reversibly from a liquid with one set of properties to another that has very different properties, upon application or removal of a trigger (typically CO_2). They can be categorized as switchable polarity solvents, SHSs, and switchable water. These solvents are switchable between polar and non-polar, between hydrophilic and hydrophobic, and between high ionic strength and low ionic strength, respectively. A recent review paper describes all of the different kinds of switch-able solvents.[68]

In a recent study SHSs were applied to pyrolysis liquids made from kraft lignin through microwave pyrolysis. The SHS extraction method should be general to all pyrolysis methods such as fluidized bed pyrolysis, because the derived bio-oils are similar in chemical composition; microwave pyrolysis oil was used just as an example.[69] A method for using SHSs to extract phenols as a mixture from the lignin microwave pyrolysis oil was successfully developed and scaled up to 10 g of bio-oil. The original oil and its fractions after separation were analyzed by quantitative NMR spectroscopy, which showed that the phenols were concentrated in a single fraction, that contained, for example,

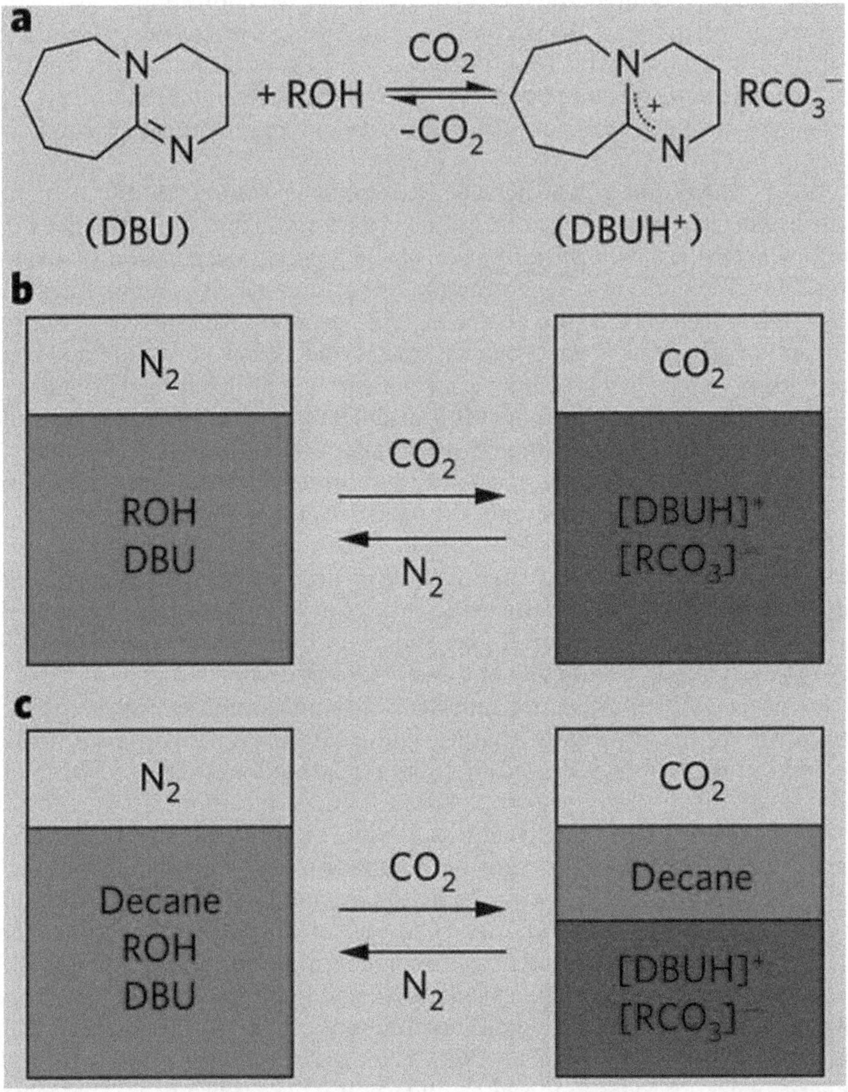

Figure 9.7 The "switching" of a switchable solvent SHS.[67] (Reproduced from Ref. 67 with permission from The Royal Chemistry Society). **a:** Protonation of DBU (1,8-diazabicyclo-[5.4.0]-undec-7-ene) in the presence of an alcohol and carbon dioxide is reversed when CO_2 is removed. **b:** Polarity switching in the reaction shown in **a**, in which CO_2 causes a non-polar liquid (shown in medium grey) mixture of hexanol and DBU to change over one hour into a polar, ionic liquid (shown in dark grey); nitrogen gas reverses the process by stripping out CO_2 from the reaction. **c:** The different polarity of each liquid under the two conditions is illustrated by the miscibility of decane with the hexanol/DBU mixture under nitrogen, before exposure to CO_2: however, decane separates out once the mixture becomes polar in the presence of CO_2. Again, N_2 reverses the process.

72% of the guaiacol and 70% of the 4-methylguaiacol present in the original unfractionated oil. The SHS (91%) was recovered after extraction.

9.2.2.3 Separations Based Upon Salt Formation

Typical technologies for the separation of phenolic compounds from bio-oil are not very different from the separation technologies presented before. Phenolic compounds of pyrolytic oils are usually extracted with basic aqueous solutions of 8–15% wt% NaOH in soluble polar organic solvent media such as ethyl acetate and methyl isobutyl ketone. In general, sodium phenolates are formed which can easily be separated. An example at laboratory scale can be found elsewhere.[70,71]

Gallivan and Matschei[72] from American Can. Co. investigated a method of fractionating pyrolysis oils to obtain a phenolic fraction which was suitable as a total or partial substitute for phenol in making phenol formaldehyde (PF) resins. Pyrolysis oils were obtained from the pyrolysis of wastes such as bark, sawdust, tree-tops and limbs. The fractionation method (see the schematic diagram in Figure 9.8) comprises mixing the oil with NaOH to a pH level at

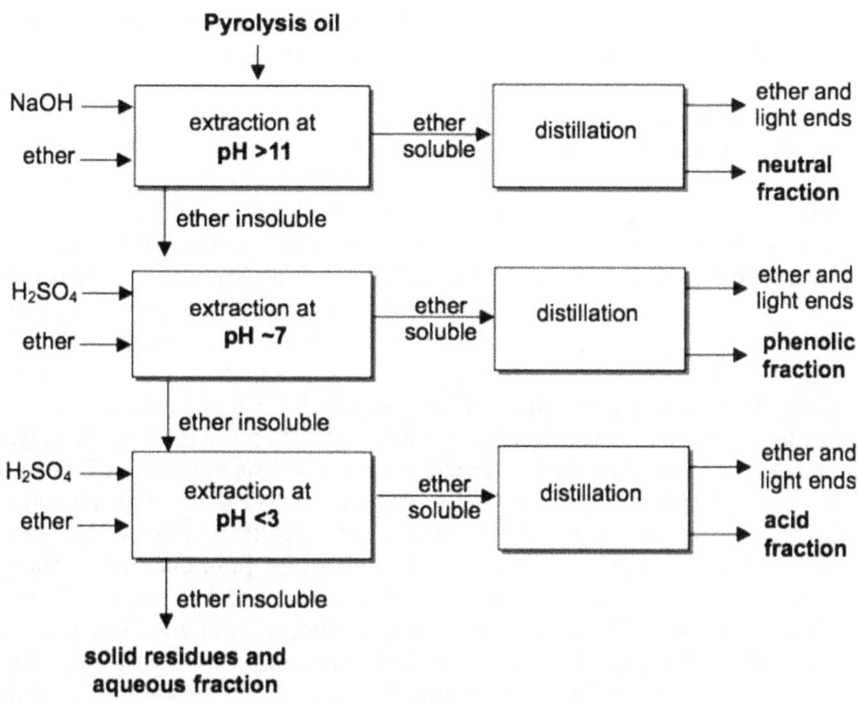

Figure 9.8 Schematic diagram of fractionation steps for isolation of phenolic compounds.[72] Reprinted from Renewable and Sustainable Energy Reviews, **12**(8), A. Effendi, H. Gerhauser, A. V. Bridgewater, Production of renewable phenolic resins by thermochemical conversion of biomass: A review, 2092–2116, Copyright (2008) with permission from Elsevier.[73]

which the neutral fraction of the oil is selectively soluble in a solvent such as methylene chloride or ether. The mixture is extracted with the solvent to obtain a first extract containing the solvent and the neutral fraction, and a first raffinate containing the remaining fractions of the oil, *i.e.* phenolics, organic acid fractions and amorphous residues. The neutral fraction is recovered by distillation and the first raffinate is mixed with sulphuric acid to lower its pH to a level at which the phenolic fraction is selectively soluble in the solvent.

This raffinate is extracted with the solvent to obtain a second extract containing the solvent and the phenolic fraction and a second raffinate containing the organic acids and the residues. The phenolic fraction is recovered by distillation and the second raffinate is mixed with sulphuric acid to lower its pH to a level at which the organic acids are selectively soluble in the solvent. After separating the residues, the second raffinate is extracted with the solvent to obtain a third extract, which is distilled to recover the organic acid fraction of the oil. The complexity of solvent extraction routes limits the industrial application of such a process. From the method applied roughly 33% phenolic, 32% neutral and 24% acid fractions could be recovered. The phenolic fraction may be used for partial or total replacement of pure phenol in making PF resins. A resin was formulated by Gallivan and Matschei using the whole phenolic fraction (25 g), 37% formaldehyde (87 g), water and NaOH catalyst. The mixture was refluxed for less than 1 h. The resulting adhesive was used to laminate veneer panels (plywood), which were tested for wood failure. The phenolic-fraction-based adhesive showed comparable results with those made from petroleum PF resin.[73]

A complex liquid–liquid partition scheme of bio-oils from slow and fast pyrolysis was developed by Maggi & Delmon.[74] They separated various bio-oils from fast and slow pyrolysis processes into four main fractions: phenols and acidic compounds (I), bases (II), "polar" neutrals (III), and hydrocarbon or "non-polar" neutrals (IV). In their partition scheme, the cumulated yields of the various fractions did not reach 100%. The difference corresponds to very polar molecules retained in the aqueous layer obtained during the separation of the acidic fraction (I). This aqueous layer was re-extracted with chloroform and named the "aqueous fraction" (V).[74] The yields obtained after the extraction procedure are presented in Table 9.4. The efficiency of the partition was controlled by thin layer chromatography (TLC) and Fourier transform infrared spectroscopy (FTIR). Most of the GC-detectable components were found in the acidic fraction (I). They contain phenols and methoxy derivatives. The basic fraction (II) is small and contained N-containing aromatic compounds such as quinoline and aminonaphthalene. The neutral fraction (III) represents also a small fraction. Here, cyclic esters (*e.g.* γ-butyrolactone) was detected. The "non-polar" fraction (IV) contained mainly aromatic and cyclic compounds of 2–4 rings. The aqueous fractions (V) contain mainly carboxylic acids and esters, ethers and alcohols.

Shriner *et al.*[75] report general procedures and schemes for isolation of water-soluble and water-insoluble mixtures of any kind and Mohan *et al.*[76] adopted their procedures to bio-oils.

Table 9.4 Yields obtained after liquid–liquid partition. Reprinted from Fuel, 73(5), R. Maggi, B. Delmon, Comparison between 'slow' and 'flash' pyrolysis oils from biomass, 671–677, Copyright (1994) with permission from Elsevier.[74]

	Oil (wt%)				
Fraction	Acacia	Oak	Eucalyptus	Pine bark	RTP[a]
I (acidic)	40.0	43.08	26.6	21.8	34.5
II (basic)	0.7	0.39	0.57	0.8	0.4
III (neutral)	6.7	4.96	6.82	5.6	6.0
IV (hydrocarbon)	10.0	11.02	7.95	11.9	2.7
V (aqueous)	22.7	12.0	[b]	[b]	17.0
Precipitates	13.3	9.12	13.57	11.3	18.9
Yield	93.4	80.57	55.51	51.4	79.5

[a]Rapid Thermal Processing.
[b]Not recovered.

Further practical and useful details on laboratory extraction procedures and the theory behind them can be found elsewhere.[77] However, several authors have reported many practical problems associated with the extraction of phenols from pyrolysis oil such as the redistribution of phenols in both phases and precipitation between the organic phases.[78]

An important extraction scheme was developed by Chum *et al.* aimed at producing phenolic resins from bio-oil.[79–81] First, the whole bio-oil (1 kg) was dissolved in ethyl acetate on a 1:1 (wt:wt) basis. The oil was then vacuum-filtered through filter paper to remove char. Upon standing, the EA/pyrolysis oil separated into two phases: an organic-rich, EA-soluble (top) phase and an EA-insoluble (bottom) phase. Most of the water formed during pyrolysis was contained in the EA-insoluble phase. The EA-soluble portion of the oil was washed with water (2 × 75 mL) to remove the remaining water-soluble products. The EA-soluble phase was then extracted with aqueous $NaHCO_3$ (5% w/w, 10 × 200 mL) and the basic aqueous layer was saved for isolation of the acidic organic fraction, which contained the phenolic and neutral (P/N) fractions. The yield of the phenolic and neutral fractions from the extraction procedure is about 30% of the fast pyrolysis oil derived from sawdust and about 50% of the oil derived from bark. Phenolic/neutral (P/N) fractions have molecular weights of 100–800 g mol^{-1}. The fraction also contains many compounds with aldehyde groups. This is advantageous, as less formaldehyde is needed for resin formulations.[73] EA was evaporated from the EA extraction layer. Because EA was not dried prior to evaporation, water formed an azeotrope during distillation. The scheme of the procedure was discussed by Mohan *et al.*[76] (see Figure 9.9).

Wang *et al.*[82] explored the multistep separation of mono-phenols and pyrolytic lignin from the water-insoluble phase of bio-oil. They employed acid and alkaline solutions combined with organic solvents for the separation. The relative contents of total phenolic components in alkaline solutions amounted to 94.3%, the content of guaiacols reached 48.3%.

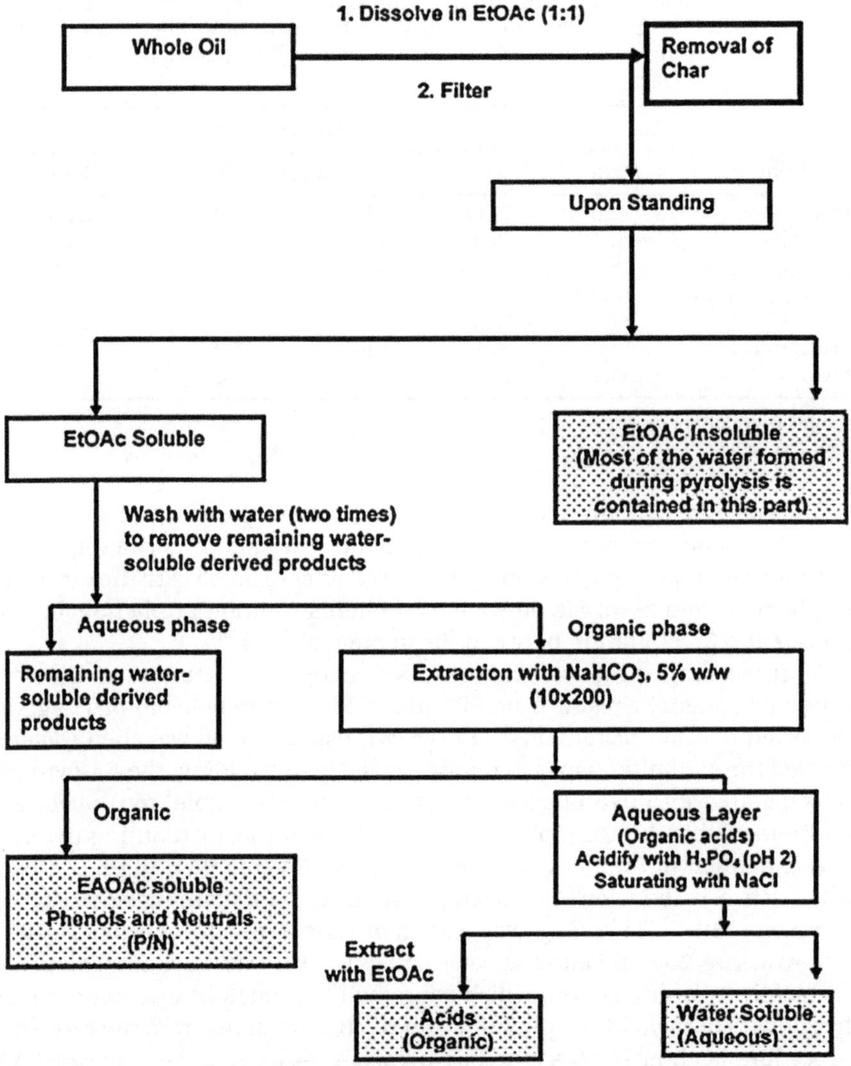

Figure 9.9 Midwest Research Institute (MRI) Ventures Inc. extraction scheme.[76,79,80] Reprinted with permission from ref. 76, Copyright (2006) American Chemical Society.

The addition of pure salts and salt solutions for inducing phase separation of bio-oil was investigated by Song *et al.*[83] Samples of bio-oil (10 mL) were placed into glass tubes (15 mm in diameter, with a capacity of 15 mL) and 0.3 g salt (LiCl, $FeCl_3$, $(NH_4)_2SO_4$, $Fe(NO_3)_3$) or 1 mL of 30% salt aqueous solution (LiCl, $CaCl_2$, $(NH_4)_2SO_4$, K_2CO_3), were added with stirring and sonication in a water-cooled bath below 15 °C. Afterwards, the tubes were sealed and stored for 10 h at room temperature, forming two phases (upper/bottom layer). The upper layer was decanted and the two phases were weighed and characterized. The upper layer was further extracted with diethylether and the lower phase was extracted with dichloromethane.

The addition of a little salt (3 wt% of bio-oil) or a solution (10 wt% of bio-oil) into bio-oil resulted in a quick phase separation (40–80 wt% of top phase, 20–60 wt% of bottom phase). The ratio of the two phases depends on salt added and its dosage. Some compounds with similar polarity are concentrated in different phases, such as acetic acid, alcohols, and other water-soluble compounds in upper layers, and lignin-pyrolysis compounds in the bottom layers. The upper layers exhibit high contents of water, acetic acid, and water-soluble compounds, low density and viscosity, low calorific values, and high distillable substances (up to 65%); and the bottom layers have contrary properties: low water contents, high amounts of lignin-derived compounds, high viscosity and calorific values, and low amount of distillable substances (<10%). The nature of salt and its dosage has an influence on the physicochemical properties and composition of components of the two phases. Addition of neat water (10 wt% of bio-oil) results in a very low portion of the bottom phase (5 wt%). Addition of aqueous salt solutions destroys hydrogen bonds and enhances polarity of aqueous phase, causing agglomeration and separation of the lignin micelles.[83]

9.3 Affinity-Based Separation

9.3.1 Adsorption – Desorption

9.3.1.1 Introduction

Purification of most of the products produced in biorefineries cannot be achieved by standalone conventional technologies and requires the use of combinations of separation and purification technologies. An attractive alternative is adsorption, which can perform many selective separations that are not feasible by conventional techniques alone, such as distillation or extraction.[84] In the adsorption process, molecules are preferentially bound to a solid surface; these surface phenomena can occur at any solid–liquid interface. Adsorption occurs due to the attractive forces existing between the adsorbent surface and the adsorbate molecules. The basic types of attractive forces are van der Waals forces, electrostatic forces, and chemical bonding.[84] Therefore, the type of adsorbent plays a significant role in the separation processes. The adsorbents have a broad range of chemical and geometrical structural forms. For bio-oils the main adsorbent is silica in different forms.

9.3.1.2 Open-Column Chromatography (Sequential Elution by Solvents Chromatography, SESC)

The principle of column chromatography is that substances are separated based on their different adsorption capabilities on a stationary phase. Thus, the process of column chromatography involves adsorption, desorption, re-adsorption, and re-desorption. For very complex mixtures such as coal or biomass-derived liquids mainly normal and reverse phase chromatographic methods have been applied. In normal phase chromatography polar silica gel is commonly used as the stationary phase, and an eluent is selected according to the polarity of the components. The less polar components will be eluted faster than the more polar ones. Paraffin eluents, such as hexane and pentane, are used to separate aliphatic compounds. Aromatic compounds are usually eluted with benzene or toluene. Highly polar compounds are obtained by elution with methanol.[85,86]

In reverse phase chromatography, the polarities of the mobile and stationary phases are opposite to what they were when performing normal phase chromatography. Instead of choosing a non-polar mobile phase solvent, a polar solvent will be chosen. Moreover, the stationary phase is based on modified silica, *e.g.* with alkyl side chains (C18). Thus, if the separation requires a solvent polarity gradient, the gradient must be carried out with the most polar solvent first and the least polar solvent last (reverse order of normal phase chromatography).

Farcasiu[87] was the first to develop a method based on silica gel to fractionate complex coal-derived liquids from direct coal liquefaction processes by sequential elution with different solvents (SESC = sequential elution by solvents chromatography). The SESC method uses a liquid chromatographic separation methodology. Silica gel was chosen as stationary phase because its mildly acidic properties are favorable to the separation of coal liquids, and because fewer of the substances present in the mixture are irreversibly adsorbed by silica gel than by other adsorbents. The criteria used in choosing the sequence of the solvents for eluting the polar materials are based on the Hildebrand solubility parameters. In principle, the sequence of solvents for polar compounds (see below) changes systematically from solvents with dipole moment but no proton-acceptor properties (as in chloroform) to those that are highly polar and have good acceptor properties (such as pyridine). The ratio of sample : silica gel was 1 : 25 (by weight). Silica gel was washed with methanol and dried, and the water content was adjusted to 4%. The fractions were eluted successively with the following solvents:

1. Hexane
2. Hexane–15% benzene
3. Chloroform
4. Chloroform–10% diethyl ether
5. Diethyl ether–3% ethanol
6. Methanol
7. Chloroform–3% ethanol
8. Tetrahydrofuran–3% ethanol
9. Pyridine–3% ethanol

Separation into chemically different fractions was obtained and confirmed by various analytical methods.

Bio-oil fractionation using the SESC methodology was first reported by Davis *et al.*[88] They separated wood oils from direct liquefaction into fractions according to chemical functionalities. Meier *et al.*[78] also applied SESC for oils obtained from direct biomass liquefaction. SESC has proved to be fairly simple, reproducible, and time-saving. An open glass chromatography column (length = 48 cm, internal diameter = 8 mm) was packed with 12 g silica gel.[60] The oil (200 mg) dissolved in dichloromethane was adsorbed on 800 mg of silica gel; after evaporating the solvent this silica gel was transferred to the column. The elution was carried out using seven freshly distilled solvents in order of increasing polarity:

I. 1-Chlorobutane
II. Chloroform
III. Chloroform + 6% diethylether
IV. Diethylether + 4% ethanol
V. Methanol
VI. Acetic acid
VII. Water

Complete elution of each fraction was obtained using 150 mL of solvent. The solvent was removed by rotary evaporation and the weight of the residue was determined. Yields are presented in Table 9.5. Fraction I, which has the lowest polarity, is a lightly colored oil of very low viscosity. All of the feedstock oils examined contained between 2 and 8% of this fraction. The more

Table 9.5 Percentage yield of SESC fractions. Reprinted from Fuel, **65**(7), D. Meier, D. R. Larimer, O. Faix, Direct liquefaction of different lignocellulosics and their constituents 1. Fractionation, elemental composition, 910–915, Copyright (1986) with permission from Elsevier.[78]

	Fractions (%)							
Feedstocks	I	II	III	IV	I–IV	V	VI–VII	Total
Spruce wood	7.0	27.0	8.0	9.0	51.0	22.5	19.0	92.5
Birch wood	3.5	6.5	22.5	29.0	61.5	15.5	14.5	91.5
Sugar cane bagasse	3.5	6.5	14.5	26.5	51.0	18.5	12.5	82.0
Barley straw	2.7	6.7	16.7	28.0	54.1	20.0	8.7	82.8
Pine bark	8.0	5.0	24.5	41.5	79.0	11.0	8.0	98.0
Cellulose	5.3	5.0	17.7	31.3	59.3	17.3	8.3	84.9
Spruce holocellulose	3.5	7.0	21.0	30.0	61.5	10.5	13.5	91.5
Birch holocellulose	3.5	7.0	21.0	30.0	61.5	10.5	13.5	85.5
Spruce organosolv lignin	5.0	17.5	9.0	14.0	45.5	19.0	14.0	78.5
Bagasse organosolv lignin	3.5	12.5	14.5	27.0	57.5	24.5	10.0	92.0
Birch Willstätter lignin	2.0	10.0	14.7	31.7	58.4	18.7	8.3	85.4
Mean standard deviation of replicates	±1.3	±2.5	±2.3	±5.9	±5.9	±1.9	±3.4	±4.8

polar fractions II and III showed increased viscosity. Fraction IV did not flow at room temperature. The highly polar SESC fractions V, VI, and VII were practically insoluble in acetone and methanol.

The principal advantage of the SESC fractionation is the separation of the bio-oil into distinguishable classes according to chemical functionality. Another advantage over the LLE procedure is minimal loss of the bio-oil during fractionation. On an average only approximately 12% of the oil was irreversibly adsorbed onto the silica gel.[78]

Achladas[89] presented a method for separating the phenolic fraction from biomass pyrolysis liquids by combining silica gel open-column chromatography with alkaline extraction through salt formation (sodium phenolates) with NaOH.

9.3.1.3 *Solid-Phase Extraction*

SPE is an increasingly useful sample preparation technique. With SPE, many of the problems associated with LLE can be prevented, such as incomplete phase separations, less-than-quantitative recoveries, use of expensive, breakable specialty glassware, and disposal of large quantities of organic solvents. SPE is more efficient than LLE, yields quantitative extractions that are easy to perform, is rapid, and can be automated. Solvent use and lab time are reduced. SPE is used most often to prepare liquid samples and extract semi-volatile or non-volatile analytes, but also can be used with solids that are pre-extracted into solvents. SPE products are excellent for sample extraction, concentration, and clean-up. They are available in a wide variety of chemistries, adsorbents, and sizes. Selecting the most suitable product for each application and sample is important.[90] SPE has been developed to separate chemical compounds or groups from a wide variety of matrices. Especially in the area of pharmacy (*e.g.* purification) and sport medicine (*e.g.* doping tests) this method is used successfully, but in the area of bio-oil only limited information about utilization of SPE is available.[91–94]

SPE has become a useful tool in bio-oil characterization and separation of distinct fractions. This is because SPE features a set of special separation functions for bio-oils. In broad terms, functionality in bio-oil characterization can be divided into two categories: polar derived products and non-polar remaining structures from lignin (pyrolytic lignin). The affinity of dissolved analytes for a solid phase through which the sample is passed addresses the general principle of effect; whereby compounds can be categorized according to their specific properties.

Numerous types of SPE-cartridges are available, and new types are continually being introduced. A comprehensive summary of applicable SPE materials for bio-oils was compiled by Brodzinski,[95] who devoted her thesis to the analysis of bio-oil characterization with SPE. The use of reversed phase C18 columns was recommended. Generally, the columns are conditioned first with methanol followed by water. The sample amount is around 10 wt% of the column's bed material weight. The bio-oil sample is placed on top of the

Table 9.6 Elution of chemical groups with C18 cartridge. Reproduced with permission from ref. 95.

Solvent	Groups
100% Water	Anhydrosugars, acids, furans
15% Methanol	Furan- and pyran derivatives
35% Methanol	Aromatic ketones and aldehydes
50% Methanol	Alkyl-derivatives from syringol
100% Methanol	Alkene-derivatives from von guaiacyl- and syringyl- moieties

column and—as the samples are often viscous—distributed through application of a slight vacuum. This helps to distribute the oil evenly over the cross section of the SPE cartridge. Subsequently, dedicated solvents are used to elute corresponding functional groups.

This approach can serve two different objectives: (1) clean-up to obtain a desired target fraction; (2) fractionation into distinctive groups. Brodzinski[95] studied the latter case taking whole bio-oil. Separations with water and increasing amounts of methanol led to the following results (see Table 9.6). However, the results also revealed incomplete separation and insofar the use of the method is restricted.

In addition, the clean-up approach with SPE seems to be promising for the separation and enrichment of sugars in bio-oil. In this case the aforementioned pyrolytic lignin precipitation method B is the starting procedure: After separation of the pyrolytic lignin in the syringe filter, the aqueous filtrate contains all water solubles in a very diluted form. The filtrate is passed through a C18 SPE cartridge. Very polar components like sugars remain unretained whereas less polar pyrolysis products are retained and can be completely eluted with organic solvents like acetone. Thus, all water-soluble bio-oil components are separated into only two solvents: water and acetone. The water phase can be freeze-dried to get a sugar enriched fraction. The acetone fraction is enriched with lignin-derived components (phenolics and aromatics).

9.4 Fractional Condensation

One of the options to steer the bio-oil composition is by controlling the pyrolysis conditions (reactor temperature and residence time of vapors and solids). Downstream fractionation of the whole pyrolysis oil obtained after condensation by for example, distillation, but the high reactivity of some bio-oil fractions typically results in the formation of cross-linked solid carbonaceous residues when bio-oil is heated.[96] Consequently, fractional condensation has been investigated increasingly as an alternative means of fractionating bio-oil both for reducing the water content or to enrich chemcial families such as sugars or phenolics.[96–101]

It is vital that bio-oil be separated into less complex fractions/mixtures for successful upgrading to liquid drop-in transportation fuels and/or valuable chemicals. There are several attempts described in literature to fractionate the hot vapors from biomass fast pyrolysis right after their generation prior to condensation. The product stream exiting a fast pyrolysis reactor is a mixture of vapors and aerosols.[102] Many of the compounds in bio-oil are reactive, viscous, and/or non-volatile, which complicates their recovery. Generally, bio-oil recovery systems are of two types: spray towers and condensing heat exchangers. Based on good experiences with the so-called "Hamburg-Process" for fast pyrolysis of plastic waste[103] Meier *et al.*[35] included an electrostatic filter and Gerdes *et al.*[104] added a spray quench tower in their recovery units to trap aerosols from the fast pyrolysis vapors. A spray quench system utilizes a re-circulated spray of bio-oil or an immiscible hydrocarbon solvent to cool and collect bio-oil. Many variations are possible.[105]

When reviewing the bio-oil collection systems in the literature two themes are found: First, most systems make little effort to recover bio-oil as distinct fractions. Even if several stages are used, the fractions are recombined to form a "whole bio-oil", probably because the separate fractions are not distinctive enough to preserve. Most systems also employ ESPs at the end of the process train to recover aerosols formed during pyrolysis or droplets of bio-oil or solvent generated in quench vessels that escape mist eliminators.[105]

The most sophisticated bio-oil recovery system employing fractional condensation is established at Iowa State University. It was developed to recover bio-oil as distinctive stage fractions while eliminating many of the problems associated with traditional condenser-based bio-oil recovery systems. Fractionation is achieved by replacing the chilled water traditionally used in bio-oil condensers with coolants operated at temperatures carefully prescribed to encourage selective condensation of classes of molecules according to their dew points in the pyrolysis vapor stream. More details can be found elsewhere.[105,106] By careful control of the temperatures of coolants, heat exchange surfaces, and the gas stream, it is possible to recover bio-oil as stage fractions with distinctive physical and chemical characteristics.

As illustrated in Figure 9.10, the PDU consisted of a biomass feed system, a fluidized bed pyrolysis reactor, two hot gas cyclones for particulate matter removal, and a five-stage bio-oil recovery system.

This bio-oil collection system was developed on the premise that both vapors and aerosols are generated during fast pyrolysis. Vapors were presumed to be predominantly decomposition products of carbohydrate polymers while aerosols were the non-volatile, lignin-derived oligomers often described as "pyrolytic lignin" in bio-oil.[11,40] The condenser stages were designed to recover vapors according to condensation temperatures. Electrostatic precipitators were used to effectively recover aerosols. Five stages of bio-oil recovery allows separate collection of both vapors and aerosols from the oligomeric rich "heavy-ends", a middle portion of monomeric phenols and furans, and an aqueous phase containing most of the "light oxygenates".[105]

SF1 is designed and operated to capture levoglucosan and other compounds with high dew points. The temperature of the coolant water is controlled to 85 °C, which kept the condensed bio-oil warm enough to readily

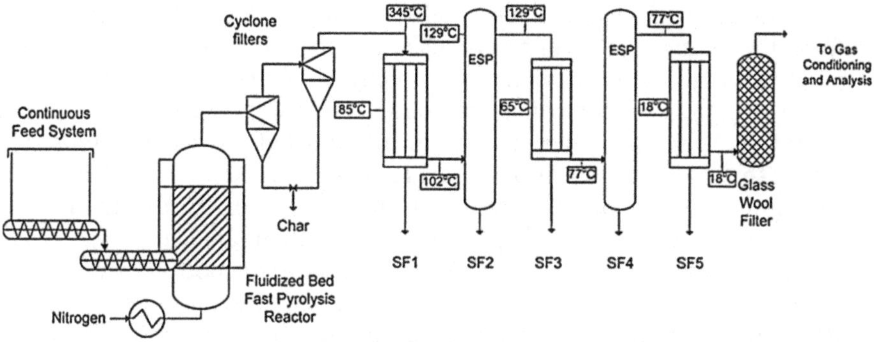

Figure 9.10 Schematic of the pyrolysis process development unit used to evaluate the fractionating bio-oil recovery system.[105,107] Reprinted from Journal of Analytical and Applied Pyrolysis, **93**, A. S. Pollard, M. R. Rover, R. C. Brown, Characterization of bio-oil recovered as stage fractions with unique chemical and physical properties, 129–138, Copyright (2012) with permission from Elsevier.[105]

flow by gravity into a collection pot at the bottom of the condenser. Stage 2, consisting of an electrostatic precipitator (ESP) operated at 40 kVDC, was designed to collect aerosols produced in the pyrolysis reactor or stage 1. The ESP was heat traced to 129 °C to prevent condensation of vapors in this stage since the goal was to segregate volatile and non-volatile molecules. Stage 3, designed to capture compounds with dew points close to that of phenol, was also a shell-and-tube heat exchanger. It is operated with gas inlet and outlet temperatures of 129 °C and 77 °C, respectively. The coolant water was controlled to 65 °C. Stage 4 was an ESP and insulated rather than heat traced, which resulted in an operating temperature of about 77 °C. Stage 5, designed to remove water and light oxygenated compounds, such as acetic acid, was a shell-in-tube heat exchanger using water at 18 °C as a coolant. Residence times in the individual units of the collection system range from 1 s to 10 s. These five stages were followed by a packed column of glass wool to trap any remaining aerosols although in practice only negligible quantities of oil were collected at this point.[105] The bio-oil recovery system was successful in producing bio-oil fractions of low water content (7–15 wt%) and decreased acidity (reduced by up to 50%). The concentration of water-soluble levoglucosan and water-insoluble "pyrolytic lignin" in the first two stage fractions suggest their separation by water washing, ultimately yielding an aqueous stream of fermentable sugar.[105]

9.5 Industrial Processes

The German company "ProFagus" (Bodenfelde) is producing approximately 35 000 tons of barbeque charcoal and in addition about 25 000 tons of a crude wood tar, which is extracted to a large extent to obtain various chemical products. The carbonization process is based on the "Reichert retort" also known as the DEGUSSA process which started 110 years ago. The DEGUSSA retort is different from most of other retorts used in charcoal making, mainly in the type of heating. The retort is heated directly with pre-heated entrained tail gas exploiting the exothermicity of wood. In this way less heat is required. Only residual, pre-dried beech wood from the wood processing industry is used for the carbonization process. Six retorts are operated in series (25 t of wood input each) at a pyrolysis temperature of 450–500 °C for approximately 13–20 hours.[108] A schematic diagram is depicted in Figure 9.11.

The tar coming out of the process flows down into the bottom of the retort. This tar fraction, known as “pot tar”, is separated and later used as the tar fraction for liquid smoke production. Part of the crude tar is cooled and used in a scrubber. The “scrubber tar” is stored in a separate tank and used as fuel in the boiler system. The vapors leaving the scrubber are cooled down and condensed by heat exchangers using water from a river close by. The temperature of the wood gas after the coolers is about 40–50 °C. The condensate from the coolers is the base material of the intermediate product "raw pyroligneous acid". The gas is further cleaned in another scrubber and then heated to 550 °C and recycled to the retorts. Part of the gas is burned in the gas heating systems to achieve the process heat. The rest of energy is produced using natural gas. All liquid intermediate products are stored to have a buffer for subsequent processing. The acid content in the scrubber tar is reduced by distillation and reused as fuel for the steam production to provide the heat for the distillation columns.

The raw pyroligneous acids are freed from methanol through distillation. The waste from the distillation is wood spirit going over the top of the columns. It has a methanol content of about 60% and is burned in the boiling system. The remaining tar is very aromatic, almost insoluble in water

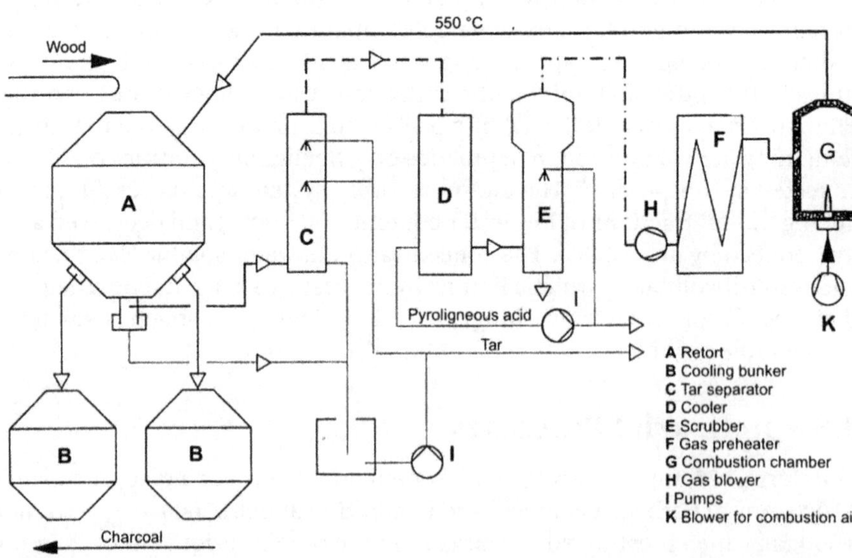

Figure 9.11 Schematic of the carbonizing plant using the DEGUSSA process.

and contains mostly lignin-derived components. It is subjected to fractional vacuum distillation under steam addition. Hence, light, middle, and heavy oils are generated which are further rectified to obtain creosote with pharmaceutical quality for example.

After distillation, the raw pyroligneous acid exhibits a methanol content of only 0.03% and an average acetic acid content of about 11%. Acetic acid is further concentrated by continuous extraction with ethylacetate. The remaining tar ("ex-tar") from the raw pyroligneous liquid is separated from the crude acetic acid by vacuum distillation and used as raw material for the production of liquid smoke aromas. In several distillation stages a very pure acetic acid is obtained suitable for the food industry and the chemical industry.[109]

9.6 Conclusions

The growing interest in bio-oil production for energy and bio-based materials (*e.g.* commodity chemicals) has intensified research in developing economic-environmental models[110] as well as separation processes and technologies. Fractional condensation and solvent extraction remain the most important separation technologies for bio-oils. They are established and have been applied from laboratory to industrial scale not only for pyrolysis liquids from biomass but also for coal tars.[111] Other than coal tars, fast pyrolysis bio-oils are difficult to distill due the existence of highly reactive molecules which are prone to condensation reactions. As a consequence, innovative but costly methods are under investigation such as molecular distillation, extraction with supercritical media, and SHSs. Affordable and reliable separation technologies for bio-oils are indispensable to meet the economic requirements for a future bio-based economy.

References

1. S. Black and J. R. Ferrell, *Energy Fuels*, 2016, **30**, 1071–1077.
2. Q. Lu, Z. Zhang, C. Zhang, S. Su, W. Li and C. Dong, *Appl. Mech. Mater.*, 2012, **130–134**, 422–425.
3. M. Stas, D. Kubicka, J. Chudoba and M. Pospisil, *Energy Fuels*, 2014, **28**, 385–402.
4. N. Charon, J. Ponthus, D. Espinat, F. Broust, G. Volle, J. Valette and D. Meier, *J. Anal. Appl. Pyrolysis*, 2015, **116**, 18–26.
5. D. Meier and M. Windt, in *Transformation of Biomass - Theory to Practice*, ed. A. Hornung, John Wiley & Sons Ltd, Chichester, UK, 2014, pp. 227–256.
6. Z. Lei and B. Chen, in *Separation and Purification Technologies in Biorefineries*, ed. S. Ramaswamy, H.-J. Huang and B. V. Ramarao, John Wiley & Sons Ltd, Chichester, UK, 2013, pp. 39–60.
7. H. J. Huang, S. Ramaswamy, U. W. Tschirner and B. V. Ramarao, *Sep. Purif. Technol.*, 2008, **62**, 1–21.

8. A. K. Jana, *Appl. Energy*, 2010, **87**, 1477–1494.
9. J. G. Speight, *The Chemistry and Technology of Petroleum*, CRC Press, 2014.
10. Y. Elkasabi, C. A. Mullen, A. L. M. T. Pighinelli and A. A. Boateng, *Fuel Process. Technol.*, 2014, **123**, 11–18.
11. D. S. Scott, *Chemicals and Fuels from Biomass Flash Pyrolysis–Part of the Bioenergy Development Program (BDP)*, Univ. Waterloo, 1988.
12. S. Czernik and A. V. Bridgwater, *Energy Fuels*, 2004, **18**, 590–598.
13. S. Wang, in *Biomass Now–Sustainable Growth and Use*, ed. M. D. Matovic, InTech, 2013, pp. 401–418.
14. J. A. Capunitan and S. C. Capareda, *Fuel*, 2013, **112**, 60–73.
15. Y. Elkasabi, C. A. Mullen and A. A. Boateng, *ACS Sustainable Chem. Eng.*, 2014, **2**, 2042–2052.
16. X. S. Zhang, G. X. Yang, H. Jiang, W. J. Liu and H. S. Ding, *Sci. Rep.*, 2013, **3**, 1120.
17. Z. G. Guo, S. R. Wang, G. H. Xu and Q. J. Cai, *BioResources*, 2011, **6**, 2539–2550.
18. J. Lutišan and J. Cvengroš, *Chem. Eng. J.*, 1995, **56**, 39–50.
19. J. Lutišan, J. Cvengroš and M. Micov, *Chem. Eng. J.*, 2002, **85**, 225–234.
20. X. J. Guo, S. R. Wang, Z. G. Guo, Q. A. Liu, Z. Y. Luo and K. F. Cen, *Appl. Energy*, 2010, **87**, 2892–2898.
21. Z. G. Guo, S. R. Wang, Y. L. Gu, G. H. Xu, X. Li and Z. Y. Luo, *Sep. Purif. Technol.*, 2010, **76**, 52–57.
22. S. R. Wang, Y. L. Go, Q. Liu, Y. Yao, Z. G. Guo, Z. Y. Luo and K. F. Cen, *Fuel Process. Technol.*, 2009, **90**, 738–745.
23. X. Y. Guo, Y. Zheng, B. H. Zhang and J. Y. Chen, *Biomass Bioenergy*, 2009, **33**, 1469–1473.
24. Z.-g. Guo, S.-r. Wang, Y.-y. Zhu, Z.-y. Luo and K.-f. Cen, *J. Fuel Chem. Technol.*, 2009, **37**, 49–52.
25. Y. Elkasabi, B. M. E. Chagas, C. A. Mullen and A. A. Boateng, *Energy Fuels*, 2016, **30**, 4925–4932.
26. B. M. E. Chagas, C. A. Mullen, C. Dorado, Y. Elkasabi, A. A. Boateng, M. A. F. Melo and C. H. Ataide, *Ind. Eng. Chem. Res.*, 2016, **55**, 6734–6741.
27. J. S. Kim, *Bioresour. Technol.*, 2015, **178**, 90–98.
28. R. J. M. Westerhof, N. J. M. Kuipers, S. R. A. Kersten and W. P. M. van Swaaij, *Ind. Eng. Chem. Res.*, 2007, **46**, 9238–9247.
29. J. Asikkala and A. Gutierrez, USA Pat. US 9187444 B2, 2015.
30. D. Radlein, J. Piskorz and P. Majerski, EU Pat. EP 0718392 A1, 1996.
31. B. Schuur, *Chem. Pap.*, 2015, **69**, 245–253.
32. H.-J. Huang and S. Ramaswamy, in *Separation and Purification Technologies in Biorefineries*, ed. S. Ramaswamy, H.-J. Huang and B. V. Ramarao, John Wiley and Sons, Chichester, 2013, pp. 1–36.
33. I. A. Vasalos, M. C. Samolada and G. E. Achladas, in *Res. Thermochem. Biomass Convers., [Ed. Rev. Pap. Int. Conf.]*, ed. A. V. Bridgwater and L. J. Kuester, Elsevier, London, 1988, pp. 251–263.
34. J. N. Murwanashyaka, H. Pakdel and C. Roy, *Sep. Purif. Technol.*, 2001, **24**, 155–165.

35. D. Meier, S. Wehlte, P. Wulzinger and O. Faix, in *Bio-Oil Production and Utilization, Proceedings of the 2nd EU-Canada Workshop on Thermal Biomass Processing*, ed. A. V. Bridgwater and E. N. Hogan, CPL Press, Newbury, UK, 1996, pp. 102–112.
36. J. Zhang and B. Hu, in *Separation and Purification Technologies in Biorefineries*, John Wiley & Sons, Ltd, 2013, pp. 61–78.
37. X. Yang, L. Hang, K. Chen, X. Zhu, S. Zhang and J. Chen, *BioResources*, 2014, **9**, 5219–5233.
38. M. Garcia-Perez, A. Chaala, H. Pakdel, D. Kretschmer and C. Roy, *Biomass Bioenergy*, 2007, **31**, 222–242.
39. K. Sipilä, E. Kuoppala, L. Fagernäs and A. Oasmaa, *Biomass Bioenergy*, 1998, **14**, 103–113.
40. B. Scholze and D. Meier, *J. Anal. Appl. Pyrolysis*, 2001, **60**, 41–54.
41. Y. Feng and D. Meier, *Biomass Bioenergy*, 2016, **85**, 346–354.
42. A. Oasmaa, E. Kuoppala, A. Ardiyanti, R. H. Venderbosch and H. J. Heeres, *Energy Fuels*, 2010, **24**, 5264–5272.
43. A. Oasmaa, E. Kuoppala and D. C. Elliott, *Energy Fuels*, 2012, **26**, 2454–2460.
44. A. Oasmaa, E. Kuoppala and Y. Solantausta, *Energy Fuels*, 2003, **17**, 433–443.
45. N. M. Bennett, S. S. Helle and S. J. B. Duff, *Bioresour. Technol.*, 2009, **100**, 6059–6063.
46. C. R. Vitasari, G. W. Meindersma and A. B. de Haan, *Bioresour. Technol.*, 2011, **102**, 7204–7210.
47. C. R. Vitasari, G. W. Meindersma and A. B. de Haan, *Chem. Eng. Res. Des.*, 2015, **95**, 133–143.
48. C. R. Vitasari, G. W. Meindersma and A. B. de Haan, *Sep. Purif. Technol.*, 2012, **95**, 39–43.
49. M. Nakagawa, Y. Sakai and T. Yasui, *J. Ferment. Technol.*, 1984, **62**, 201–203.
50. X. L. Zhuang, H. X. Zhang and J. J. Tang, *Biomass Bioenergy*, 2001, **21**, 53–60.
51. I. L. Chien, K.-L. Zeng, H.-Y. Chao and J. Hong Liu, *Chem. Eng. Sci.*, 2004, **59**, 4547–4567.
52. I. L. Chien and C.-L. Kuo, *Chem. Eng. Sci.*, 2006, **61**, 569–585.
53. D. Radlein, J. Piskorz and D. S. Scott, *J. Anal. Appl. Pyrolysis*, 1991, **19**, 41–63.
54. E. Reverchon and I. De Marco, *J. Supercrit. Fluids*, 2006, **38**, 146–166.
55. T. Cheng, Y. Han, Y. Zhang and C. Xu, *Fuel*, 2016, **172**, 49–57.
56. L. M. Petkovic and D. M. Ginosar, *Abstr. Pap. Am. Chem. Soc.*, 2015, **250**, I+EC-102.
57. S. Naik, V. V. Goud, P. K. Rout and A. K. Dalai, *Bioresour. Technol.*, 2010, **101**, 7605–7613.
58. P. K. Rout, M. K. Naik, S. N. Naik, V. V. Goud, L. M. Das and A. K. Dalai, *Energy Fuels*, 2009, **23**, 6181–6188.
59. J. Wang, H. Cui, S. Wei, S. Zhuo, L. Wang, Z. Li and W. Yi, *Smart Grid Renewable Energy*, 2010, **1**, 98–107.

60. R. N. Patel, S. Bandyopadhyay and A. Ganesh, *Energy*, 2011, **36**, 1535–1542.
61. R. N. Patel, S. Bandyopadhyay and A. Ganesh, *Energy Convers. Manage.*, 2010, **52**, 652–657.
62. S. Rudyk, S. Hussain and P. Spirov, *J. Supercrit. Fluids*, 2013, **78**, 63–69.
63. A. W. Francis, *J. Phys. Chem.*, 1954, **58**, 1099–1114.
64. Y. Feng and D. Meier, *J. Anal. Appl. Pyrolysis*, 2015, **113**, 174–185.
65. P. G. Jessop, L. Kozycz, Z. G. Rahami, D. Schoenmakers, A. R. Boyd, D. Wechsler and A. M. Holland, *Green Chem.*, 2011, **13**, 619–623.
66. P. G. Jessop, L. Phan, A. Carrier, S. Robinson, C. J. Durr and J. R. Harjani, *Green Chem.*, 2010, **12**, 809–814.
67. P. G. Jessop, D. J. Heldebrant, X. Li, C. A. Eckert and C. L. Liotta, *Nature*, 2005, **436**, 1102.
68. P. G. Jessop, S. M. Mercer and D. J. Heldebrant, *Energy Environ. Sci.*, 2012, **5**, 7240–7253.
69. D. B. Fu, S. Farag, J. Chaouki and P. G. Jessop, *Bioresour. Technol.*, 2014, **154**, 101–108.
70. C. Roy, L. Calve, X. Lu, H. Pakdel and C. Amen-Chen, *Biomass: A Growth Opportunity in Green Energy and Value-Added Products, Oakland, Calif., Aug. 29–Sept. 2, 1999*, 1999.
71. C. Amen-Chen, H. Pakdel and C. Roy, *Biomass Bioenergy*, 1997, **13**, 25–37.
72. R. M. Gallivan and P. K. Matschei, USA Pat. US 4209647, 1980.
73. A. Effendi, H. Gerhauser and A. V. Bridgwater, *Renewable Sustainable Energy Rev.*, 2008, **12**, 2092–2116.
74. R. Maggi and B. Delmon, *Fuel*, 1994, **73**, 671–677.
75. R. L. Shriner, C. K. F. Hermann, T. C. Morrill, D. Y. Curtin and R. C. Fuson, *The Systematic Identification of Organic Compounds*, John Wiley & Sons, Inc., 1997.
76. D. Mohan, C. U. Pittman Jr and P. H. Steele, *Energy Fuels*, 2006, **20**, 848–889.
77. D. Harvey, *Modern Analytial Chemistry*, Mc Graw-Hill Higher Education, 1st edn, 2000.
78. D. Meier, D. R. Larimer and O. Faix, *Fuel*, 1986, **65**, 910–915.
79. H. L. Chum, S. K. Black, J. P. Diebold and R. E. Kreibich, USA Pat. US 5,091,499, 1993.
80. H. Chum, J. Diebold, J. Scahill, D. Johnson, S. Black, H. Schroeder and R. E. Kreibich, *ACS Symp. Ser.*, 1989, **385**, 135–151.
81. H. L. Chum and S. K. Black, USA Pat. US 4942269 A, 1990.
82. S. R. Wang, Y. R. Wang, Q. J. Cai, X. Y. Wang, H. Jin and Z. Y. Luo, *Sep. Purif. Technol.*, 2014, **122**, 248–255.
83. Q. H. Song, J. Q. Nie, M. G. Ren and Q. X. Guo, *Energy Fuels*, 2009, **23**, 3307–3312.
84. S. Venkatesan, in *Separation and Purification Technologies in Biorefineries*, ed. S. Ramaswamy, H.-J. Huang and B. V. Ramarao, John Wiley & Sons, Ltd, Chichester, UK, 2013, pp. 101–148.
85. M. Ertas and M. H. Alma, *J. Anal. Appl. Pyrolysis*, 2010, **88**, 22–29.

86. O. Onay, A. F. Gaines, O. M. Kockar, M. Adams, T. R. Tyagi and C. E. Snape, *Fuel*, 2006, **85**, 382–392.
87. M. Farcasiu, *Fuel*, 1977, **56**, 9–14.
88. H. G. Davis, M. A. Eames, C. Figueroa, R. R. Gansley, L. L. Schaleger and D. W. Watt, *The Products of Direct Liquefaction of Biomass*, Lawrence Berkeley Laboratory, University of California, 1982.
89. G. E. Achladas, *J. Chromatogr.*, 1991, **542**, 263–275.
90. Sigma Aldrich: Guide to solid phase extraction, Bulletin 910, http://www.sigmaaldrich.com/Graphics/Supelco/objects/4600/4538.pdf, date accessed: 2 Aug 2016.
91. C. Brage and K. Sjöström, *J. Chromatogr.*, 1991, **538**, 303–310.
92. C. Brage, Q. Z. Yu, G. X. Chen and K. Sjostrom, *Fuel*, 1997, **76**, 137–142.
93. I. Brodzinski and D. Meier, *Proceedings 2nd World Biomass Conference*, Rome, 2004.
94. D. Fabbri, A. Adamiano and C. Torri, *Anal. Bioanal. Chem.*, 2010, **397**, 309–317.
95. I. Brodzinski, *Dissertation*, University of Hamburg, 2006.
96. R. J. M. Westerhof, D. W. F. Brilman, M. Garcia-Perez, Z. Wang, S. R. G. Oudenhoven, W. P. M. van Swaaij and S. R. A. Kersten, *Energy Fuels*, 2011, **25**, 1817–1829.
97. P. Kim, S. Weaver, K. Noh and N. Labbe, *Energy Fuels*, 2014, **28**, 6966–6973.
98. V. S. K. K. Palla, K. Papadikis and S. Gu, *Biomass Bioenergy*, 2015, **74**, 180–192.
99. A. Tumbalam Gooty, D. Li, F. Berruti and C. Briens, *J. Anal. Appl. Pyrolysis*, 2014, **106**, 33–40.
100. R. Yin, R. Liu, Y. Mei, W. Fei and X. Sun, *Fuel*, 2013, **112**, 96–104.
101. A. Tumbalam Gooty, D. Li, C. Briens and F. Berruti, *Sep. Purif. Technol.*, 2014, **124**, 81–88.
102. A. V. Bridgwater, D. Meier and D. Radlein, *Org. Geochem.*, 1999, **30**, 1479–1493.
103. W. Kaminsky, *Angew. Makromol. Chem.*, 1995, **232**, 151–165.
104. C. Gerdes, C. M. Simon, T. Ollesch, D. Meier and W. Kaminsky, *Eng. Life Sci.*, 2002, **2**, 167–174.
105. A. S. Pollard, M. R. Rover and R. C. Brown, *J. Anal. Appl. Pyrolysis*, 2012, **93**, 129–138.
106. R. C. Brown, S. T. Jones and A. Pollard, USA Pat. US 8,476,480 B1, 2013.
107. M. R. Rover, *Dissertation*, Iowa State University, 2013.
108. H.-G. Brocksiepe, *Ullmann's Encyclopedia of Industrial Chemistry*, Wiley-VCH Verlag GmbH & Co. KGaA, 2000, vol. A 6, pp. 157–162.
109. H.-G. Brocksiepe, in *Ullmanns Encyclopedie der Technischen Chemie*, ed. E. Bartholomé, E. Biekert, H. Hellmann, H. Ley and W. M. Weigert, Verlag Chemie, Weinheim, 1976, vol. 12, pp. 703–708.
110. Y. N. Zhang, G. P. Hu and R. C. Brown, *Bioresour. Technol.*, 2014, **157**, 28–36.
111. T. Jiao, X. Zhuang, H. He, C. Li, H. Chen and S. Zhang, *Ind. Eng. Chem. Res.*, 2015, **54**, 2573–2579.

CHAPTER 10

Catalytic Fast Pyrolysis Over Zeolites†

GÜRAY YILDIZ*[a], FREDERIK RONSSE[a] AND WOLTER PRINS[a]

[a]Department of Biosystems Engineering, Ghent University, Coupure Links 653, B-9000 Ghent, Belgium
*E-mail: guryildiz@gmail.com

10.1 Introduction

Renewable energy sources typically used for heat and power production (*i.e.* solar, tidal, wind, hydropower, *etc.*), are not suitable for the production of transportation fuels and raw materials (*e.g.* petroleum-based chemicals) used in the current industries.[1] To contribute to the sustainability of the current petrochemical-based economy and to continue the production of petroleum-like products, chemicals and (transportation) fuels, biomass should be utilized. Due to its abundance and fast turnover in organic carbon and hydrogen, there is a growing interest in biomass as an alternative for the currently-used non-renewable fossil resources. Biomass can be converted *via* thermo-chemical processes (*i.e.* pyrolysis, torrefaction, gasification) to fuel intermediates and chemicals in liquid, solid, and/or gaseous form and potentially replace fossil fuels, either fully or in blends of various fractions. This makes biomass the most promising renewable source for chemicals production with an increasing potential for the worldwide biofuels and biochemicals market.[2,3]

†Electronic supplementary information (ESI) available. See DOI: 10.1039/9781788010245.

Green Chemistry Series No. 50
Fast Pyrolysis of Biomass: Advances in Science and Technology
Edited by Robert C. Brown and Kaige Wang

Published by the Royal Society of Chemistry, www.rsc.org

Ideally, the production of fuels and/or chemicals from solid biomass materials would include a single-step process with a high degree of conversion. Flexibility in using various types of biomass feedstock and short conversion times would be an asset to the process. Pyrolysis has been considered as a viable biomass conversion technology for the production of fuel intermediates, chemicals, and heating sources in the form of liquid, solid, and gaseous product fractions in various proportions. The outcome of a pyrolysis process depends on several factors related to the type and the properties of the biomass feedstock and to the reactor, process conditions, and intended application of the final products.[4] In order to maximize the yield of the liquid product, pyrolysis has to be conducted at temperatures of *ca.* 500 °C, and at higher biomass heating rates (>500 W m^2 K^{-1}).[5] Residence times of 2 to 20 seconds are needed for complete devolatilization of small biomass particles (of a few millimeters in diameter) followed by rapid quenching of the released pyrolysis vapors (short vapor residence times of *ca.* 1 s). This type of pyrolysis process is called fast pyrolysis. The main product, bio-oil (referring to crude, non-catalytically produced single-phase pyrolysis liquid), can be obtained in yields of 60 to 70 wt% (on a dry-feed basis) in continuously operated bench- and pilot-scale process units.

Crude bio-oil is a complex and homogeneous mixture of water and a large number of mainly oxygenated chemicals (30–40 wt%[6]). The molecular weights of the organic macromolecular lignocellulosic degradation products range from that of light hydrocarbons up to 5000 g mol^{-1} or more (pyrolytic lignins, sugar oligomers).[7–9] The organic chemicals in crude bio-oil can be classified into four main categories: (1) anhydrosugars; (2) furan derivatives; (3) low molecular weight oxygenates (*e.g.* carboxylic acids, ketones, aldehydes); and (4) phenols and aromatics.[10] Without any modification, this mixture of chemicals could be used as feed for boilers, turbines, engines, and gasifiers. However, some adverse properties of crude bio-oil including its water content (15–30%), the high oxygen content (35–40 wt%), its corrosiveness (pH of 2–3), the low heating value if compared to fossil fuels (lower heating value of approximately 17 MJ kg^{-1}), poor volatility, and the presence of small solid particulates (*i.e.* char, sand) due to the inefficient filtration, makes it unsuitable for the direct use as a transportation fuel or as a source for the production of fuels and/or chemicals.[11,12] Its immiscibility with hydrocarbon fuels and its chemical/thermal instability under storage and heating conditions further obstruct its direct use in almost any application.[11–13] Although the applications are still not all fully developed, unmodified bio-oil could in principle be used as a fuel for gasifiers, boilers, and turbines.[14] For specific applications of bio-oil, like in diesel or gasoline engines, or as an intermediate bio-based feed material in the existing petroleum refineries, a certain degree of upgrading including the (partial) removal of oxygen and cracking of the oligomers present in bio-oil is required. However, the precise way of upgrading is not evident beforehand, because it is closely related to the actual end-application.[15] In the case of engines, whether stationary (diesel generators) or mobile (car engines), a certain degree of bio-oil upgrading has to be carried out with the primary objectives to:

- reduce the viscosity and instability
- reduce the acidity, allowing the use of conventional injectors and pumps
- prevent the formation of coke (avoiding plugging of injectors)
- increase the cetane number; which is an indicator for ignition behavior in a diesel engine
- improve the miscibility with conventional fossil fuels

To reach these objectives, the quality of bio-oil can be improved through actions (individual or in combination) taken before, during, or after the fast pyrolysis step. These can be the demineralization of the feed by an acid wash (leaching), staged condensation to lower the acid and water content of crude bio-oil, catalytic upgrading of the produced bio-oil (*e.g.* hydrogenation, hydrodeoxygenation), or the application of heterogeneous catalysis in the pyrolysis process (*i.e.* catalytic pyrolysis).[16] The common target of all possible bio-oil upgrading techniques must be to favor the pathways towards reducing the oxygen content (*i.e.* deoxygenation) while cracking the large aromatic structures to smaller ones.[12,17] This results in an upgraded bio-oil with increased stability, higher heating value, and miscibility with petroleum refinery feedstock and/or petroleum-based fuels.

Early on it was recognized that the application of catalysis could be of major importance in controlling the quality and the chemical composition of the bio-oil from fast pyrolysis.[18] In the last decade, many pyrolysis researchers have focused their attention to two main technologies for improving the quality of crude bio-oil.[19] Here, the quality improvement refers to the production of certain chemicals (*i.e.* hydrocarbons, phenols, aromatics) and/or liquid fuels or fuel intermediates similar in physical and chemical properties to current gasoline and diesel fuels. Hydrodeoxygenation (HDO) is a downstream upgrading process for pyrolysis liquids which is based on the hydrotreatment of bio-oil over heterogeneous catalysts (*e.g.* CoMo- and NiMo-based catalysts) under high hydrogen pressure (mostly in the range of 30–140 bar[20]) and at elevated temperatures (330–450 °C).[21–23] However, the economic viability of the HDO process strongly depends on the amount of hydrogen consumed and requires control of the composition and the amount of the high molecular mass fraction (*e.g.* sugars) in the bio-oil. For the latter, catalytic fast pyrolysis (CFP) of biomass seems to be an attractive alternative. In terms of capital and operational costs, CFP seems more economical since it is a single-step 'biomass-to-liquids' process and does not require the supply of expensive high pressure hydrogen gas.[24,25] Although both techniques exhibit major operating problems (*e.g.* reactor plugging, rapid catalyst deactivation) and low yields of liquid fuels,[26] it was stated that CFP could have several advantages over other techniques.[27] The claimed advantages are in the possibilities to complete all of the targeted chemistry in a single reactor while using relatively inexpensive aluminosilicate catalysts, to process a wide range of different lignocellulosic feedstock materials after a simple pretreatment (*e.g.* grinding, drying), and to blend the liquid product as a feedstock (or co-feedstock) into the existing gasoline infrastructure.

In CFP, the heat transfer material, generally assumed inert in conventional fast pyrolysis, is (partially) replaced by a solid catalyst (*i.e. in situ* CFP, see Figure 10.1). Besides, the primary fast pyrolysis vapors can be catalytically reformed in a separate catalytic reactor just after the fast pyrolysis reactor (*i.e. ex situ* CFP). The presence of the catalyst favors oxygen removal *via* decarbonylation (CO rejection), decarboxylation (CO_2 rejection) and dehydration (H_2O removal) reactions. CFP induces vapor phase reactions that cause a change in oxygen functionalities and oxygen concentration, alleviate the acidity, increase the calorific value, and improve the stability of the produced liquid product (predominantly organic phase, *i.e.* CFP-oil). Hence, the CFP-oil becomes more similar in chemical composition to current gasoline and diesel fuels than the conventional crude bio-oil.[28,29]

In CFP of biomass, the selection of an appropriate catalyst is crucial for the formation of higher value compounds (phenolics, alkanes, monoaromatic hydrocarbons, *etc.*) and for the elimination of undesirable compounds (acids, aldehydes, ketones, poly-aromatic hydrocarbons, *etc.*).[30,31] Lifetime, activity at various temperatures and other process conditions, deoxygenation performance, and the capability to suppress the formation of coke are important issues in the catalyst selection process. Zeolites, and particularly ZSM-5, were shown to be effective in the conversion of biomass to reasonable yields of aromatics[32] and in the selective deoxygenation of pyrolytic vapors thereby increasing the C : O ratio.[2,33] The use of zeolite cracking catalysts also leads to smaller molecules being formed, due to cracking of the high molecular weight molecules present in the primary pyrolysis vapors.[34,35]

In recent years, CFP of biomass has been extensively studied and various literature reviews have been published.[1,2,10,19,36–41] The objective of this chapter is to focus on the research and development advances and trends in the field of CFP of woody biomass over zeolite catalysts based on a selection of studies published from 1994 up till May 2016. Only the publications reporting the *quantitative results* (which includes pyrolysis product yields, pyrolysis products' elemental composition, *etc.*) obtained from meso- and macro-level laboratory-scale setups, pilot, process development units (PDU), and (semi-) commercial process units wherein the woody biomass feedstock (in some cases also the catalyst) is fed continuously. The definition of such boundaries may help to translate laboratory results to viable process concepts and proper pilot plant trials. When zeolite-type catalysts are used, key issues involved are the most suitable processing mode and the applied reactor technology. Therefore, the spotlight will be on research aiming at optimal process conditions, reactor technologies, and the end product (*i.e.* CFP-oil) properties. The final goal is to come up with a clear picture on how to realize this technique at a commercial/industrial scale.

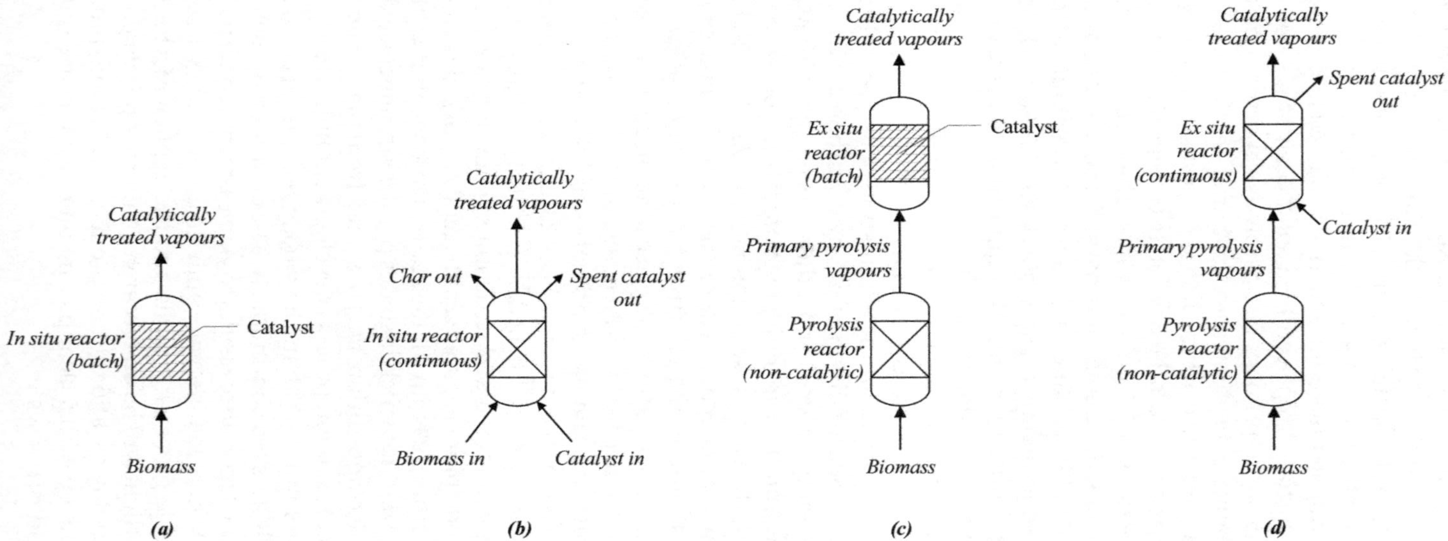

Figure 10.1 Sketch showing the difference between two operation modes in CFP: (a) *in situ*, batch operation, (b) *in situ*, continuous operation, (c) *ex situ*, batch operation, and (d) *ex situ*, continuous operation.

10.2 Process Conditions for CFP of Biomass

The efficiency of a CFP process is influenced by the type of the biomass feedstock (woody biomass, forestry or agricultural residues, *etc.*), operating temperature of the catalyst, residence time and the heating rate of the biomass, the catalyst-to-biomass ratio (or weight hourly space velocity, WHSV), vapor residence time, type of the operation mode (*i.e. in situ* or *ex situ*), type of the reactor (bubbling/circulating fluidized beds, auger reactors, *etc.*), and the type of the catalyst (including pore size, acidity, nature of active sites, and the presence of metals). The success of the process is closely related to the precise optimization of these factors which would yield a maximum quantity of catalytic fast pyrolysis oil (CFP-oil) with the desired quality, while the amounts of by-products (*i.e.* aqueous liquid phase, char, coke on catalyst, and non-condensable gases) are minimized.

10.2.1 Choice of the Biomass Feedstock

Lignocellulosic biomass, referring to most plants or plant-based materials, is composed of three main structural constituents (*i.e.* cellulose, hemicellulose and lignin), and some non-structural components (minerals, extractives, moisture), *viz.* in varying weight percentages depending on the type of biomass.[38,42] The relative proportions of cellulose, hemicellulose, and lignin in the biomass feedstock have a distinct effect on the quality and the quantity of the produced bio-oil.[43] Especially during the screening of potential catalysts for CFP, it is essential to reveal the thermal degradation mechanisms associated with the conversion of these constituents individually.[39] This would help to also understand the complex nature of the bio-oils. When the target of CFP is the production of certain chemicals (*i.e.* aromatics, phenols, olefins) in high yields, model compounds can be studied to unravel the reaction mechanisms and develop an understanding of any catalytic factors which may influence the overall reactions.[27,44] This may eventually help in designing appropriate CFP catalysts.

According to the U.S. Department of Energy,[45] biomass feedstock can be classified into three families: forestry, agriculture, and municipal. Hundreds of biomass materials have been classified[7,46] *viz.* as: woody (soft and hard woods, forestry residues, sawdust), agricultural residues (husks, stalks, stovers, straws), industrial by-products (fruit skins and seeds, lignin), non-woody (grass), waste products (sewage sludge, animal litter) and marine (seaweeds).[43] In case of conventional non-catalytic pyrolysis, different types of biomass feedstock result in varying bio-oil yields[24,39] depending on their cellulose, hemicellulose, lignin, and ash contents. In case of CFP, however, high lignin (because of its lower oxygen content and the presence of aromatic moieties in it) and low ash containing biomass feedstock may be preferred. Compared to many other types, woody biomasses have relatively low ash content and a higher fraction of lignin[47] and thus yield a bio-oil with higher heating value. On the other hand, agricultural residues and grasses have higher ash contents. These high-ash-containing biomass feedstocks are less suitable for CFP because of the undesirable catalytic effect of ash under pyrolysis conditions.[11,48] Ash is known to favor the formation of water and gas at the expense of liquid organics and inhibit the activity of the catalyst by poisoning it. Moreover, due to their higher hemicellulose content compared to that of woody feedstocks, agricultural and non-woody biomass produce more non-condensable gases in pyrolysis.

Generally, woody biomass is seen to produce the best quality bio-oil in terms of C, H, and water content.[43] Indeed, low ash containing woody biomass seems to be the ideal feedstock for industrial scale production of high quality bio-oil *via* CFP. The literature studies[2,19,37,40,43,49,50] also show that woody biomass has been used in most studies due to the fact it has a vast availability on the international market, and is easy to process (*e.g.* handling and feeding), relatively clean, and well characterized (especially for some specific wood types, such as pine and beech).

10.2.2 Key Process Parameters

10.2.2.1 Temperature

In pyrolysis, the temperature strongly influences the yield distribution and the physical/chemical properties of the products (*e.g.* carbon content, oxygen content, pH, and viscosity of bio-oils).[51] Since the thermal degradation reactions of biomass occurring during pyrolysis are endothermic, the decrease in pyrolysis temperature implies a decrease of the input energy for the system, which is beneficial from an economic point of view. In CFP, the reaction temperature is a most critical process parameter determining the relative rate of the catalyzed vapor phase reactions.[1] When embedded or impregnated in the biomass, catalysis in CFP is known to have a lowering effect on the reaction temperature; however, the precise mechanism is still unclear at present.[10]

10.2.2.2 Heating Rate and Residence Time of Biomass

Being closely related to the biomass particle size, the heating rate and residence time of biomass together affect the degree of biomass devolatilization significantly. They play a major role in defining the quality, composition, and distribution of the products.[41] In CFP of biomass, biomass residence time simply refers to the run time (t_{run}) for batch reactors where the external bed-to particle heat transfer rates (>500 W m^2 K^{-1})[5] are achieved by submerging the biomass particles quickly in a bed of hot, fine particles of inert solid material (*e.g.* sand).[52] However, in flow-through or continuous reactors, biomass residence time refers to the time that the biomass particles spend in the hot reactor zone. In order to achieve complete devolatilization in such types of reactors within the available retention time, the biomass particles should be smaller than 2–3 mm.[11] In the presence of a catalyst, high biomass heating rates (>500 °C s^{-1}) are needed also to avoid coke formation by homogeneous thermal decomposition reactions, and to produce high-grade gasoline range compounds. On the other hand, for heating rates above 50 °C s^{-1} in the presence of a zeolite catalyst, the increased production of aromatics was shown not to be a function of the heating rate but for the lower values it was.[53]

10.2.2.3 Catalyst-to-Biomass Ratio

If the catalyst in the reaction zone is deactivated (*e.g.* coking, fouling, attrition, *etc.*) before all the released primary pyrolysis vapors are being reformed by it, the produced bio-oil will be a mixture of non-catalytic bio-oil and CFP-oil. This is also true for very short vapor residence times and, correspondingly, an insufficient contact with the catalyst surface. Indeed, optimal catalyst-to-biomass ratios are required to ensure the adsorption of the primary pyrolysis vapors on the catalyst surface. In this way, these vapors can interact with the catalyst before they thermally decompose to coke *via* the secondary reactions or before being quenched in the condenser system. High catalyst-to-biomass ratios are reported to suppress coke and polycyclic aromatic hydrocarbon-forming polymerization reactions to a higher degree in case of catalysts with high acidity.[54,55] Another term frequently used in CFP is the WHSV h^{-1}, defined as the ratio of the mass flow rate of feed (in g h^{-1}) to the mass of catalyst (in g). It is one of the most important process variables in CFP. Any variations in WHSV may affect the yield of CFP-oil and the formation of water and other by-products.[51]

10.2.2.4 Vapor Residence Time

Another important parameter in fast pyrolysis (either catalytic or non-catalytic) is the vapor residence time. In CFP, the hot vapor residence time in and after the reactor needs to be carefully controlled because the desired products may undergo secondary reactions on the catalyst surface to form heterogeneous coke. Undesirable consecutive reactions can also be catalyzed by biomass originated alkali and alkaline earth metals, or take place simply as uncatalyzed homogeneous reactions in vapor phase.[1,16] Hence, short vapor residence times (<2 s)[56] in the heated zone (including the reactor zone) followed by rapid quenching of these vapors is essential. When utilized, careful optimization of the carrier gas (*i.e.* pure inert gas, recycled oxygen-free flue gas and/or non-condensable gases, or the mixture of these in any combination[41]) flow rate is also necessary. Such an optimization assists to achieve the balance between the time necessary for the completion of catalytic reactions and the suppression of unwanted secondary vapor phase reactions.

Overall, for an optimal CFP process resulting in sufficient yields of CFP-oil with improved quality, carefully optimized reaction temperatures, high heating rates for biomass, a biomass residence time sufficient for a complete devolatilization, optimal catalyst-to-feed ratios or WHSV levels (>1 h^{-1}), fast separation of the solid by-products (*i.e.* char), short hot vapor residence times (<2 s), and fast quenching of the produced vapors are all necessary.[33,51,57,58]

10.2.2.5 In situ vs. Ex situ *Catalytic Pyrolysis*

Based on the location of the catalyst in the process, catalytic pyrolysis can be categorized as either *in situ* and *ex situ*[39] (Figure 10.1). In the *in situ* process mode, a pure catalyst or a catalyst-sand mixture replaces the heat carrier which is typically pure sand in a traditional, non-CFP process. Thus, CFP occurs in a single reactor where biomass feedstock and the catalysts are mixed intensively to enable a good contact between the two. Generated primary pyrolysis vapors instantly diffuse into the catalyst pores where the catalytic cracking/reforming reactions occur in batch (Figure 10.1a) or in continuous modes (Figure 10.1b). In small-scale laboratory reactors, such as micro-pyrolysis units and fixed beds, catalyst and powdered biomass are mixed and the mixture is placed in a tube with the aid of ceramic (or quartz) wool, or a sintered disc support.[1,21] When *in situ* processing is applied in continuously operated larger scale units (circulating fluidized bed, auger reactor, *etc.* See Section 10.2.3.2) it is advisable to separate the char from the catalyst mixture before the regeneration step.[41] Although practically hard to implement, the reason for this crucial separation is to prevent the accumulation of char-bound ash (minerals) on the active surface of the catalyst material. In CFP of biomass, negative effects of accumulating biomass ash on the bio-oil yield and quality have been observed.[11,48]

In the *ex situ* mode, pyrolysis vapors produced in the primary (non-catalytic pyrolysis) reactor are carried through a secondary (fixed bed, moving bed, or circulated transported bed) reactor where they are contacted with a batch of catalyst (Figure 10.1c) or with continuously fed, fresh (or regenerated) catalyst (Figure 10.1d). One of the biggest advantages of the *ex situ* processing is that the temperatures of both reactors are independently adjustable, which helps in controlling the product distribution and selectivity.[39] Moreover, *ex situ* processing allows the possible secondary input of gas (*e.g.* steam, hydrogen) to the *ex situ* reactor, which might improve the efficiency of the catalytic processing.[1] The contact time between the primary pyrolysis vapors and the catalyst should be carefully controlled for a complete reforming of these vapors. *Ex situ* catalysis can be applied in a separate fixed/packed bed in the form of pellets or monoliths. This avoids the damage due to catalyst particle attrition that may readily occur in a circulating transported bed; particle attrition can severely limit the catalyst lifetime.[1] The char formed inside the pyrolysis reactor can be further combusted in order to provide the heat needed for the pyrolysis process.[59] Moreover, the periodic regeneration of the spent catalysts (burning off the coke on catalyst) can be conducted in the absence of biomass char, and thus the accumulation of biomass ash (mainly within the char structure) in the catalytic reactor can be prevented. Adversely, the need for an additional catalytic reactor (or two or more additional reactors operating sequentially in operation and regeneration modes) increases the capital and operating costs to a great extent, which can be considered as the main drawback of the *ex situ* processing.[60] Moreover, the additional vapor residence time in such a system may cause further secondary reactions and promote the undesired water and coke production. This could decrease the quality and the quantity of the liquid product. For more detailed information regarding the process modes in CFP of biomass the reader is referred to the recent publication of Yildiz *et al.*[41]

10.2.3 Reactors and Setups Used for CFP of Biomass

In pyrolysis processes, the reactor is the core and generally the most researched part of the fast pyrolysis technology, in particular from an engineering point of view. A large body of literature is available for CFP of biomass carried out both at analytical scale (*e.g.* analytical pyrolysis setups or py-GC-MS, packed (fixed) bed, tubular quartz micro reactor) and laboratory, bench, and pilot-scale (*i.e.* bubbling and circulating fluidized beds, auger reactor, conical spouted bed, entrained flow reactor) reactors. In these studies, the performance of various zeolite catalysts on product yields and properties have been investigated. There are also PDUs and (semi-)commercial units; for the latter, most of the quantitative information remains obscure. Nevertheless, it is crucial to note here that the results obtained are difficult to compare because of the variety in the reactor types, the scales of the reactors, and the process parameters used. Besides, the performance analysis is often incomplete.

10.2.3.1 Analytical Reactors (μg- and mg-Scale)

Analytical pyrolysis, with a micro-pyrolyzer directly connected to a GC-MS system (py-GC-MS), is a well-established technique to screen various catalyst/biomass combinations by performing simultaneous separation and identification of produced compounds in the GC-detectable vapor range. This helps to identify some of the main reactions that occur during CFP and to understand how the catalysts behave on a molecular basis leading to the production of desirable hydrocarbons and other chemicals. In this way, promising catalyst/biomass candidates can be identified and selected for larger scale experimentation.[61,62] Various biomass feedstock materials and individual biomass constituents (*e.g.* cellulose, hemicellulose, and lignin) have been tested batch-wise by analytical pyrolysis, *viz.* in packed bed and tubular quartz micro reactors.[61,63] The results show good selectivity for target compounds (*e.g.* aromatics, C_1–C_4 hydrocarbons) even though the absolute quantification of pyrolysis products is often missing. Besides, it is difficult to have an accurately closed mass balance. These types of reactors are not the real representatives of industrial/commercial scale reactors because they cannot be scaled up.[64] They are designed specifically for laboratory screening experiments. The differences in product yields and distribution between analytical and large scale reactor configurations can be attributed to the difference in heat and mass transfer[64] which are not limiting in py-GC-MS, the much shorter vapor residence times and the fact that no bio-oil condensation takes place as only the composition of the evolved vapors are considered in analytical pyrolysis. Hence, the results obtained from py-GC-MS systems are likely not comparable with the real bio-oil samples that are collected from the condensers of larger scale units.[15]

10.2.3.2 Bench- and Pilot-Scale Reactors (g to kg Scale)

Fluidized bed reactors are widely used for biomass fast pyrolysis. They are applied also in a vast number of industrial processes due to their excellent mass and heat transfer characteristics, scalability and simplicity of operation.[36,64] In laboratories, they have been frequently used for non-CFP and CFP of biomass due to their ability to provide high biomass heating rates, high catalyst-to-feed ratios,[53] and high bio-oil yields. In the case of CFP, biomass is fed into a hot fluidized bed reactor, where it thermally decomposes to form primary pyrolysis vapors. These vapors are directly contacted with the catalysts, either *in situ* or *ex situ*, and then condensed to form an upgraded liquid product (*i.e.* CFP-oil).

According to the CFP literature, two types of fluidized bed reactors have been used in laboratories, namely bubbling and circulating fluidized beds. Bubbling fluidized beds (BFBs) can be considered as 'semi-batch' reactors in which all the solid bed material (*i.e.* generated char, catalyst, sand) stays together during the operational run time without being transported outside the reactor bed. While being quite suitable for conventional non-catalytic biomass pyrolysis, BFBs are less ideal for CFP because the bed material needs to contain a certain amount of catalyst. With the start of continuous feeding of the biomass feedstock, the deactivation of the catalyst takes place within a short period while the WHSV (h^{-1}) increases over time. Consequently, the catalyst behaves as an inert material for the rest of the operation. A BFB reactor could also be used for the *ex situ* upgrading of non-catalytic pyrolysis vapors if placed after the primary pyrolysis reactor.[65] However, catalyst deactivation would again be a serious issue requiring the catalyst to be refreshed continuously, either by feeding fresh catalyst or by recycling a regenerated catalyst.[26] This can be solved by the application of two or more parallel *ex situ* reactors operating sequentially in an active and in a regeneration mode.

Another drawback of BFBs in case of CFP is the change in the fluidization behavior when a catalyst is added to the bed material (*e.g.* sand–catalyst mixture). Typical inert solid materials such as Al_2O_3, SiO_2, and sand used in BFB reactors have higher densities and larger particle sizes than zeolite catalysts (*e.g.* Fluid Catalytic Cracking, ZSM-5). A bed material mixture composed of Geldart's class B and C particles[66] cannot be fluidized together. It would quickly result in segregation and excessive entrainment of the fine catalyst particles. Because relatively high fluidization velocities are required to limit the vapor residence times, the BFB can be considered unsuitable for CFP. Commercially available catalysts have high porosities and small particle diameters; they are designed particularly for the applications in circulating fluidized beds. To overcome this problem, Yildiz *et al.* developed a research setup with a reactor (similar in design to the conventional BFB reactors) in which a mechanical mixer continuously agitates the bed contents.[48,67] This way, the bed contents remain well mixed without any risk of particle segregation. However, since the bed material is not continuously replaced, catalyst deactivation, as previously explained, still occurs. Recently, Jae *et al.* developed a unit with continuous catalyst addition and removal based on bubbling fluidized bed technology.[68]

Circulating fluidized bed (CFB) and riser reactors, which include continuous catalyst regeneration, seem the most appropriate technologies for CFP of biomass. In these systems, char and the bed material (*i.e.* catalyst) coming from reactor bed are separated from the pyrolysis vapors and gases by cyclones, and transferred to a combustion reactor (*e.g.* regenerator).[69] In contrast to laboratory-scale bubbling fluidized bed reactors where the catalyst stays in contact with the biomass originated char (containing most of the biomass ash) during the operational run time without being regenerated, process systems based on solids recirculation offer a few seconds of contact (retention) time (<1 s)[13] between the catalyst and the biomass. By frequent regeneration, the catalyst should in theory retain most of its original activity and contact the freshly-fed biomass always in an active state.[36,54] Unlike in BFB reactors, high fluidization velocities in CFB's are not a problem for the small catalyst particles, as both the char and the catalyst are meant to be passed through the cyclones. A serious concern is the available devolatilization time for biomass particles. For a full devolatilization, only very small biomass particles (perhaps far below 1 mm) can be fed. Circulating fluid beds with continuous catalyst regeneration system have been used for CFP of woody biomass by Lappas and co-workers at CPERI[13,21] and by the scientists at RTI.[65]

Currently, a considerable interest exists in using auger reactors both for research and industrial use.[70] Auger reactors offer some distinct advantages such as high heating rates when heated by an externally supplied heat carrier, being compact and robust, less consumption of carrier gas, feedstock flexibility and anticipated processing capacities from 50 to 100 tons day^{-1}.[70–72] Practically, the biomass feedstock and the heat carrier stay in contact on the auger screw for a time necessary to complete the biomass devolatilization. Pyrolysis vapors are then physically separated from the solids (heat carrier and char) along the entire length or at the end of the screw. Continuously fed biomass feedstock is always brought into contact with fresh (or regenerated) catalyst, which makes the auger reactor technology quite suitable for CFP. Veses *et al.*,[73–75] Yildiz *et al.*,[15,76] and Li *et al.*[77] conducted *in situ* and *ex situ* CFP of woody biomass (pine wood) based on auger reactor technology.

Besides the fluidized beds and auger reactors, conical spouted bed and entrained flow reactors have been applied for CFP of biomass. Olazar and co-workers reported the catalytic pyrolysis of sawdust in a conical spouted bed reactor by using a H-ZSM-5 catalyst.[78] In this reactor setup, the bottom section of the reactor is conical in shape and a high velocity stream of gas (the spout) induces circulation within the catalyst bed. Imran *et al.* performed *in situ*[79] and *ex situ*[80] CFP of wood fibers in a continuously operated entrained flow reactor. In this system, the feedstock (biomass or biomass/catalyst) is fed to the electrically heated entrained down flow reactor from the top position. The primary devolatilization of biomass particles takes place in a few seconds (size limited) under inert atmosphere where the solids and the vapors are separated by a cyclone located at the bottom the reactor.

10.2.3.3 PDU's and (Semi-)Commercial Process Units (kg to t Scale)

Although laboratory studies regarding the CFP of biomass have been carried out for more than two decades, the technology development is still in its infancy. The most serious commercial attempt up to date was by KiOR Inc. The company built the first commercial scale CFP facility in Columbus (Mississippi, USA), which started production in 2012. The selected process technology was based on existing Fluid Catalytic Cracking (FCC) technology, used for crude oil refining. The reactor was composed of a riser, a solid separator, and a catalyst regenerator, continuously operating in *in situ* mode. The facility was designed as an initial scale commercial facility; processing 500 bone-dry tons of woody biomass per day. It should have produced around 50 million liters of gasoline, diesel, and fuel oil blend-stocks annually, enough to fuel 25 000 cars. The company was next looking to constructing its "flagship" facility in Natchez, Mississippi; it was designed to process approximately 1500 bone-dry tons of feedstock per day—three times the size of its Columbus facility—to take advantage of economies of scales.[81] However, in November 2014, KiOR Inc. filed for bankruptcy protection with a plan to sell its assets.[82] The main reason for this was the insufficient production yields of the marketable product, which could not sustain the overall process economy.

Anellotech is a spin-off company and was founded in 2008. The company has the exclusive license from the University of Massachusetts for the core CFP process technology developed in the George Huber's laboratory. The first application of the technology, Biomass to Aromatics™ (BTA), is said to produce "green" benzene, toluene, and xylenes (BTX) in the future. The reactor technology is similar to that of KiOR's and composed of a fluidized bed reactor operating in combination with a catalyst separator and a catalyst regenerator.[83] Although not mentioned anywhere, the process mode is supposed to be *in situ*. With the successful start-up of its Pearl River pilot plant (erected in March 2013), Anellotech is now able to meet its commitment to provide BTX developmental quantities for evaluation by a number of strategic partners who invested in the development of the process. At this pilot plant, the company is currently producing kilogram-scale BTX directly from non-food biomass such as palm wastes, bagasse, corn stover, and even from wood feedstocks. If all goes according to schedule, Anellotech expects to start up its first commercial plant in 2019–2020.

Being a semi-commercial application of a catalytic hydropyrolysis process, the IH^2 technology is meant to provide a cost-effective route, ~\$0.53 liter^{-1} in 2012 at a scale of 2000 ton dry feed day^{-1}, to produce fungible liquid hydrocarbon transportation fuels from renewable resources. This process was developed by the Gas Technology Institute (GTI) of Des Plaines, IL. GTI has licensed the IH^2 technology to CRI Catalyst Company for exclusive worldwide deployment. The process involves the hydrodeoxygenation of the volatilized biomass to produce a raw hydrocarbon product over proprietary CRI catalysts (non-zeolite) in the presence of low-pressure hydrogen, followed by a fixed bed hydrotreater, which uses other proprietary non-zeolite CRI catalysts to polish the first-stage product and transform it into a finished hydrocarbon fuel or blend stock. As a step in preparation of the commercialization process, GTI commissioned a new 50 kg day^{-1} Pilot-Scale IH^2 Plant in the first quarter of 2012. The goal of the pilot plant studies is to provide validation of the operational and performance factors that are keys to achieving commercial deployment.[84–87] These efforts led Shell India Markets Pvt. Ltd. (SIMPL) to proceed with the installation of a 5 tons day^{-1} IH^2 technology demonstration plant on the site of SIMPL's new Technology Centre in Bangalore, India.[88]

Unfortunately the precise performance of the aforementioned (semi-) commercial plants in terms of product yields, the information regarding the product quality (*e.g.* acidity, viscosity, chemical composition), and total operation time remains obscure. Websites of private enterprises do not mention such data although it appears that the companies are sending out larger samples for closer inspection by future clients.

10.2.4 Catalysts Used in CFP of Biomass

In the petrochemical industry, heterogeneous catalysis are widely used for the conversion of heavy oil fractions into lighter fuels and chemicals. Heterogeneous catalysts are those where the catalyst and the reactants are in different phases, most usually a solid catalyst and gas phase reactants. This phase difference between the two allows the gaseous reactants to react on the large surface area of the porous solid catalyst and be easily separated from each other afterwards.[38] In the last decades, improving the quality of bio-oil using heterogeneous catalysis has received considerable attention and a similar approach is carried over to biomass fast pyrolysis where the heavy oxygenated volatiles from the decomposition of biomass are cracked/deoxygenated and converted to lighter compounds when in contact with a suitable heterogeneous catalyst.[36,89] The challenges in developing this technology involve the development/optimization of advanced porous materials as efficient mono- and bi-functional catalysts for the production of high-grade bio-oils from biomass. To this end, cracking catalysts from different classes can be used. The reader is referred to some extensive reviews covering the types of catalysts used in CFP research.[19,49,50]

The selection of appropriate catalysts for CFP is one of the key factors affecting the outcome of the process. Apart from any cost consideration, a successful catalyst is characterized by three factors: activity, referring to the rate of reaction; selectivity, defined as the percentage of the spent reactant that forms the desired product; and stability, which is determined by how quickly the catalytic activity declines and determines how often the catalyst has to be replaced.[38] In CFP of biomass, an ideal catalyst should produce high yields of liquid product with a reduced oxygen content (by favoring CO, CO_2 and H_2O production), suppress the formation of coke, and minimize the amount of undesirable compounds present in the product oil (*e.g.* aldehydes, ketones, acids, poly-aromatics). The latter are known to be detrimental for the direct use or further co-processing of bio-oil.[36] Moreover, the catalysts need to be both mechanically and thermally stable (resistance to deactivation) and robust (the behavior of the catalyst upon successive regeneration cycles) to improve overall lifetimes. To evaluate the catalytic activity and/or to compare different catalysts, turnover frequency, defined as the number of molecules that react per active site per unit time (in units of reciprocal time), can be used.[38]

In commercial CFP processes, all the details of catalyst performance, deactivation modes, and regeneration protocols must be carefully defined. Increasing the catalytic conversion by improving the solid/vapor interface chemistry will be the key for catalyst improvements. By using a combination of computational and experimental techniques, rational design and development of catalysts can be accomplished.[90]

10.3 Reported Results of Continuous CFP of Woody Biomass Over Zeolites

Existing CFP research spans a wide range of catalyst and feedstock combinations tested over a range of temperatures and with various process parameters in an effort to develop fundamental knowledge on the effects of the catalysts. As candidate catalysts for CFP of biomass, micro- and meso-porous acidic zeolites, mesoporous materials with uniform pore size distribution, microporous/mesoporous hybrid materials doped with noble and transition metals, basic (metal oxides), and metal catalysts have been investigated in the literature.[36] Zeolites are known to be effective in cracking, deoxygenation, and synthesis of aromatic hydrocarbons[31] The reaction chemistry for the CFP of biomass on a solid acid catalyst is shown in Figure 10.2. The reason this catalyst has been selected was that the majority of the previous studies concerning the upgrading of biomass pyrolysis vapors were performed with this type of catalyst (see Tables ESI-1 and ESI-2 in the Electronic Supplementary Information). Most of these studies were carried out in semi-continuous mode mainly by using bubbling fluidized bed reactor technology. Here, semi-continuous operation refers to the continuous biomass feeding into a stagnant bed of catalyst (or catalyst-sand mixture).

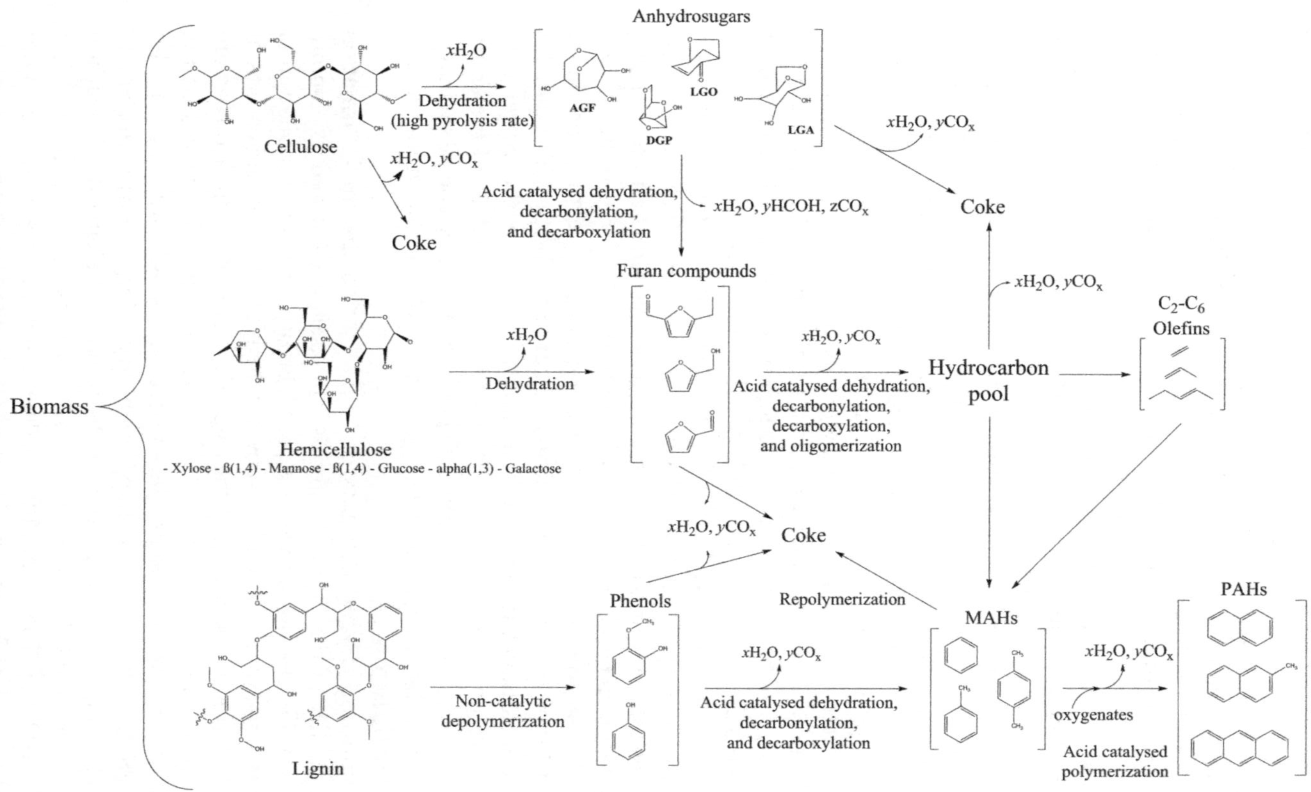

Figure 10.2 Reaction chemistry for the CFP of biomass on solid acid catalyst. Scheme constructed from previously published suggestions.[28,39,55,64,91] AGF: 1,6-Anhydro-β-D-glucofuranose; DGP: 1,4:3,6-dianhydro-α-D-glucopyranose; LGO: levoglucosenone; LGA: levoglucosan; MAHs: mono aromatic hydrocarbons; PAHs: poly-aromatic hydrocarbons. Reprinted with permission from ref. 67. Copyright (2014) American Chemical Society.

As discussed in Section 10.2, the product distribution in CFP of biomass and the quality of the resulting CFP-oil are likely dependent on several factors and parameters. At the end, process reliability, catalyst consumption rate, and the price and quality of the CFP-oil will be decisive for the success of a technology developer. A comparison of results obtained from different studies could be misleading due to the differences in the applied parameters and the scales at which such data have been obtained. One of the indications of a successful process unit for woody biomass thermal processing could be the performance of the said unit under the non-catalytic conditions. The main performance criterion could be specified as the maximum achievable liquid yield obtained in non-catalytic pyrolysis, while the formation of by-products (*i.e.* water, char, and non-condensable gases) is minimized. For example, continuously or semi-continuously operated laboratory, bench, pilot, and PDU scale process units reporting non-catalytic organic liquid yields below a threshold of 40 wt% (based on biomass feed) could be recognized as inefficient ones (with respect to the yield in pyrolysis vapors which would subsequently be subjected to heterogeneous catalysts in CFP), because these yields fall into the intermediate and/or slow pyrolysis categories. In the literature for CFP of woody biomass, only a limited number of studies report yields above this threshold (Table 10.1). In the following sections of this chapter, only the results provided by these studies will be considered.

Tables ESI-1 (for *in situ* studies) and ESI-2 (for *ex situ* studies) in the Electronic Supplementary Information show the process conditions, *i.e.* the type and operating temperatures of reactors, type of catalyst (including Si : Al ratio) used, and the given catalyst-to-biomass ratio or WHSV of the studies shown in Table 10.1. The amount of catalyst consumed per unit weight of produced liquid and CFP-oil (organics) were calculated based on the provided data for each individual study. The decrease in organics, decrease in the oxygen content of the organics, and increase in the higher heating values of the liquid organic phase were included, as well. The latter are listed as a percent change in the results of the catalytic pyrolysis relative to the corresponding results for non-catalytic pyrolysis.

The carbon and oxygen contents of CFP-oil reveal useful information about the quality of the liquid. In almost all cases, carbon and oxygen are rejected from the CFP-oil to various extents. Figure 10.3 is a compilation of data from *in situ* and *ex situ* studies showing the biomass-to-liquid carbon or oxygen yield ratios as a function of the catalytic over non-catalytic oil (organics) yield ratio. The data points with same shape and color represent the results for various process parameters tested in a single study (see Tables ESI-1 and ESI-2 in the Electronic Supplementary Information). The elemental yields $(\eta_i)_{CFP}$ and $(\eta_i)_{FP}$ are defined by the fraction of element i (*i.e.* carbon or oxygen) in the feed that is recovered in the dry-ash-free organic product (*i.e.* CFP-oil or fast pyrolysis oil, FP-oil). The reduction in a certain element (η_C for carbon and η_O for oxygen) is based on the carbon or oxygen content of CFP-oil compared with that obtained by the non-catalytic fast pyrolysis

Table 10.1 Obtained non-catalytic product yields from the selected literature studies.

	Non-catalytic yields (wt% on biomass fed)						
Biomass type (intake, kg h^{-1})	Organics	Aqueous phase	H_2O-KF	Char	NCG's	Total	Ref.
Hybrid poplar wood (0.1)	(63.3)[a]	-	N/A	12.7	24	100	42
Pine wood (0.015)	41.7	-	6	24.2	9.3	81.2	92
Wood mixture (0.22)	43.3	-	24.1	17.2	14.3	98.9	93
Mixed wood (0.22)	40.4	14.8	-	17.1	15.2	87.5	94
Mixed wood (0.22)	40.4	14.8	-	17.1	15.2	87.5	95
Beech wood (0.33)	53.6	-	16.9	7.1	11.5	89.1[b]	13
	54.7	-	17.6	8.9	9.5	90.5[c]	
Wood fibers, Lignocel (1)	47.2	21.7	-	16.2	11.6	96.8	80
Sawdust (3)	(63.4)[a]	-	N/A	18	18	99.4	3
Hybrid poplar wood (N/A)	51	-	15	15.1	15.1	96.2	96
White oak (2.22)	54	-	7.8	18	13	92.8	97
Pine wood (0.12 to 0.36)	55.4	-	14.3	12.1	18.2	100	78
European spruce (0.7–0.8)	59	-	11	13	10	93	54
Pine sawdust (20)	63	-	10	18	9	100	98
Radiata pine (0.15)	(60.3)[a]	-	N/A	20	19.8	100.1	99
Oak wood (0.04)	47	-	13.2	7	24.3	91.5	100
Pine wood (0.4)	45.8	-	21.1	12	17.9	96.8	101
Hybrid poplar wood (N/A)	51	-	15	15	18.9	99.9	102
Wood waste (0.22)	48.2	17.7	-	17.1	15.2	98.2	103
Wood fibers, Lignocel (1)	47.2	-	21	16	11.6	95.8	79
Southern yellow pine (200 ml)	45.8	-	21.1	12	17.9	96.8	104
Wood fibers, Lignocel (1)	43.6	20	-	16.3	11.6	91.5	105

[a]Total liquid yield, water content is not available. Provided total liquid yields above 60 wt% without any water content information are also included, exceptionally.
[b]Performed at a reactor temperature of 460 °C.
[c]Performed at a reactor temperature of 457 °C. H_2O-KF: Water content of the bio-oil determined by Karl-Fischer Titration method; N/A: not available.

reaction (Eqn (10.1)). In eqn (10.1), $(m_i)_{\text{CFP-oil,daf}}$ and $(m_i)_{\text{FP-oil,daf}}$ represent the amount of carbon and oxygen in wt% of the organic fraction while Y represents the yields of catalytic and non-catalytic organic fraction, as derived from each individual study.

$$\eta_i = \frac{(\eta_i)_{\text{CFP}}}{(\eta_i)_{\text{FP}}} = \frac{(m_i)_{\text{CFP-oil,daf}} * Y_{\text{CFP-oil}}}{(m_i)_{\text{FP-oil,daf}} * Y_{\text{FP-oil}}}, i: \text{C or O} \tag{10.1}$$

In Figure 10.3, the carbon and oxygen yield ratios on y-axes were combined in a graph with a common x-axis representing the yield ratio. In both graphs, linear trendlines were generated using the given data points. According to the example shown by the set of lines with arrows, for a case in which the organic oil yield decreases with 61% (x-axis), a decrease of 70% in oxygen yield and a corresponding decrease of 55% in carbon yield is observed when going from

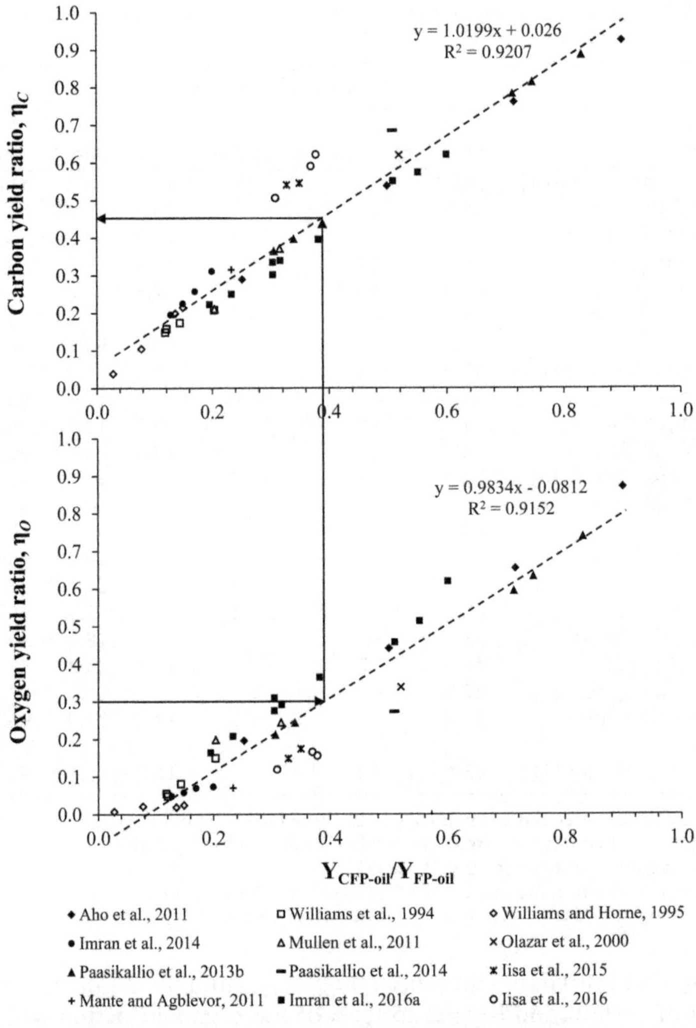

Figure 10.3 Compilation of literature data showing the carbon and oxygen yield ratios (see eqn (10.1)), related to the ratio of CFP-oil yields and fast pyrolysis oil (FP-oil) yields reported in each individual study. The dashed lines represent the trend lines. The lines with arrows show example values for the change in the organic liquid yields and the carbon yield for a given oxygen yield. The process conditions for each individual study can be seen in Tables ESI-1 and ESI-2 in the Electronic Supplementary Information.

non-catalytic to catalytic pyrolysis. Carbon is lost together with the process by-products; *i.e.* coke on catalyst, char, carbon containing non-condensable gas molecules, and organic molecules solved in the aqueous liquid phase. The arrows can be further shifted to find other values for the oxygen and carbon yield reduction for various organic oil yield ratios. As an extreme

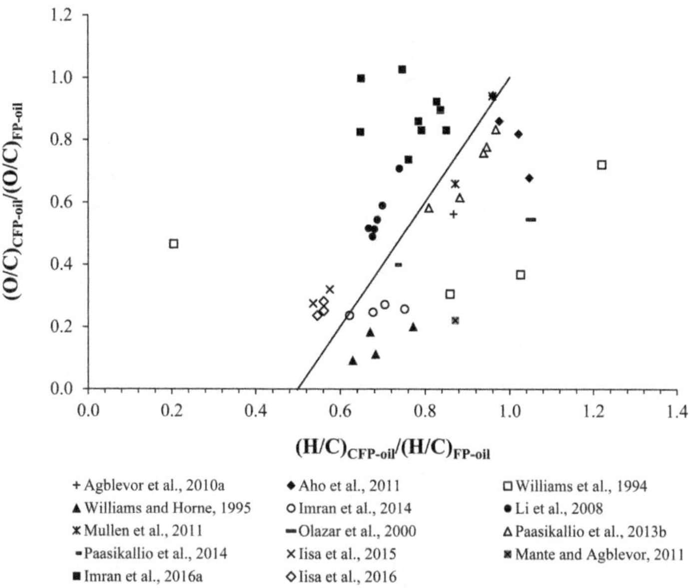

Figure 10.4 Comparison of O/C and H/C ratios for CFP-oils (dry-ash-free basis), related to the identical ratios reported for non-catalytic pyrolysis oil. The line represents the dehydration line. The process conditions for each individual study can be seen in Tables ESI-1 and ESI-2 in the Electronic Supplementary Information.

example, a fully deoxygenated CFP-oil ($\eta_O = 0$) yields a very viscous (or solid) product in very low yields reduced by a factor more than 10 (see ref. 93 and ref. 95 in Table ESI-2 in the Electronic Supplementary Information).

The target of a CFP process should be to produce a maximum amount of CFP-oil with minimum oxygen content, while keeping its H:C ratio as high as possible. Additionally, lower O:C ratios are preferable in terms of the energy content of the CFP-oil. A useful representation providing insight into the effects of zeolite catalysts and process conditions on the elemental composition of the CFP-oil is a van Krevelen plot.[106] In the studies of which the details are shown in Tables ESI-1 and ESI-2 (see the Electronic Supplementary Information), various types of feedstocks, catalysts, and process parameters were used. Comparing the O:C and H:C ratios of produced organic liquid fractions could further demonstrate the dehydration performances based on these differences. The results are provided in Figure 10.4, where the ratios of O:C and H:C, for the CFP-oil and the FP-oil respectively, are shown.

Although the collected data points are widely spread, they are clustered around the dehydration line. In the studies where a single type of catalyst tested under different process conditions (*e.g.* temperature) some trends can be observed. For example, in the studies of Imran *et al.*, 2014[80] and Williams *et al.*[93] it is seen that increasing the catalytic reaction temperature decreases the H:C ratio of the resulting CFP-oil. On the other hand, in the study of

Li *et al.*,[3] O : C ratios decrease with increasing reactor temperature. This information shows that in CFP of biomass over zeolites, higher temperatures favor the cracking and deoxygenation reactions. The promotion of cracking and deoxygenation mechanisms in CFP is also possible with the changes in catalyst-to-biomass ratio and/or in WHSV. Under the same process conditions, increasing the catalyst-to-biomass ratio was shown to decrease the amount of organics[13] (see Table ESI-1 in the Electronic Supplementary Information) and the O : C ratio.[92] Besides, Paasikallio *et al.* showed that decreasing WHSV favored the dehydration mechanism.[54] The degree of dehydration varies with the type of catalyst, as shown by Williams and Horne[95]; Na-ZSM-5 appeared to favor dehydration more than that of Y-zeolite. On the other hand, Ca–Y-zeolite was shown to be more effective in hydrogenation than β-zeolite.[97] These results show that advancements can be achieved in tailored catalyst design by impregnating metals in them.

As shown in Figure 10.2, the primary fast pyrolysis vapors from woody biomass undergo various reactions that result in the production of hydrocarbons on the active surface of a zeolite catalyst. These complex set of reactions include deoxygenation reactions (*i.e.* dehydration, decarboxylation, decarbonylation), but also result in additional heterogeneous coke generated on the catalyst surface. In CFP of woody biomass, the water, CO_x (the sum of CO and CO_2 yields), and coke yields could reveal valuable information regarding the performance of a zeolite catalyst. Considering the deoxygenation reactions, decarboxylation (CO_2 rejection) is more desirable than dehydration (H_2O removal) because it minimizes the loss of hydrogen which is important for the hydrocarbon-forming reactions. Moreover, compared to decarbonylation (CO rejection) wherein a single oxygen atom is removed per carbon atom, decarboxylation maximizes oxygen removal with minimal carbon loss (two oxygen atoms per carbon atom) and thus allows maintaining high H : C ratio and low O : C ratio for the bio-oil.[76,107]

Figure 10.5a shows the comparison of the water yield (wt% on feed basis) related to the CO : CO_2 ratios reported in the selected CFP studies. Although the desired deoxygenation mechanism is decarboxylation (the region at the left bottom in Figure 10.5a), the generation of water and CO is unavoidable in CFP. Obviously, significant amounts of water are produced, ranging from 8 to 36 wt%. Besides, the decarbonylation mechanism dominates the decarboxylation in most of the studies. Decreasing the catalyst-to-biomass ratio from 18 to 2.9 (open spheres in Figure 10.5a) with ZSM-5 based FCC (10 wt% USY) resulted in suppressed dehydration while the CO : CO_2 ratio was constant.[13] A similar result was obtained in the work of Williams *et al.*[93] when the temperature of the *ex situ* reactor was increased. In the work of Paasikallio[54] the increase in WHSV suppressed decarbonylation while an increase in acidity of ZSM-5 catalyst favored it. Decarboxylation was dominant when 15% vegetable oil was impregnated to wood fibers while this showed an overall decrease in the production of CO_x compounds.[105] A remarkable result was achieved by Olazar *et al.*;[78] pine wood was catalytically pyrolyzed in the presence of H-ZSM-5 catalyst in a conical spouted bed reactor operating at 450 °C. At this conditions, a considerable degree of dehydration occurred while the CO_2 removal was boosted.

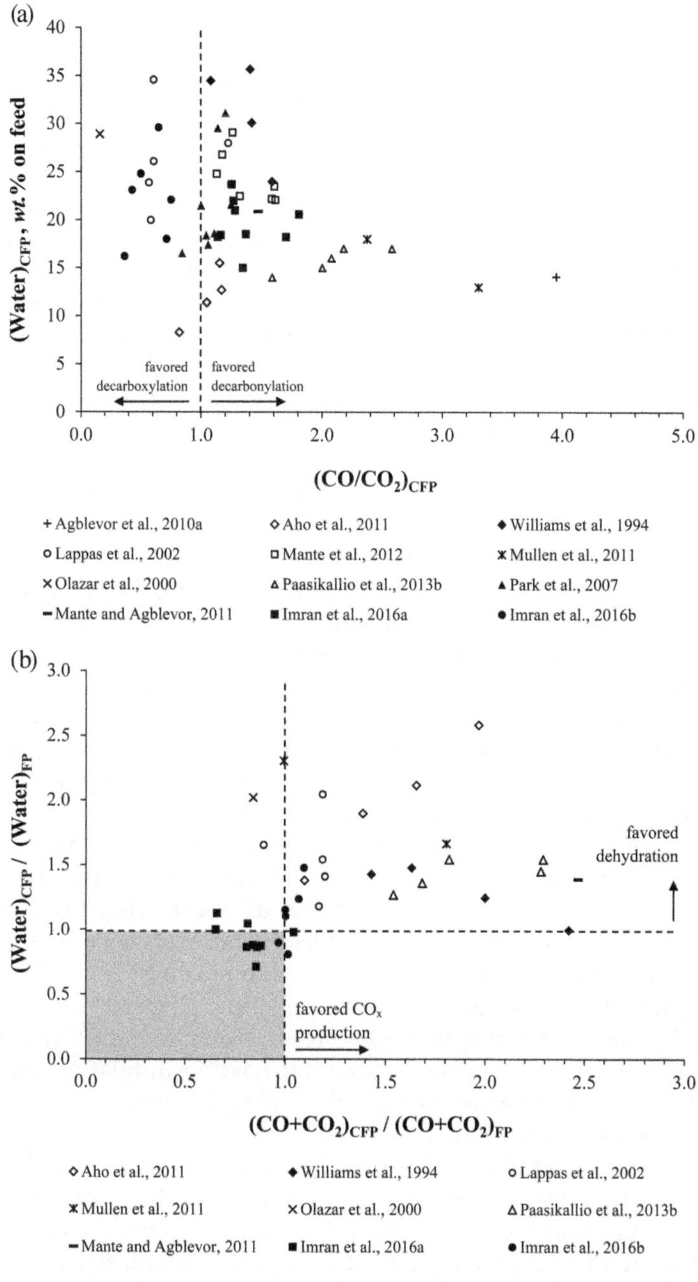

Figure 10.5 (a) Comparison of water yields (wt% on feed basis) related to the $CO:CO_2$ ratios obtained in CFP. The process conditions for each individual study can be seen in Tables ESI-1 and ESI-2 in the Electronic Supplementary Information. (b) Comparison of water yield ratios and $CO + CO_2$ ratios obtained in CFP related to the same ratios reported for non-catalytic pyrolysis obtained in the same setup. The region outside the grey zone depicts the region of deoxygenation. The process conditions for each individual study can be seen in Tables ESI-1 and ESI-2 in the Electronic Supplementary Information.

As a complementary figure, Figure 10.5b shows the comparison of water yield ratios and $CO + CO_2$ ratios obtained in CFP related to the same ratios reported for non-catalytic pyrolysis obtained in the same setup. In principle, all data points should be bigger than one (outside the grey zone) disclosing the deoxygenation in the case of CFP. This figure reveals that the CO_x production (the sum of CO and CO_2 yields) is favored in most of the studies, while the main mechanisms seem to be the dehydration and decarbonylation. The exception is the work of Imran *et al.* 2016a, where H-ZSM-5 catalyst does not contribute to the deoxygenation (black squares in the grey zone).

Together with the deoxygenation capability, the ability to suppress the formation of coke is a critical parameter that values a zeolite catalyst in CFP conditions. Figure 10.6a compares the coke yield (wt% on feed basis) with the change in $CO:CO_2$ ratios obtained in CFP. Such a comparison could reveal valuable hints regarding in which way the biomass carbon is consumed if not converted to liquids. Similar to the Figure 10.5b, Figure 10.6b shows the coke yield related to the $CO + CO_2$ ratios obtained in CFP and non-catalytic pyrolysis.

As seen in Figure 10.6a and Figure 10.6b, coke yield varies remarkably. The main reasons for this variation are the used of different type of catalysts and process parameters. In all studies, CO_x production is favored when compared with the non-catalytic case (Figure 10.6b). As a general trend, it can be concluded that when the decarbonylation mechanism is dominant, coke yield increases. In the presence of an FCC catalyst, the combined effect of decreasing the catalyst-to-biomass ratio from 22 to 3.5 and the riser temperature from 468 °C to 421 °C resulted in a shift through CO_2 production without causing any change in coke yield.[13] On the other hand, at a fixed riser temperature with ZSM-5 based FCC (10 wt% USY), decreasing the catalyst-to-biomass ratio promoted the coke yield while keeping the $CO:CO_2$ ratio almost constant.[13] Similarly, Aho *et al.* found that increasing the catalyst-to-biomass ratio (while other process parameters were kept constant) favored CO_2 production while suppressing the coke yield.[92] Williams *et al.* reported that increasing the reactor temperature favored the CO production but did not make a reasonable change in the coke yield.[93] H-ZSM-5 and Na-ZSM-5 were shown to be more resistant to coke accumulation and favored decarboxylation mechanism compared to Y-zeolite and activated alumina under same process conditions.[95]

A continuously operated CFP process unit with a catalyst recycling system requires a certain amount of fresh catalyst input to the system.[41] Together with the energy and biomass feedstock inputs, the catalyst is considered as a serious item of expenditure for the overall process. Hence, the target should be to achieve a marketable product with desired quality and quantity while keeping the catalyst consumption at its minimum. Figure 10.7 shows the deoxygenation of CFP-oil in relation with the catalyst consumed per unit weight of CFP-oil produced in a given process unit. Based on the data obtained from the selected literature studies, the oxygen reduction comes with a high price; with decreasing oxygen content the catalyst consumption

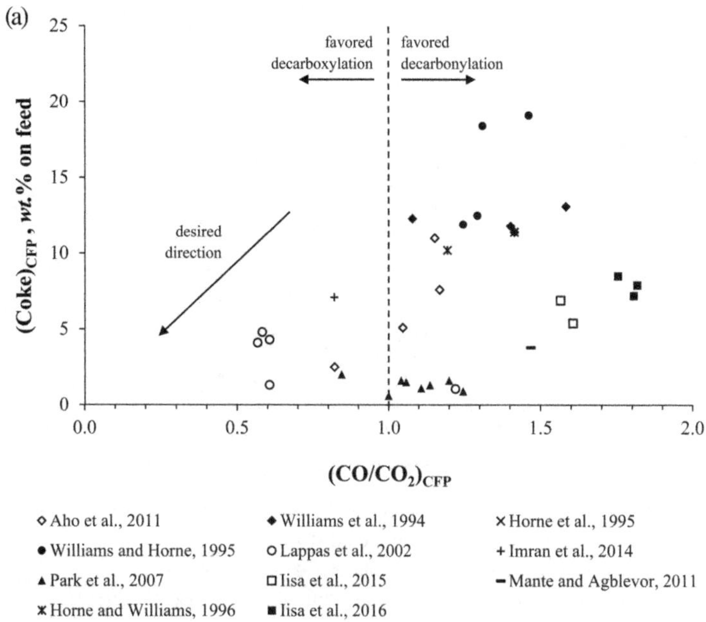

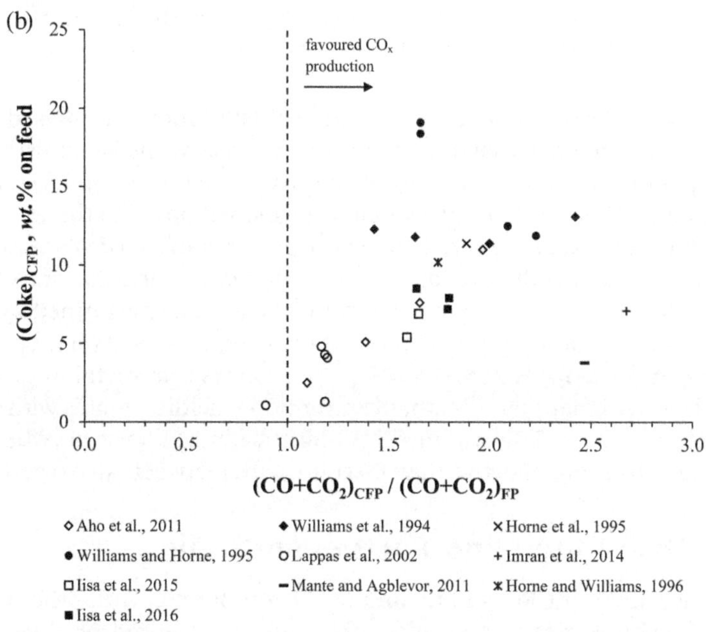

Figure 10.6 (a) Comparison of coke yield (wt% on feed basis) related to the $CO:CO_2$ ratios obtained in CFP. The process conditions for each individual study can be seen in Tables ESI-1 and ESI-2 in the Electronic Supplementary Information. (b) Comparison of coke yield (wt% on feed basis) related to the $CO + CO_2$ ratios obtained in CFP and fast pyrolysis (FP). The process conditions for each individual study can be seen in Tables ESI-1 and ESI-2 in the Electronic Supplementary Information.

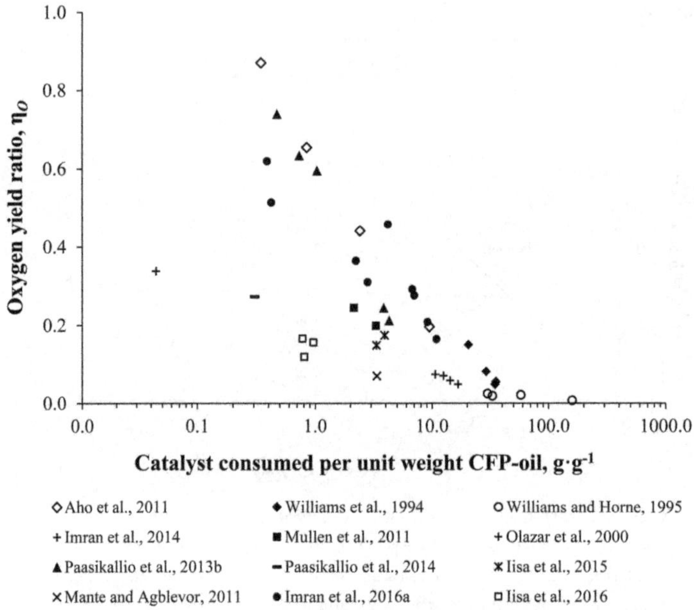

Figure 10.7 The oxygen yield ratio of CFP-oil (dry-ash-free basis) *versus* the amount of catalyst consumed per unit weight of CFP-oil produced.

(defined by catalyst-to-biomass ratio and/or WHSV) increases. Based on Figure 10.7, it is concluded that a temperature increase, while keeping the other process parameters the same, favors the oxygen removal while increasing the catalyst-to-CPF-oil ratio (black diamond shaped boxes to the +x,−y direction).[93] Paasikallio *et al.* reported that in the presence of a basic ZSM-5 catalyst (Si : Al ratio = 140), a reduction of the WHSV increased the catalyst-to-CPF-oil ratio and decreased the oxygen yield ratio.[54] In a study performed by Imran *et al.*, the deoxygenation performances of the used catalysts were reported as Ca–Y > Mg–Y > Na–Y = H-ZSM-5 > H–Y.[79] This shows that metal impregnated zeolites have a higher potential than conventional acidic zeolites with respect to the deoxygenation. This was further confirmed by Mullen *et al.*, where Ca–Y was found to be more effective than β-zeolite with respect to deoxygenation.[97]

10.4 Discussion and Further Outlook

The development of CFP is in its infancy in comparison with other thermochemical technologies (*e.g.* gasification), and the required product specifications are still fairly undefined. The selection of a catalyst for instance, is determined in the first place by the requirements imposed by any future user of the liquid product. Such requirements could be assigned according to targeting on, for example, the degree of deoxygenation and/or the production of target chemical compounds in high yields. In order to obtain target products in CFP of biomass with desired yields and compositions, the influence of the reactor type, the operation mode (*e.g. in situ*, *ex situ*), and the type and the performance of a catalyst (*e.g.* resistance to deactivation) in the specified process conditions should be well tuned.

The main target of CFP of lignocellulosic biomass is to deoxygenate the liquid product (CFP-oil) as much as possible while limiting the loss of carbon and hydrogen. The degree of coke formation on the zeolite catalyst should also be minimized. Complete deoxygenation, however, would result in producing only hydrocarbons (or just carbon in the worst case) at a very low yield. Hence, catalysis should be applied rather to steer the oxygen functionalities in the pyrolysis vapors, resulting in the production of certain fuel compounds or chemicals (aromatics, phenols, alcohols, furans, *etc.*) in the produced liquid. Catalyst design should therefore focus on the production of cheap catalysts with improved abilities for cracking, deoxygenation (mainly *via* decarboxylation), and the resistance against deactivation.

In CFP of biomass, the main problem to be dealt with is the presence of the biomass originated alkaline ash which destroys and physically swamps any catalyst utilized in the process. Practically, the direct physical contact between the catalyst and the biomass must be prevented in order to achieve a successful operation. Hence, prior to scaling up this technology to a commercial level, this particular problem should be clearly identified and solved. A solution may be to separate the biomass fast pyrolysis from the catalytic treatment of the vapors (*i.e. ex situ* processing mode). Although technically challenging and costly, *ex situ* operation provides some more flexibility in selecting the optimal conditions for catalysis. In this way, the periodic regeneration of the spent catalysts by coke combustion can be conducted in the absence of char which contains the vast majority of the biomass ash. Hence, the physical contact between the biomass minerals and the catalyst is excluded (no risk of catalyst poisoning by minerals) so that the activity of the catalyst is eventually conserved over a much longer time period. This allows the catalyst to be re-used for larger number of reaction/regeneration cycles than in an *in situ* operation. However, the quality and quantity of the CFP-products obtained in the *ex situ* processing may be different from those of *in situ* processing. The reason for this difference could be that the catalyst is not contacted directly with the primary fast pyrolysis vapors, but with the vapors that may have changed their composition during the transfer from the pyrolysis reactor to the catalytic reactor. This must be investigated more thoroughly.

CFP process yields a liquid product composed of two phases, *viz.* an aqueous phase (rich in water and acids) and a heavy phase rich in (partially) deoxygenated organic compounds (*i.e.* CFP-oil). In principle, the latter can be used to co-feed a FCC unit in existing petroleum oil refineries; further (co-)processing in such a unit would at least require a significant degree of deoxygenation and a good miscibility with the regular refinery feedstock. That would enable the production of drop-in hydrocarbon fuels without building a separate bio-oil refinery. When the CFP-oil is co-fed to a crude oil refinery, it is being processed in the existing refinery units together with the crude oil and then ends up in the various product fractions in various proportions. It is not clearly known in which fraction(s) the CFP-oil ends up in the end. If it ends up in asphaltene fraction for example, the economical contribution of the CFP-oil to the overall economy of the refinery is low. In order to be considered economical, the CFP-oil should have a composition allowing it to be converted high value products in the refineries. Overall, in a CFP process, consideration should be given to the production of liquid intermediates with improved physical and chemical properties instead of finished fuels.

Most of the literature studies concerning CFP of biomass deal with the catalyst screening in small laboratory systems. However, the efficient and the economic use of all products of the process (*i.e.* utilization of by-products for process heat generation within the system), and the operational barriers related to the implementation of CFP technology at a large scale, are hardly discussed. This chapter, therefore, focused on various aspects of continuous operation. Keeping in mind that the catalyst screening in small-scale laboratory setups (*e.g.* analytical pyrolysis devices, *mg*-scale batch reactors) is not the ultimate target but a tool for the development of this technology, the best performing catalysts in such studies should be tested in larger and continuous setups allowing for longer operations. For that purpose, the priority should be given to the design of novel catalytic systems and new reactor technologies, based on the specific process requirements for CFP rather than that of conventional fast pyrolysis.

References

1. M. A. Morris, in *Handbook of Biofuels Production*, ed. R. Luque, J. Campelo and J. Clark, Woodhead Publishing Limited, 2011, Production of bio-oils *via* catalytic pyrolysis, pp. 349–389.
2. H. J. Park, J.-K. Jeon, D. J. Suh, Y.-W. Suh, H. S. Heo and Y.-K. Park, *Catal. Surv. Asia*, 2011, **15**, 161.
3. H. Li, Y. Yan and Z. Ren, *J. Fuel Chem. Technol.*, 2008, **36**, 666.
4. T. Bridgwater, *J. Sci. Food Agric.*, 2006, **86**, 1755.
5. R. H. Venderbosch and W. Prins, *Biofuels, Bioprod. Biorefin.*, 2010, **4**, 178.
6. D. L. Compton, M. A. Jackson, D. J. Mihalcik, C. A. Mullen and A. A. Boateng, *J. Anal. Appl. Pyrolysis*, 2011, **90**, 174.
7. D. Mohan, C. U. Pittman and P. H. Steele, *Energy Fuels*, 2006, **20**, 848.
8. P. K. Kanaujia, Y. Sharma, M. Garg, D. Tripathi and R. Singh, *J. Anal. Appl. Pyrolysis*, 2014, **105**, 55.
9. R. French and S. Czernik, *Fuel Process. Technol.*, 2010, **91**, 25.
10. C. Hu, Y. Yang, J. Luo, P. Pan, D. Tong and G. Li, *Front. Chem. Sci. Eng.*, 2011, **5**, 188.
11. A. V. Bridgwater, *Biomass Bioenergy*, 2012, **38**, 68.
12. S. Czernik and A. V. Bridgwater, *Energy Fuels*, 2004, **18**, 590.
13. A. A. Lappas, M. C. Samolada, D. K. Iatridis, S. S. Voutetakis and I. A. Vasalos, *Fuel*, 2002, **81**, 2087.

14. *Heat and Power from Pyrolysis Oil*, https://www.btg-btl.com/en/applications/heat-power, May 2015.
15. G. Yildiz, F. Ronsse, J. Vercruysse, J. Daels, H. E. Toraman, K. Van Geem, G. Marin, R. van Duren and W. Prins, *Fuel Process. Technol.*, 2016, **144**, 312.
16. S. Kersten and M. Garcia-Perez, *Curr. Opin. Biotechnol.*, 2013, **24**, 414.
17. D. A. Bulushev and J. R. Ross, *Catal. Today*, 2011, **171**, 1.
18. J. Soltes and T. A. Milne, Pyrolysis Oils From Biomass, *ACS Symposium Series 376*, American Chemical Society, Washington, DC, 1988.
19. P. S. Rezaei, H. Shafaghat and W. M. A. W. Daud, *Appl. Catal., A*, 2014, **469**, 490.
20. E. Furimsky, *Appl. Catal., A*, 2000, **199**, 147.
21. A. Lappas, S. Bezergianni and I. Vasalos, *Catal. Today*, 2009, **145**, 55.
22. A. R. Ardiyanti, S. A. Khromova, R. H. Venderbosch, V. A. Yakovlev, I. V. Melian-Cabrera and H. J. Heeres, *Appl. Catal., A*, 2012, **449**, 121.
23. A. Kloekhorst, J. Wildschut and H. J. Heeres, *Catal. Sci. Technol.*, 2014, **4**(8), 2367.
24. X. Wu, J. Markham, X. S. Sun and D. Wang, *Trans. ASABE*, 2012, **55**(5), 1879.
25. M.-J. Jeon, J.-K. Jeon, D. J. Suh, S. H. Park, Y. J. Sa, S. H. Joo and Y.-K. Park, *Catal. Today*, 2013, **204**, 170.
26. M. C. Samolada, A. Papafotica and I. A. Vasalos, *Energy Fuels*, 2000, **14**, 1161.
27. A. J. Foster, J. Jae, Y.-T. Cheng, G. W. Huber and R. F. Lobo, *Appl. Catal., A*, 2012, **423–424**, 154.
28. T. Mochizuki, S.-Y. Chen, M. Toba and Y. Yoshimura, *Appl. Catal., A*, 2013, **456**, 174.
29. A. Pattiya, J. O. Titiloye and A. V. Bridgwater, *J. Anal. Appl. Pyrolysis*, 2008, **81**, 72.
30. M. A. Jackson, D. L. Compton and A. A. Boateng, *J. Anal. Appl. Pyrolysis*, 2009, **85**, 226.
31. A. Pattiya, J. O. Titiloye and A. V. Bridgwater, *Fuel*, 2010, **89**, 244.
32. J. Jae, G. A. Tompsett, A. J. Foster, K. D. Hammond, S. M. Auerbach, R. F. Lobo and G. W. Huber, *J. Catal.*, 2011, **279**, 257.
33. T. R. Carlson, T. P. Vispute and G. W. Huber, *ChemSusChem*, 2008, **1**, 397.
34. H. Zhang, Y.-T. Cheng, T. P. Vispute, R. Xiao and G. W. Huber, *Energy Environ. Sci.*, 2011, **4**, 2297.
35. L. Qiang, L. W. Zhi, Z. Dong and Z. X Feng, *J. Anal. Appl. Pyrolysis*, 2009, **84**, 131.
36. A. A. Lappas, K. G. Kalogiannis, E. F. Iliopoulou, K. S. Triantafyllidis and S. D. Stefanidis, *Wiley Interdiscip. Rev.: Energy Environ.*, 2012, **1**, 285.
37. T. Dickerson and J. Soria, *Energies*, 2013, **6**, 514.
38. Y. C. Lin and G. W. Huber, *Energy Environ. Sci.*, 2009, **2**, 68.
39. S. Tan, Z. Zhang, J. Sun and Q. Wang, *Chin. J. Catal.*, 2013, **34**, 641.
40. R. H. Venderbosch, *ChemSusChem*, 2015, **8**(8), 1306.

41. G. Yildiz, F. Ronsse, R. van Duren and W. Prins, *Renewable Sustainable Energy Rev.*, 2016, **57**, 1596.
42. F. A. Agblevor, S. Beis, O. Mante and N. Abdoulmoumine, *Ind. Eng. Chem. Res.*, 2010, **49**, 3533.
43. E. Butler, G. Devlin, D. Meier and K. McDonnell, *Renewable Sustainable Energy Rev.*, 2011, **15**, 4171.
44. K. Wang, K. H. Kim and R. C. Brown, *Green Chem.*, 2013, **16**, 727.
45. *Feedstock supply*, http://energy.gov/eere/bioenergy/feedstock-supply#-Feedstock_Types, December 4, 2014.
46. A. V. Bridgwater, IEA Bioenergy Update 27, *Biomass Bioenergy*, 2007, **31**, VII.
47. D. J. M. Hayes, in *The Role of Catalysis for the Sustainable Production of Bio-Fuels and Bio-Chemicals*, ed. K. S. Triantafyllidis, A. A. Lappas and M. Stöcker, Elsevier B.V., Biomass composition and its relevance to biorefining, 2013, pp. 27–65.
48. G. Yildiz, F. Ronsse, R. Venderbosch, R. van Duren, S. R. A. Kersten and W. Prins, *Appl. Catal., B*, 2015, **168**, 203.
49. M. Asadieraghi, W. M. A. W. Daud and H. F. Abbas, *RSC Adv.*, 2015, **5**, 22234.
50. C. Liu, H. Wang, A. M. Karim, J. Sun and Y. Wang, *Chem. Soc. Rev.*, 2014, **43**, 7594.
51. O. D. Mante, F. Agblevor and R. McClung, *Fuel*, 2013, **108**, 451.
52. W. Prins, G. J. Harmsen, P. de Jong and W. P. M. van Swaaij, *Presented in part at 6th Int. Conf. on Fluidization*, Banff, Canada, 1989.
53. T. R. Carlson, G. A. Tompsett, W. C. Conner and G. W. Huber, *Top. Catal.*, 2009, **52**, 241.
54. V. Paasikallio, C. Lindfors, J. Lehto, A. Oasmaa and M. Reinikainen, *Top. Catal.*, 2013, **56**, 800.
55. Z. Ma, E. Troussard and J. A. van Bokhoven, *Appl. Catal., A*, 2012, **423–424**, 130.
56. D. S. Scott, J. Piskorz, M. A. Bergougnou, R. Graham and R. P. Overend, *Ind. Eng. Chem. Res.*, 1988, **27**(1), 8.
57. H. Zhang, R. Xiao, H. Huang and G. Xiao, *Bioresour. Technol.*, 2009, **100**, 1428.
58. Q. Zhang, J. Chang, T. J. Wang and Y. Xu, *Energy Convers. Manage.*, 2007, **48**, 87.
59. A. Aho, N. Kumar, K. Eranen, T. Salmi, M. Hupa and D. Y. Murzin, *Fuel*, 2008, **87**, 2493.
60. B. Li, L. Ou, Q. Dang, P. Meyer, S. Jones, R. Brown and M. Wright, *Bioresour. Technol.*, 2015, **196**, 49.
61. D. J. Mihalcik, C. A. Mullen and A. A. Boateng, *J. Anal. Appl. Pyrolysis*, 2011, **92**, 224.
62. C. Torri, M. Reinikainen, C. Lindfors, D. Fabbri, A. Oasmaa and E. Kuoppala, *J. Anal. Appl. Pyrolysis*, 2010, **88**, 7.
63. B. Zhang, Z. Zhong, M. Min, K. Ding, Q. Xie and R. Ruan, *Bioresour. Technol.*, 2015, **189**, 30.

64. T. R. Carlson, Y. T. Cheng, J. Jae and G. W. Huber, *Energy Environ. Sci.*, 2011, **4**, 145.
65. D. C. Dayton, J. R. Carpenter, A. Kataria, J. E. Peters, D. Barbee, O. D. Mante and R. Gupta, *Green Chem.*, 2015, **17**(9), 4680.
66. J. V. Valverde Millan, in *Fluidization of Fine Powders*, ed. J. M. Valverde Millan, SpringerLink, Netherlands, 2013, Introduction, 1–6.
67. G. Yildiz, T. Lathouwers, H. E. Toraman, K. M. Van Geem, G. B. Marin, F. Ronsse, R. van Duren, S. R. A. Kersten and W. Prins, *Energy Fuels*, 2014, **28**, 4560.
68. J. Jae, R. Coolman, T. Mountziaris and G. W. Huber, *Chem. Eng. Sci.*, 2014, **108**, 33.
69. H. Zhang, R. Xiao, B. Jin, D. Shen, R. Chen and G. Xiao, *Bioresour. Technol.*, 2013, **137**, 82.
70. S.-S. Liaw, Z. Wang, P. Ndegwa, C. Frear, S. Ha, C.-Z. Li and M. Garcia-Perez, *J. Anal. Appl. Pyrolysis*, 2012, **93**, 52.
71. J. Brown and R. Brown, *Bioresour. Technol.*, 2012, **103**, 405.
72. N. Puy, R. Murillo, M. V. Navarro, J. M. Lopez, J. Rieradevall, G. Fowler, I. Aranguren, T. Garcia, J. Bartroli and A. M. Mastral, *Waste Manage.*, 2011, **31**, 1339.
73. A. Veses, M. Aznar, I. Martínez, J. D. Martínez, J. M. López, M. V. Navarro, M. S. Callén, R. Murillo and T. García, *Bioresour. Technol.*, 2014, **162**, 250.
74. A. Veses, M. Aznar, J. M. López, M. S. Callén, R. Murillo and T. García, *Fuel*, 2015, **141**, 17.
75. A. Veses, M. Aznar, M. S. Callén, R. Murillo and T. García, *Fuel*, 2016, **181**, 430.
76. G. Yildiz, M. Pronk, M. Djokic, K. M. Van Geem, F. Ronsse, R. van Duren and W. Prins, *J. Anal. Appl. Pyrolysis*, 2013, **103**, 343.
77. B. Li, W. Lv, Q. Zhang, T. Wang and L. Ma, *Fuel*, 2014, **129**, 61.
78. M. Olazar, R. Aguado, J. Bilbao and A. Barona, *AIChE J.*, 2000, **46**, 1025.
79. A. Imran, E. A. Bramer, K. Seshan and G. Brem, *Energies*, 2016, **9**(3), 187.
80. A. Imran, E. A. Bramer, K. Seshan and G. Brem, *Fuel Process. Technol.*, 2014, **127**, 72.
81. KiOR Inc, http://www.kior.com/, December 4, 2014.
82. Kior Inc, Biofuel Company, Files Bankruptcy, Plans Sale, http://www.bloomberg.com/news/articles/2014-11-10/kior-inc-biofuel-company-files-bankruptcy-plans-sale, December 4, 2014.
83. Anellotech, http://anellotech.com/, December 2014.
84. CRI/Criterion Inc, IH2 technology, http://s04.static-shell.com/content/dam/shell-new/local/business/cri-catalyst/downloads/pdf/cri-ih2.pdf, 2012.
85. *New IH2® Advanced Biofuels Plant Broadens Options for Converting Biomass into Transportation Fuels*, http://www.gastechnology.org/news/Pages/New-IH2-Advanced-Biofuels-Plant-Broadens-Options-for-Converting-Biomass-into-Transportation-Fuels.aspx, December 2014.

86. T. L. Marker, L. G. Felix, M. B. Linck and M. J. Roberts, *Environ. Prog. Sustainable Energy*, 2012, **31**(2), 191.
87. T. L. Marker, L. G. Felix, M. B. Linck, M. J. Roberts, P. Ortiz-Toral and J. Wangerow, *Environ. Prog. Sustainable Energy*, 2014, **33**(3), 762.
88. *IH2 Technology Demonstration Plant to be Build*, http://www.cricatalyst.com/cricatalyst/en_us/catalysts/renewables/in-the-news/ihtwo-technology-demonstration-plant-to-be-build.html, December 2014.
89. A. V. Bridgwater, *Appl. Catal., A*, 1994, **116**, 5.
90. *In-situ Catalytic Fast Pyrolysis Technology Pathway*, http://www.nrel.gov/docs/fy13osti/58056.pdf, December 2014.
91. C. A. Mullen and A. A. Boateng, *Fuel Process. Technol.*, 2010, **91**, 1446.
92. A. Aho, A. Tokarev, P. Backman, N. Kumar, K. Eranen, M. Hupa, B. Holmbom, T. Salmi and D. Yu. Murzin, *Top. Catal.*, 2011, **54**, 941.
93. P. T. Williams and P. A. Horne, *Biomass Bioenergy*, 1994, **7**, 223.
94. P. A. Horne, N. Nugranad and P. T. Williams, *J. Anal. Appl. Pyrolysis*, 1995, **34**, 87.
95. P. T. Williams and P. A. Horne, *J. Anal. Appl. Pyrolysis*, 1995, **31**, 39.
96. O. D. Mante, F. A. Agblevor, S. T. Oyama and R. McClung, *Appl. Catal., A*, 2012, **445–446**, 312.
97. C. A. Mullen, A. A. Boateng, D. J. Mihalcik and N. M. Goldberg, *Energy Fuels*, 2011, **25**(11), 5444.
98. V. Paasikallio, C. Lindfors, E. Kuoppala, Y. Solantausta, A. Oasmaa, J. Lehtoa and J. Lehtonen, *Green Chem.*, 2014, **16**, 3549.
99. H. J. Park, J.-I. Dong, J.-K. Jeon, K.-S. Yoo, J.-H. Yim, J. M. Sohn and Y.-K. Park, *Ind. Eng. Chem. Res.*, 2007, **13**, 182.
100. S. Wan, T. Pham, S. Zhang, L. Lobban, D. Resasco and R. Mallinson, *AIChE J.*, 2013, **59**, 2275.
101. K. Iisa, R. J. French, K. A. Orton, S. Budhi, C. Mukarakate, A. R. Stanton, M. M. Yung and M. R. Nimlos, *Top. Catal.*, 2015, **59**, 94.
102. O. D. Mante and F. A. Agblevor, *Biomass Convers. Biorefin.*, 2011, **1**, 203.
103. P. A. Horne and P. T. Williams, *Fuel*, 1996, **75**, 1043.
104. K. Iisa, R. J. French, K. A. Orton, M. M. Yung, D. K. Johnson, J. ten Dam, M. J. Watson and M. R. Nimlos, *Energy Fuels*, 2016, **30**(3), 2144.
105. A. Imran, E. A. Bramer, K. Seshan and G. Brem, *J. Anal. Appl. Pyrolysis*, 2016, **117**, 236.
106. D. W. Van Krevelen, *Fuel*, 1950, **29**, 269.
107. M. Zabeti, T. S. Nguyen, L. Lefferts, H. J. Heeres and K. Seshan, *Bioresour. Technol.*, 2012, **118**, 374.

CHAPTER 11

Simulating Biomass Fast Pyrolysis at the Single Particle Scale

PETER N. CIESIELSKI*[a], GAVIN M. WIGGINS[b], JOSEPH E. JAKES[c] AND C. STUART DAW[b]

[a]Biosciences Center, National Renewable Energy Laboratory, 15013 Denver W. Parkway, Golden, CO 80401, United States; [b]Oak Ridge National Laboratory, 2360 Cherahala Blvd., Knoxville, TN 37932, United States; [c]Forest Biopolymers Science and Engineering, USDA Forest Service Forest Products Laboratory, One Gifford Pinchot Drive, Madison, Wisconsin, United States 53726
*E-mail: peter.ciesielski@nrel.gov

11.1 Introduction

Biomass holds tremendous potential as a renewable feedstock for the production of fuels and chemicals. However, significant technological advancement is required before production of biofuels and bio-based chemicals will become widespread and economically self-sustaining at the industrial scale. Many of the greatest challenges surrounding biomass conversion stem from the complex nature of the feedstock. Biomass consists of the remains of once-living plant tissue, and therefore retains many of the characteristics of the original organism. These characteristics, such as microstructure, biopolymer composition, and mineral content, are species-specific and can vary

Green Chemistry Series No. 50
Fast Pyrolysis of Biomass: Advances in Science and Technology
Edited by Robert C. Brown and Kaige Wang

Published by the Royal Society of Chemistry, www.rsc.org

substantially between feedstock types. Furthermore, the commoditization of biomass feedstocks will likely result in the distribution of feedstock "blends", or combinations of several feedstock species (*e.g.* pine wood, switchgrass, and poplar wood), the proportions of which will typically be determined by economic factors such as harvesting, preparation, transportation costs, supply levels and market demand for the various constituents.

Ideally, biomass conversion processes should be robust to changing economic conditions and thereby able to maintain acceptable product yields and quality for a wide range of potential feedstocks. However, the inherent variability of biomass feedstocks presents significant process development challenges. This is especially true for thermochemical conversion processes such as fast pyrolysis, where feedstock variations can have a major impact on process performance and economics.

While feasibility studies of biomass fast pyrolysis typically focus on process simulations at the reactor scale, intra-particle processes can often become rate limiting. Thus particle-scale modeling has indeed received considerable attention from the scientific community in recent years. The vast majority of these studies have attempted to couple various kinetic models with highly simplified particle geometries along with estimates for the time and temperature histories experienced by individual biomass particles.[1–11] While these approaches can provide good agreement with the trends observed in specific experiments, their predictive utility is limited since feedstock-specific effects, such as variations of intra-particle transport due to species-specific characteristics, are typically lumped together with the intrinsic reaction kinetics in the form of rate parameters that do not resolve differences between structural and molecular effects.

Recent attempts to more effectively address the impact of biomass particle properties have included models that account for realistic particle shapes[12] as well as the anisotropic, intra-particle transport behavior that arises from the highly directional cellular structure.[13] These efforts have certainly advanced biomass particle modeling; yet recent experience indicates that the next generation of biomass conversion modeling will need to establish even more refined relationships between feedstock-dependent physical features, such as microstructure and composition, and particle-scale transport and chemical reaction parameters. Also, to be practically useful, pyrolysis simulation models should strive to minimize computational overhead, so that it is possible to make timely investigations of how reactor design and operating changes might be used to maintain yield and quality in spite of feedstock variations. Ultimately, this might include the possibility of implementing on-line model-based process control to continuously optimize process performance. As we discuss next, it appears to us that this type of model order reduction can be achieved for fast biomass pyrolysis by combining thoughtful use of suitable approximations for key transport and reaction processes with model verification by more detailed, complex simulations. Such reduced order models for particle-scale pyrolysis will facilitate efficient integration into reactor and process-scale simulations relevant to both research and industrial interests.

In this chapter, we summarize recent advances in biomass particle-scale modeling that are relevant to fast pyrolysis simulations. We begin by describing the physical structure of biomass particles and how that structure relates to intra-particle processes during fast pyrolysis. Next, we summarize the state of the art in characterizing and predicting the pyrolysis reaction chemistry and kinetic mechanisms that are driven by the rapid heating. Furthermore, we describe approaches for addressing transport effects with even simpler models and add reaction kinetics to produce simulations that predict product compositions and yields. Finally, we summarize our view of the current limitations and discuss opportunities that remain in the area of computational particle-scale modeling of biomass fast pyrolysis.

11.2 Overview of Biomass Structure

Plant-derived biomass is a porous, biopolymer composite material with a complex hierarchical structure. This structure is inherited from the remains of once-living plant tissue, where the anatomy of the original plant organism is manifested at every length scale.[14] At the macroscale, inter-species differences such as branching patterns in trees, or stem thicknesses and internode distances in grasses, are visually obvious. At the microscale, the dominant structural feature of biomass is imparted by the cellular arrangement of the tissue. Many of these features are visually depicted in Figure 11.1 for coniferous softwood, which is a common type of feedstock for biomass fast pyrolysis. Due to the tiered structure of biomass, computational simulation of any type of thermochemical biomass conversion requires an inherently multiscale approach.

A scanning electron micrograph (SEM) showing the microstructure of yellow pine is shown in Figure 11.1c. During the life of the organism, the primary function of the tissue is to transport water and nutrients throughout the plant, giving rise to many high-aspect-ratio cells oriented parallel to the trunk or stem which strongly influences the density and thermal properties of the wood. Furthermore, transport of molecular species liberated during pyrolysis processes occurs *via* convection within these open cell lumen, which is much faster than intra-cell wall transport which is primarily limited to diffusion.

Secondary cell walls, such as that of yellow pine depicted by the transmission electron micrograph (TEM) shown in Figure 11.1d, account for the majority of the mass in wood and grasses. The biopolymer composition of these different regions is known to vary significantly; the lignin composition is typically higher in the compound middle lamella (abbreviated CML, the region between adjacent cells) than in the secondary cell wall (SCW). The impact of these different regions on thermochemical conversion processes is not entirely understood at present; however, it has been recently shown that intra-cell wall diffusion for some molecules, particularly ions, is a strong function of local moisture content and occurs at different rates through the CML than the SCW.[15] These observations suggest that the local biopolymer

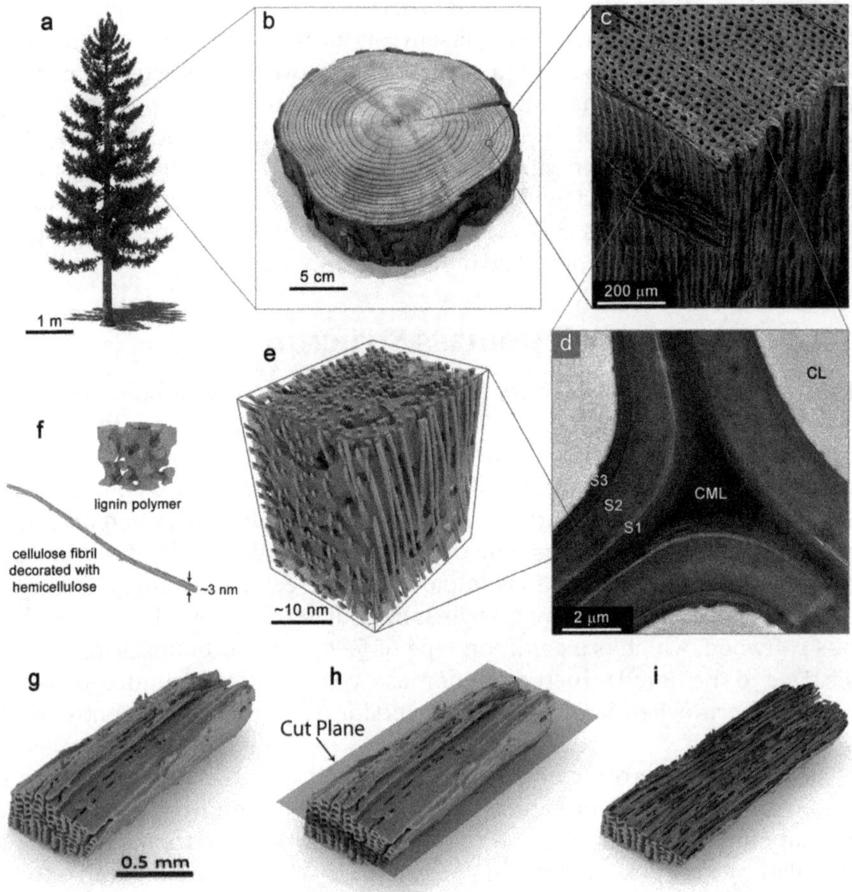

Figure 11.1 Multiscale visualization of wood structure and a typical woody feedstock. (a) Depiction of a coniferous tree. (b) Optical micrograph of section of a pine trunk. (c) Scanning electron micrograph of wood tissue showing cellular structure. (d) Transmission electron micrograph of cell wall showing various layers of the cell wall. CML: compound middle lamella; CL: cell lumen; S1, S2, and S3 denote layers of the secondary cell wall. (e) Depiction of the nanoscale arrangement of biopolymers within the cell wall. (f) Depiction of amorphous lignin polymer and a cellulose fibril decorated with hemicellulose. (g–i) X-Ray computed tomography reconstruction of a milled pine particle. The cutaway image reveals intact, directional porosity contributed by the cellular structure. Figure panels a–f reprinted with permission from ref. 14. Copyright 2016 American Chemical Society. Data in figure panels g–i are unpublished, courtesy of Joseph Jakes, USDA Forest Products Lab.

composition, which varies substantially between species and even between tissue types of the same species (particularly in grasses), can impact rates of intra-cell wall molecular transport which in turn impacts the intra-particle residence time of products formed during fast pyrolysis.

A depiction of the arrangement of nanoscale biopolymers within a secondary cell wall is shown in Figure 11.1e. Unlike conventional synthetic polymer assemblies, the nanostructure of biomass is highly ordered. Excellent, detailed discussions of the synthesis, molecular structure, and arrangement of these biopolymers are available in the literature.[16,17] In brief, cellulose nanofibrils provide the scaffolding of the cell wall; hemicellulose acts to crosslink the cellulose; and lignin, a generally amorphous polymer that imparts hydrophobicity, provides structural support, and microbial defense to the cell wall matrix. During pyrolysis, these macromolecules are thermally depolymerized to smaller, volatile compounds that must exit the remains of the cell wall and the particle.

Fast pyrolysis, like most thermochemical conversion processes, requires some form of preliminary size reduction of the raw harvested biomass. This initial step inevitably results in a range of feed particle sizes and shapes, depending on both the mechanical action of the milling process as well as the original properties of the biomass.[18] Both the size[19] and shape[12] of the reduced biomass particles can subsequently impact fast pyrolysis performance by affecting the rates of heat and mass transfer that drive the intra-particle decomposition reactions. X-ray computed tomography (XCT) reconstructions of a milled pine particle as shown in Figure 11.1g–i exemplify the non-spherical geometry that is typical of milled biomass particles. The cutaway image shown in Figure 11.1h illustrates that the internal, highly directional porosity is preserved through the milling process. All of these structural features impact the outcome of fast pyrolysis; thus the challenge of building realistic particle models with enhanced utility lies in the accurate, quantitative measurement of these structural features and subsequently incorporating them into simulations.

11.3 Representing the Microstructure, Morphology, and Material Properties of Biomass in Particle Models

Capturing the complexity of biological structures and systems *in silico* is indeed challenging in general, and biomass particles are no exception. As with most computational undertakings, increasing degrees of complexity and detail provides improved accuracy and reliability but comes at the expense of increased computational resources such as longer compute times and memory requirements. The complexity of the problem is depicted in Figure 11.2 with structural models of woody biomass particles. Various imaging techniques such as XCT and SEM provide detailed structural information that can be used to quantify key geometric features. In the case of XCT, the

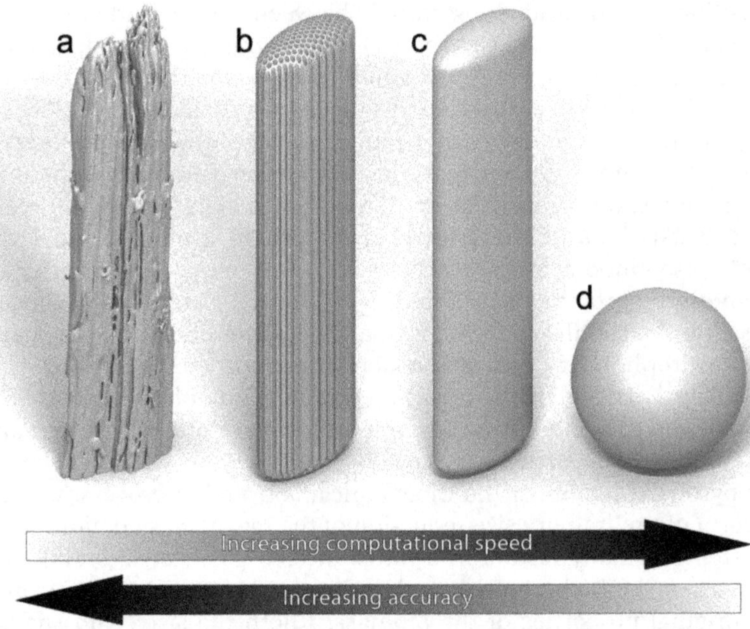

Figure 11.2 Prediction accuracy for particle models increases as more geometric details are included at the expense of reduced computational speed. (a) Full XCT model of actual wood particle. (b) Simplified geometry accounting for surface features and internal microstructure. (c) Basic geometry representing bulk surface area and volume of realistic wood particle. (d) Spherical representation of a biomass particle.

irregular geometry of actual biomass particles may be directly "mapped" into a 3-D computer modeling environment.[20] With a voxel size of ~0.5 µm, this technique provides excellent spatial resolution for resolving the microstructure of biomass and can be used to produce isosurface 3-D representations suitable for importing into computational environments such as finite element simulation software. An example of such a model is presented in Figure 11.2a. The drawback of such highly resolved particle representations is that the resulting computational analysis requires a massive number of finite elements for a particle of just a few millimeters in length. Thus for the level of detail in Figure 11.2a, computational simulations of pyrolyzing biomass particles become extremely expensive, and possibly prohibitively so, even for current high-performance computing systems. We speculate that such simulations will become more tractable as computing hardware and software continue to evolve, but no such detailed simulations based on direct XCT reconstructions have been reported to date for biomass fast pyrolysis.

Recently, we proposed an alternative method for the construction of 3-D biomass particle models that explicitly captures major structural features of the particle, such as the overall size and morphology of the particle and the

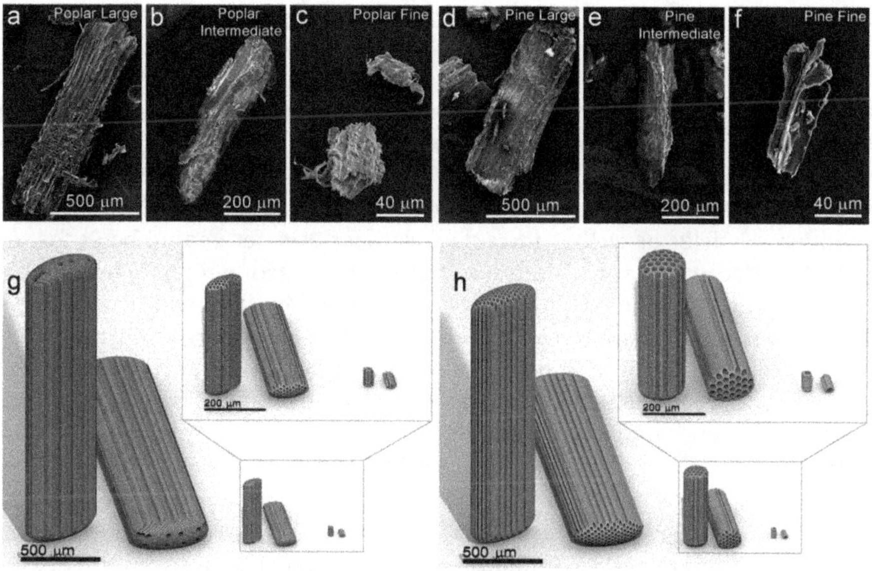

Figure 11.3 SEM and microstructure particle models of hardwood and softwood. SEM images showing representative poplar (a–c) and pine (d–f) particles. (g,h) Orthographic visualization of particle models constructed by the CSG algorithm using the dimensions and morphological parameters measured from image analysis. Inset panels show a zoom view of the intermediate and fine size classes of each feedstock. Reprinted with permission from ref. 21. Copyright 2015 American Chemical Society.

internal porosity due to the axially aligned fiber cells and vessel elements.[21] An example of one such particle model is presented in Figure 11.2b. This approach employs multiscale imaging coupled to quantitative image analysis to extract structural parameters such as the external particle size and shape from images of milled feedstock; as well as the average cell wall thickness and lumen diameters of axial tracheids and vessel elements from confocal scanning laser micrographs of particle cross-sections. These parameters are used in a custom constructive solid geometry (CSG) algorithm to build a 3-D particle model that serves as a representative surrogate of the morphological features obtained from the image analysis.

Examples of these surrogate models at various particle sizes constructed by CSG for milled pine and poplar feedstocks are presented in the lower portion of Figure 11.3. This figure demonstrates how CSG can be used to construct particle representations that account for size and shape variations along with internal features such as cell walls and axially oriented lumen. These particle models involve some loss of detailed morphological information, but the simplified structure facilitates more efficient finite element simulations of particles using present-day high-performance computing resources.

11.4 Simulating Intra-Particle Transport Phenomena

The complex internal structure of biomass provides a framework in which multiple transport processes occur during fast pyrolysis. It has been recognized that maximizing particle heating rate is critical to achieve high-yields of bio-oil.[22] Ideally, both heat transfer from the reactor environment to the biomass particles as well as intra-particle heat transfer should be as fast as possible. Because pyrolysis releases vapors and viscous liquids, heat transfer within the particle cell lumen is accompanied by fluid convection. Similarly, diffusive and convective mass transfer processes play important roles in pyrolysis as the thermal degradation products exit the particle. All these physical processes are strongly coupled to chemical reactions that produce both desired and undesired products.

11.4.1 Governing Equations for Transport

In the most general case, simulation of the intra-particle transport processes during biomass fast pyrolysis requires solving three-dimensional partial differential conservation equations for energy, mass, and momentum. If we temporarily ignore the generation terms associated with chemical reactions, the governing equations can be summarized mathematically by a series of coupled, partial differential equations (PDEs):

$$\frac{\partial \rho}{\partial t} + \nabla \cdot (\rho \mathbf{u}) = 0 \tag{11.1}$$

$$\rho \frac{\partial \mathbf{u}}{\partial t} + \rho \mathbf{u} \cdot \nabla \mathbf{u} = -\nabla p + \nabla \cdot \left[\mu \left(\nabla \mathbf{u} + (\nabla \mathbf{u})^T \right) - \frac{2}{3} \mu (\nabla \cdot \mathbf{u}) \mathbf{I} \right] \tag{11.2}$$

$$\rho C_P \left(\frac{\partial T}{\partial t} + (\mathbf{u} \cdot \nabla) T \right) = \nabla \cdot (k \nabla T) \tag{11.3}$$

where ρ is the fluid density, $\mathbf{u}$ is the velocity vector, p is the pressure, μ is the viscosity, $\mathbf{I}$ is the identity matrix, and T is the temperature (superscript T denotes the transpose operator in eqn (11.2)). In the solid domain of the biomass particle (*i.e.* within the cell walls, but not within the cell lumen), the convective component of the heat equation, $(\mathbf{u} \cdot \nabla)T$, may be omitted based on the assumption that the particle remains solid and conduction is the dominant mode of heat transfer. However, it has been demonstrated that biomass particles undergo a molten-phase transition en route to vaporization in many cases.[23] In such cases modeling the biomass particle as an extremely viscous liquid with local, temperature dependent viscosity may be more appropriate.

Pyrolyzing biomass particles that are large enough to exhibit significant spatial thermal gradients contain distinctive zones as they convert from virgin biomass to char and pyrolysis vapors within the lumens. Vapors present within the particle contain multiple components including the inert gas serving as the reactor media (typically nitrogen), condensable vapor-phase products of pyrolysis, and non-condensable light gases. In many cases, the dynamic, localized variations in the material properties can be approximated by the general rule of mixtures,[24] wherein local properties are calculated as the weighted mean of n individual components as

$$\theta_E = \sum_i^n f_i \theta_i \tag{11.4}$$

where θ_E is the ensemble material property (*e.g.* density, thermal conductivity, heat capacity, *etc.*), θ_i is the specific material property of the ith component of the ensemble, and f_i is the volume fraction of the ith material given by

$$f_i = \frac{V_i}{V_E} = \frac{V_i}{\sum_i^n V_i} \tag{11.5}$$

In the case of finite element simulations employing this strategy, all relevant material properties are evaluated by eqn (11.5) within each volume element at each time step. Because these properties are often strong functions of temperature and degree of conversion, the resulting system of PDEs can become highly non-linear, which often necessitates very small time steps, and subsequently long compute times, to achieve convergence.

Several strategies exist for solving the above transport equations computationally, the choice of which depends primarily upon the level of geometric complexity considered. Finite element methods (FEMs), which are discussed below, are typically required to simulate the most geometrically complex particle models, while less sophisticated PDE or ordinary differential equation (ODE) solvers are suitable for evaluating models with simplified geometry or reduced dimensionality. Regardless of the computational methods used, the utility of single particle simulations can be greatly enhanced by performing ensemble calculations to model the behavior of real feedstocks that contain a distribution of particle sizes, shapes, and biomass species.

11.4.2 Finite Element Simulations

Since the FEM has the ability to represent virtually any type of complex particle geometry, simulation results generated by this method are especially good for resolving the impact of detailed morphological complexity on transport processes. For this reason, FEM also provides a reliable point of reference for assessing the accuracy of less spatially resolved models. FEM subdivides the simulation geometry into smaller domains, or elements, over which boundary-value PDEs are solved. A detailed discussion of the mathematical fundamentals of this approach is outside the scope of this chapter; however, the interested reader is pointed to several excellent texts on the topic.[25,26]

As mentioned above, the geometry of biomass particles often departs significantly from simple shapes such as spheres or cylinders and contains highly variable internal porosity. Many types of biomass particles contain pores with diameters ranging from 50–100 μm (*e.g.* vessel elements in hardwoods and

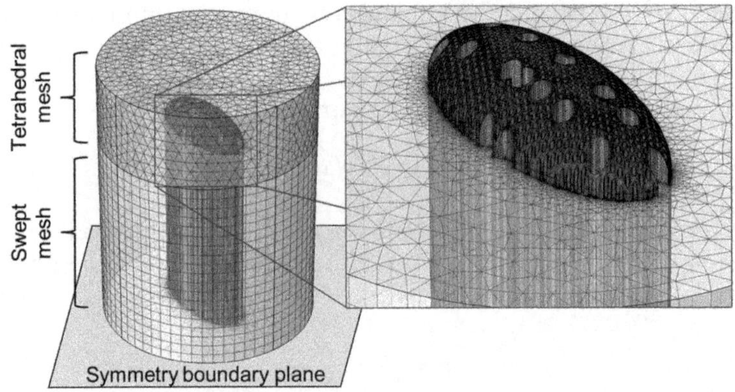

Figure 11.4 Finite element mesh and symmetry plane of a 2 mm aspen particle. Symmetry and variable mesh sizes can be utilized to reduce simulation time. Reprinted with permission from ref. 21. Copyright 2015 American Chemical Society.

vascular tissue in grasses), which can be on the same order of the dimensions of the particle exterior especially for high-aspect-ratio particles. Therefore, FEM affords the ability to explicitly account for not only the external morphology of biomass particles but also their internal porosity when necessary. However, the primary drawback to this method is the large computational expense associated with simulating geometries that require a large number of elements.

Figure 11.4 illustrates an example mesh used for FEM simulations of a ~2 mm aspen particle model which explicitly accounts for the distribution of vessel cells and axial tracheids within the particle. Even after applying applicable symmetry planes and meshing techniques such as swept prismatic meshing to reduce the number of elements, a suitable mesh for this geometry still requires ~4.8 million elements. In FEM simulations, the number of degrees of freedom that must be solved numerically scale roughly as the product of the number of elements and the number of dependent variables, which can make simulations of the geometry shown in Figure 11.4 extremely memory intensive and require long compute times even when solving for just a few dependent variables. Advances in computing hardware and solver methods such as domain decomposition will undoubtedly facilitate increasingly larger simulations in the future; however, in some cases suitable low-order approximations can be employed. Considering the constraints of current computing capabilities, the use of such high-resolution FEM models is probably most useful for identifying how and when low-order approximations are applicable to facilitate efficient use of computational resources.

11.5 Simulating Particle-Scale Reactions

Accurately predicting yields and compositions of the products from biomass fast pyrolysis also requires basic information about the rate at which chemical species are consumed and generated. Additional source and sink terms are needed in the mass and energy equations to account for these reactions. As an initial step, it is important to recognize that the structural geometry of lignocellulosic biomass is typically formed from a complex matrix of polymers with monomers consisting of cellulose ($C_6H_{10}O_5$), hemicellulose ($C_5H_8O_4$), and lignin (primarily $C_{22}H_{28}O_9$, $C_{20}H_{22}O_{10}$, $C_{15}H_{14}O_4$).[27,28] The relative amounts of these different macromolecules vary significantly among different feedstock species. In addition, there are small amounts of lower molecular weight organic species, inorganic minerals, and water. The inorganic minerals make up the residual ash left after complete devolatilization. The water initially contained in the biomass feed particles can exist in three different states: bound water which is closely associated with the carbohydrate components of the cell wall, free liquid water which is present within the cell lumen, and vapor.

All of the above components can play significant roles in the reactions (which can potentially number in the hundreds or thousands) that occur during pyrolytic conversion. Taken together, explicit simulation of all the possible species and reactions during biomass fast pyrolysis is simply beyond the current state of the art and is likely to remain so for some time. However, significant progress has been made towards developing reduced reaction mechanisms that can at least make predictions about the rates of formation of lumped product classes such as light gases, char, and tar.[28–32] In most cases, the global kinetics for these reduced reaction mechanisms are represented with first-order Arrhenius expressions in which all temperature dependence is restricted to the exponential term:

$$\begin{aligned} K_i &= A_i \mathrm{e}^{-E_i/RT} \\ \frac{\mathrm{d}C_i}{\mathrm{d}t} &= C_i K_i \end{aligned} \tag{11.6}$$

where K is the rate constant (1 s^{-1}), A is the pre-factor (1 s^{-1}), E is the activation energy (kJ mol^{-1}), R is the gas constant (kJ mol^{-1} K^{-1}), T is the temperature (Kelvin), and C is typically a mass-based concentration (kg m^{-3}) representing gas, tar, char, or wood. Table 11.1 summarizes examples of some of the simplest proposed mechanisms and their associated parameters available in the literature while Table 11.2 summarizes examples of more complex proposed mechanisms and their kinetic parameters.

An important shortcoming of the currently available reaction mechanisms and kinetics is that many of these produce inconsistent predictions, even for the same reaction conditions. This is illustrated in Figure 11.5, which depicts the fractional wood conversion and tar yield *versus* time predicted by several of the kinetic schemes in Tables 11.1 and 11.2 assuming a constant temperature of 500 °C (773 K). We conjecture that a significant portion of the disagreement between these different schemes may be the result of undocumented differences in the biomass used for experimental measurements as well as the inadvertent manifestation of feedstock species-specific transport effects in the fitted kinetic parameters.

Table 11.1 Examples of simple primary and secondary reaction mechanisms and global kinetic parameters available in the current biomass pyrolysis literature. Pre-factor represented by A (1 s^{-1}) and activation energy by E (kJ mol^{-1}).

Reference	Kinetic scheme	Kinetic parameters
Di Blasi[31] Font[37] Thurner[38]	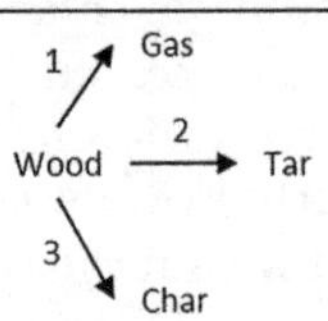	$A_1 = 1.4 \times 10^4$ to 4.4×10^9 $A_2 = 4.1 \times 10^6$ to 1.1×10^{10} $A_3 = 2.9 \times 10^2$ to 3.3×10^6 $E_1 = 88.6$ to 156 $E_2 = 112.7$ to 148 $E_3 = 61$ to 111.7
Di Blasi[39] Janse[40] Papadikis[41]	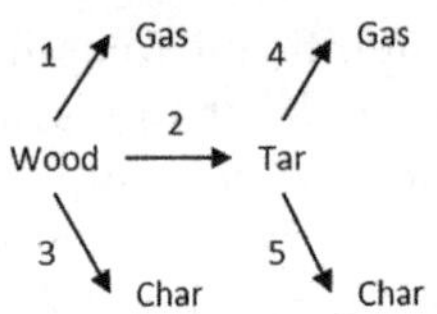	$A_1 = 5.2 \times 10^6$ to 1.1×10^{11} $A_2 = 2.0 \times 10^8$ to 1.5×10^{10} $A_3 = 1.1 \times 10^7$ to 2.7×10^{10} $A_4 = 8.6 \times 10^4$ to 4.3×10^6 $A_5 = 7.7 \times 10^4$ to 1.0×10^6 $E_1 = 88.6$ to 177 $E_2 = 112.7$ to 149 $E_3 = 106.5$ to 125 $E_4 = 87.8$ to 108 $E_5 = 87.8$ to 108
Koufopanos[42]	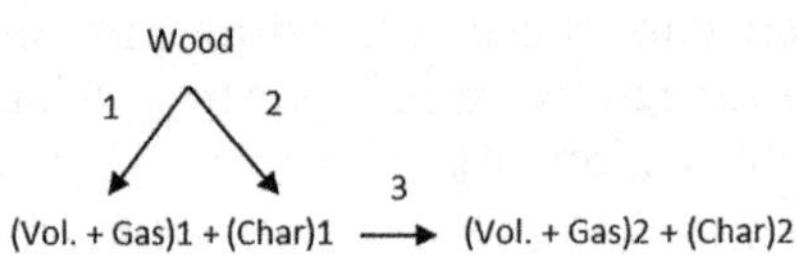	$A_1 = 9.97 \times 10^{-5}$ $G_1 = 17254.4$, $L_1 = -9061227$ $A_2 = 1.068 \times 10^{-3}$ $G_2 = 10224.4$, $L_2 = -6123081$ $A_3 = 5.7 \times 10^5$, $E_3 = 81$
Chan[43]	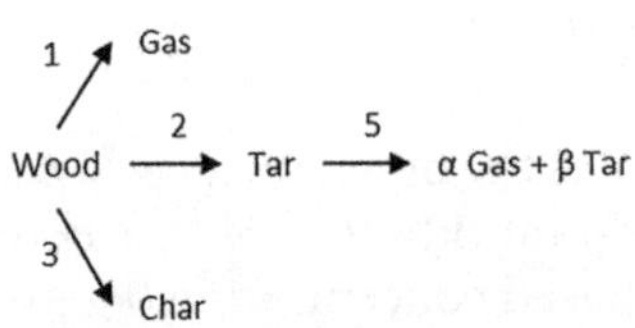	$A_1 = 1.3 \times 10^8$, $E_1 = 140$ $A_2 = 2.0 \times 10^8$, $E_2 = 133$ $A_3 = 1.08 \times 10^7$, $E_3 = 121$ $A_4 = 5.13 \times 10^6$, $E_4 = 87.9$ $A_5 = 1.48 \times 10^6$, $E_5 = 144$
Liden[44]	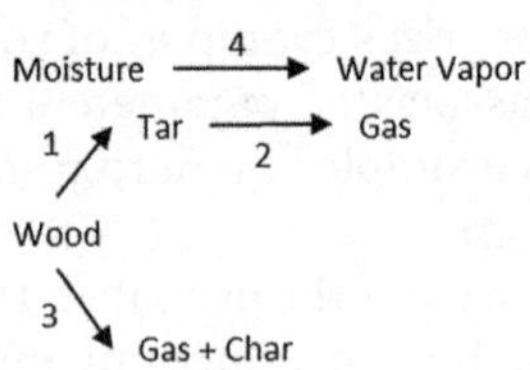	$A_2 = 4.28 \times 10^6$, $E_2 = 107.5$ $A = 1 \times 10^{13}$, $E = 183.3$ Where A and E is total wood conversion, reactions 1 and 3
Sadhukhan[45]	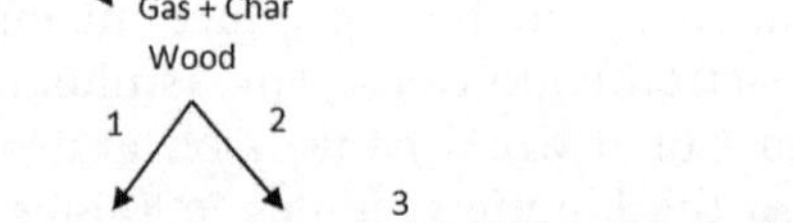	$A_1 = 168.4$, $E_1 = 51.965$ $A_2 = 13.2$, $E_2 = 45.96$ $A_3 = 5.7 \times 10^6$, $E_3 = 92.4$

Table 11.2 Examples of complex fast pyrolysis reaction mechanisms and their associated global kinetic parameters as proposed in the current literature. A is the pre-factor ($1\ s^{-1}$) while E is the activation energy (kJ mol^{-1} for the Miller and Bellan and the Anca-Couce kinetic schemes and parameters, and kcal $kmol^{-1}$ for the Ranzi kinetic scheme and parameters).

Miller and Bellan kinetic scheme and parameters[32]

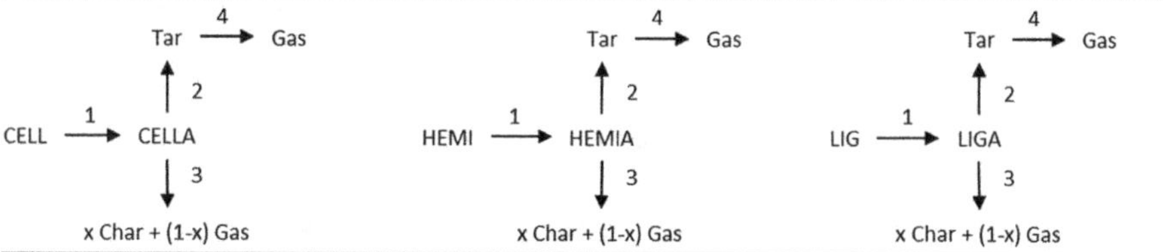

Cellulose	Hemicellulose	Lignin
$A_1 = 2.8 \times 10^{19}, E_1 = 242.4$	$A_1 = 2.1 \times 10^{16}, E_1 = 186.7$	$A_1 = 9.6 \times 10^{8}, E_1 = 107.6$
$A_2 = 3.28 \times 10^{14}, E_2 = 196.5$	$A_2 = 8.75 \times 10^{15}, E_2 = 202.4$	$A_2 = 1.5 \times 10^{9}, E_2 = 143.8$
$A_3 = 1.3 \times 10^{10}, E_3 = 150.5$	$A_3 = 2.6 \times 10^{11}, E_3 = 145.7$	$A_3 = 7.7 \times 10^{6}, E_3 = 111.4$
$A_4 = 4.28 \times 10^{6}, E_4 = 108$	$A_4 = 4.28 \times 10^{6}, E_4 = 108$	$A_4 = 4.28 \times 10^{6}, E_4 = 108$

Ranzi kinetic scheme and parameters[28]

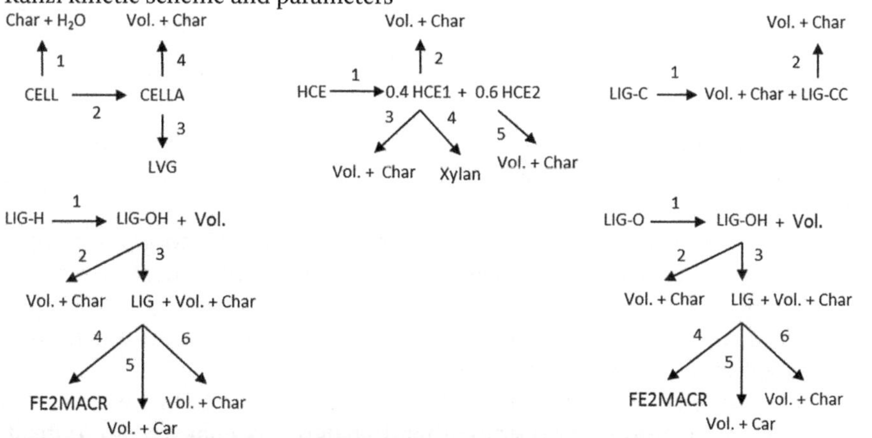

Cellulose	Hemicellulose	Lignin-C
$A_1 = 4.0 \times 10^{7}, E_1 = 31000$	$A_1 = 0.33 \times 10^{10}, E_1 = 31000$	$A_1 = 1.33 \times 10^{15}, E_1 = 48500$
$A_2 = 4.0 \times 10^{13}, E_2 = 45000$	$A_2 = 1.0 \times 10^{9}, E_2 = 32000$	$A_2 = 1.6 \times 10^{6}, E_2 = 31500$
$A_3 = 1.8 \times T, E_3 = 10000$	$A_3 = 0.05 \times T, E_3 = 8000$	
$A_4 = 0.5 \times 10^{9}, E_4 = 29000$	$A_4 = 0.9 \times T, E_4 = 11000$	
	$A_5 = 0.33 \times 10^{10}, E_5 = 33000$	
Lignin-H		**Lignin-O**
$A_1 = 0.67 \times 10^{13}, E_1 = 37500$		$A_1 = 0.33 \times 10^{9}, E_1 = 25500$
$A_2 = 33, E_2 = 15000$		$A_2 = 33, E_2 = 15000$
$A_3 = 0.5 \times 10^{8}, E_3 = 30000$		$A_3 = 0.5 \times 10^{8}, E_3 = 30000$
$A_4 = 2.4 \times T, E_4 = 12000$		$A_4 = 2.4 \times T, E_4 = 12000$
$A_5 = 0.4 \times 10^{9}, E_5 = 30000$		$A_5 = 0.4 \times 10^{9}, E_5 = 30000$
$A_6 = 0.083 \times T, E_6 = 8000$		$A_6 = 0.083 \times T, E_6 = 8000$

(*continued*)

Table 11.2 (*continued*)

Anca-Couce kinetic scheme and parameters[29]

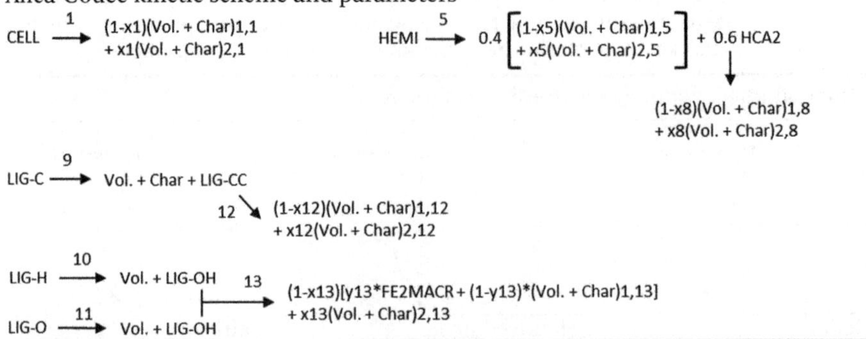

Cellulose	Hemicellulose
$A_1 = 8 \times 10^{13}$, $E_1 = 192.5$	$A_5 = 1 \times 10^{10}$, $E_5 = 129.7$
	$A_8 = 1 \times 10^{10}$, $E_8 = 138.1$
Lignin-C	Lignin-H and Lignin-O
$A_9 = 4 \times 10^{15}$, $E_9 = 202.9$	$A_{10} = 2 \times 10^{13}$, $E_{10} = 156.9$
$A_{12} = 5 \times 10^{6}$, $E_{12} = 131.8$	$A_{11} = 1 \times 10^{9}$, $E_{11} = 106.7$
	$A_{13} = 3 \times 10^{8}$, $E_{13} = 125.5$

Other important shortcomings of the currently available reaction mechanisms and kinetics in the literature are:

- There is scarce information on the catalytic effects of inorganic components such as ash (even though there is evidence that these effects can be large).[30,33–36]
- There are large inconsistencies in the experimental conditions used to obtain kinetic measurements.
- Very few mechanisms have been derived from reaction rate measurements that include product categories other than light gas, char, and tar for heating rates (500–1000 °C s^{-1}) relevant to fast pyrolysis of actual biomass.
- There are almost no mechanisms that explicitly include a role for initial particle moisture.
- There is an apparent lack of agreement on which molecular species should be included in the lumped product categories associated with "light gases", "char", and "tar".

Our review of the current pyrolysis kinetics literature reveals an imperative need to address the above shortcomings in order to develop a truly robust capability to predict product yields and compositions for industrially relevant biomass feedstocks. Otherwise, accurate simulations will only be possible for specific biomass feeds which have been previously characterized under similar experimental conditions. Even then, such simulations can probably only be expected to be interpolative rather than predictive.

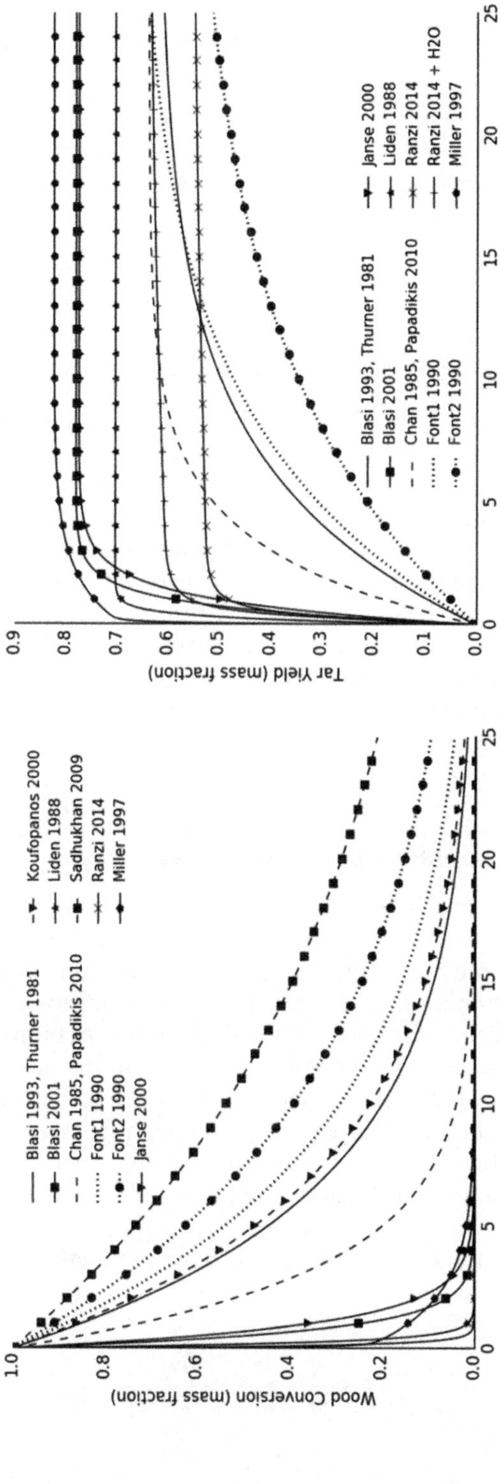

Figure 11.5 Comparison of predicted conversion and tar yield for wood pyrolyzed at 500 °C conditions based on selected kinetics from Tables 11.1 and 11.2. Left: fraction of the original wood remaining *versus* time. Right: primary tar yields *versus* time. Each line represents a particular scheme denoted by first author and year of publication.

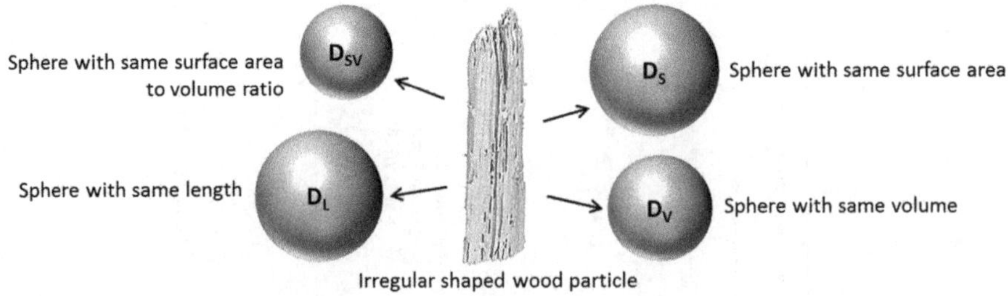

Figure 11.6 Equivalent spherical diameters to represent an irregularly shaped wood particle.

11.6 Approaches for Low-Order Particle Models

Even with the simplified 3-D geometry displayed in Figure 11.2c, it is extremely expensive to incorporate structural models with this level of detail into computational reactor-scale simulations involving thousands of biomass particles. Consequently, there is considerable motivation to develop lower-order modeling approaches that can account for the dominant particle-scale heat and mass transport effects involved in fast pyrolysis of biomass. One such approach is to approximate the multi-dimensional transport processes of biomass particles with idealized spherical particles having mathematically "similar" transport properties during rapid heat-up.[46] Figure 11.6 illustrates this concept for an irregularly shaped wood particle. We summarize an approach for utilizing this type of 1-D approximation in the following sections.

11.6.1 1-D Heat Transfer Approximations

Mathematically, approximations of 3-D transport processes are possible in 2-D and 1-D when a limited number of controlling parameters dominate the system and effectively reduce the dynamic phase space. In a recent study of particle-scale heat conduction under fast pyrolysis conditions, we demonstrated that this is typically the case for a realistic range of biomass particle sizes if the characteristic length used for 1-D simulations is based on the diameter of a surrogate spherical particle with a surface area to volume ratio (Dsv) equal to that of the original particle.[46] The significance of Dsv seems to confirm that the effective surface interface between each pyrolyzing particle and its surroundings is perhaps the most critical geometric factor controlling particle heat-up.

A widely used approach for simulating 1-D heat conduction in solid slab, cylindrical, and spherical geometries is based on solving the following transient PDE:[47]

$$\frac{1}{r^b}\frac{\partial}{\partial r}\left(kr^b\frac{\partial T}{\partial r}\right) = \rho C_p\frac{\partial T}{\partial t} \tag{11.7}$$

where r is the 1-D spatial coordinate (m), b is the shape factor (0 slab, 1 cylinder, 2 sphere), T is temperature in Kelvin (K), k is thermal conductivity (W m^{-1} K^{-1}), ρ is density (kg m^{-3}), C_p is heat capacity (kJ kg^{-1} K^{-1}), and t represents time in seconds(s). For particles with any of these shapes, spatial symmetry allows the application of a zero gradient at the particle center. The other relevant boundary condition for fast pyrolysis is the assumption that the heat flux at the particle surface can be represented by a convective heat transfer coefficient that accounts for the heat input through the external boundary layer.

As demonstrated in our particle modeling study,[46] eqn (11.7) can be successfully utilized with surrogate representations of typical biomass particles that assume a diameter (Dsv) that yields an equivalent surface area to volume ratio as the original particle. The results reported in the reference study also demonstrate that it is possible to use the bulk average thermal conductivity (k) and heat capacity (C_p) reported in standard references such as the Wood Handbook[48] for simulations. Although these bulk properties do not explicitly account for anisotropy, they effectively average the impact of the actual spatial variations. When combined with a surrogate 1-D representation of biomass particles, they appear to reasonably replicate the transient surface, center, and volume-average temperature profiles produced by the fully 3-D conductive heat transport as illustrated in Figure 11.7 for a loblolly pine particle exposed to conditions typical of fast pyrolysis.[46]

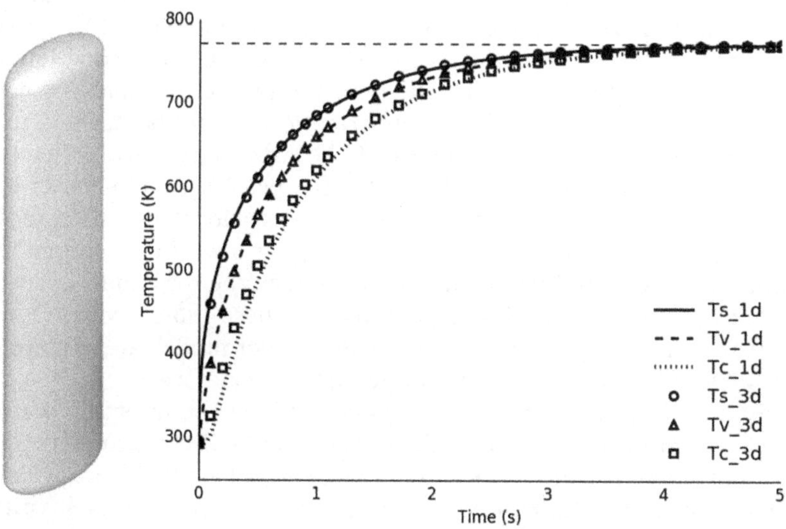

Figure 11.7 Comparison of temperature profiles from 3-D and 1-D model results for a loblolly pine particle at 500 °C (773 K). Bulk average properties of ρ = 540 kg m^{-3}, k = 0.12 W m^{-1} K^{-1}, and C_p = 103.1 + 3.867T J kg^{-1} K^{-1} provided by the Wood Handbook.[48] Particle surface area to volume diameter (Dsv) for one-dimensional model based on three-dimensional particle with a Feret diameter of 5.4 mm. Reprinted with permission from ref. 46. Copyright 2016 American Chemical Society.

11.6.2 Combining 1-D Heat Transfer and Reaction

In fast pyrolysis units, the amount of time it takes for a biomass particle to fully devolatilize is a critical parameter for reactor operation. In order to estimate this conversion, the 1-D model mentioned earlier can be coupled to a kinetic scheme to estimate pyrolysis yields and solid conversion time from wood to char. An example of combining the 1-D particle model to the kinetic scheme of Sadhukhan *et al.*[45] is shown in Figure 11.8. When the heat of reaction is included in the model the center temperature and conversion profiles match well with the experimental data for a 20 × 100 mm cylindrical wood particle. The temperature overshoot reported by the experiment at the center of the particle is also captured well with the 1-D model due to the exothermic heat of reaction. Without the heat of reaction, conversion time is prolonged and the temperature overshoot is not accounted for in the particle model. Since the model results do not account for mass transport within the particle, the effects of mass diffusion are assumed to be included to some extent *via* kinetic parameters of the reaction scheme.

11.7 Current Limitations in Particle-Scale Modeling

The recent the particle modeling efforts described above have made significant progress towards effectively capturing the complex and highly variable geometry of realistic biomass feedstocks; however, we feel that the absence of transport-independent conversion kinetics for biomass fast pyrolysis from the literature is presently the largest impediment to the development of a generalized pyrolysis model with accurate predictive capability across biomass feedstocks. In addition, more attention should be devoted to the incorporation of the catalytic effects of the ash content within biomass into kinetic schemes to accurately predict the pyrolysis products. In order to facilitate optimization of fast pyrolysis processes for the yields of desired chemical products, kinetic schemes must migrate away from lumped models and incorporate additional speciation to track the formation of specific molecules of interest. The implementation of these more detailed kinetic schemes will also require the use of reduced order models to be computationally feasible with present-day computational resources.

Additional improvements must also be made at the interface between particle modeling and reactor-scale modeling to facilitate process optimization and scale-up. Drag models have a large impact on the hydrodynamics predicted by CFD simulation software, but these models are typically established for spherical geometries which are not representative of biomass particles produced from milling and grinding processes. Improved drag models that are specific to biomass particles should be developed by combined experimental and computational efforts. These models must also be able to account for the distribution of particle shapes and sizes in order to predict particle behavior in realistic industrial scale fast pyrolysis processes. Similarly, interfacial heat transfer coefficients that

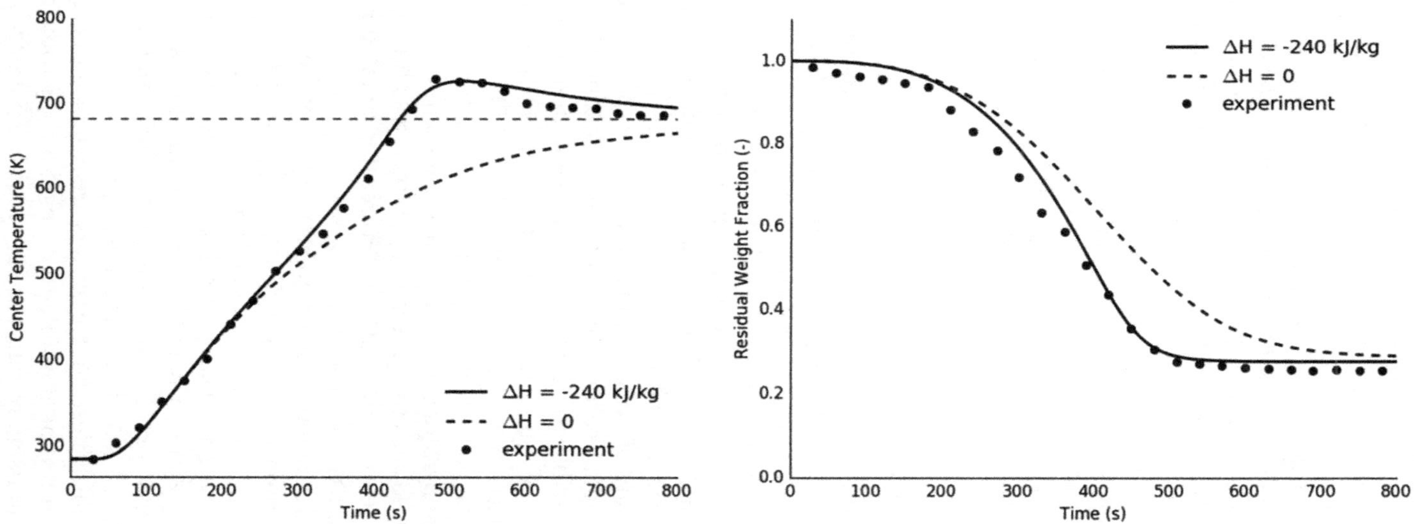

Figure 11.8 Center temperature profiles and conversion for a 20 × 100 mm wood cylinder at 683 K. Symbols represent experimental data from Sadhukhan *et al.*[45] The solid line denotes 1-D model results with $\Delta H = -240$ kJ kg^{-1} while the dashed line is with no heat of reaction. Reprinted with permission from ref. 46. Copyright 2016 American Chemical Society.

are typically used to model heat transfer from the reactor environment to the particle were developed for spherical particles. Our recent experience modeling interfacial heat transfer with realistic biomass particle models indicate that most correlations for heat transfer coefficients in the literature can provide poor agreement between simulations of conjugate heat transfer and simulations that employ interfacial heat transfer coefficients. Furthermore, we have observed interfacial heat transfer to be species-specific due to differences in particle microstructure that acts to modify the exterior geometry of the particle, and these species-specific affects are completely absent from the correlations in the current literature. In general, to accurately simulate the hydrodynamics and heating behavior of realistic biomass particles in pyrolysis reactors, many engineering correlations previously developed for other systems, such as coal pyrolysis, would need to be revisited in the context of realistic, species-specific biomass particle models.

11.8 Conclusions

Modeling fast pyrolysis at the particle scale provides the opportunity to assess the impacts of feedstock-specific parameters such as morphology, microstructure, composition, and moisture content. Since these parameters vary substantially between feedstocks, we feel that biomass particle modeling will be of increasing importance as we strive towards a renewable bioeconomy that commoditizes feedstocks and their biofuel and biochemical products provided by fast pyrolysis.

While the complexity of typical biomass feeds makes detailed computer simulations of individual particle behavior during fast pyrolysis extremely challenging, it is possible to develop 3-D representations of biomass particles that include the most important structural features revealed by advanced characterization methods such as XCT and SEM. FEM simulations using these 3-D representations can reveal important details of particle-scale processes during fast pyrolysis, but this comes at a high computational cost and thus must be used selectively. It is not currently feasible to use particle models with this level of structural detail in reactor simulations involving hundreds or thousands of particles.

Although numerous reaction mechanisms and kinetic parameters have been proposed for biomass fast pyrolysis, it appears that there remain serious shortcomings which need to be addressed. Chief among these are a lack of accounting for catalytic ash effects, inadequate separation of transport effects from intrinsic kinetics, inconsistent and poorly documented experimental protocols, inadequate differentiation of product species and associated reactions, and inadequate accounting for initial particle moisture. Until these shortcomings are resolved in the literature, we expect that it will not be possible to develop a truly robust predictive capability for an industrially relevant range of biomass feedstocks and feedstock blends.

1-D surrogate models of intra-particle conductive heat transfer can generate predictions of the transient intra-particle temperatures that are reasonable approximations of the simulation results produced by fully 3-D FEM simulations. The external surface area to volume ratio of particles is a key geometric factor, since it determines the available area per unit mass through which heat can enter the particle. Predictions from 1-D particle models combined with simplified pyrolysis kinetics generate predicted yields of char, light gas, and tar that appear to be reasonably consistent with experimental measurements. As with any modeling effort, the development of these improved models must be closely integrated with experimental results.

Overcoming the challenges described in this chapter will provide substantial benefit to the fast pyrolysis and biofuels community by enabling accurate predictions of feedstock-specific yields and optimal process conditions. This information will improve the state of technology and de-risk its commercialization, but development of these improved models will require large, coordinated efforts of computational and experimental teams.

References

1. B. V. Babu and A. S. Chaurasia, *Energy Convers. Manage.*, 2003, **44**, 2251–2275.
2. B. V. Babu and A. S. Chaurasia, *Energy Convers. Manage.*, 2004, **45**, 1297–1327.
3. C. Diblasi, *Combust. Sci. Technol.*, 1993, **90**, 315–340.
4. M. G. Gronli and M. C. Melaaen, *Energy Fuels*, 2000, **14**, 791–800.
5. A. M. C. Janse, R. W. J. Westerhout and W. Prins, *Chem. Eng. Process.*, 2000, **39**, 239–252.
6. C. A. Koufopanos, N. Papayannakos, G. Maschio and A. Lucchesi, *Can. J. Chem. Eng.*, 1991, **69**, 907–915.
7. K. Papadikis, S. Gu and A. V. Bridgwater, *Fuel Process. Technol.*, 2010, **91**, 68–79.
8. A. K. Sadhukhan, P. Gupta and R. K. Saha, *Bioresour. Technol.*, 2009, **100**, 3134–3139.
9. B. V. Babu and A. S. Chaurasia, *Chem. Eng. Sci.*, 2004, **59**, 1999–2012.
10. K. Papadikis, S. Gu and A. V. Bridgwater, *Chem. Eng. J.*, 2009, **149**, 417–427.
11. A. S. Chaurasia and B. D. Kulkarni, *Energy Convers. Manage.*, 2007, **48**, 836–849.
12. Y. Habibi, L. A. Lucia and O. J. Rojas, *Chem. Rev.*, 2010, **110**, 3479–3500.
13. C. DiBlasi, *Chem. Eng. Sci.*, 1996, **51**, 1121–1132.
14. H. Zhu, W. Luo, P. N. Ciesielski, Z. Fang, J. Zhu, G. Henriksson, M. E. Himmel and L. Hu, *Chem. Rev.*, 2016, **116**, 9305–9374.
15. S. L. Zelinka, S.-C. Gleber, S. Vogt, G. M. Rodríguez López and J. E. Jakes, *Holzforschung*, 2015, **69**, 441–448.
16. D. J. Cosgrove, *Nat. Rev. Mol. Cell Biol.*, 2005, **6**, 850–861.
17. J. H. Grabber, *Crop Sci.*, 2005, **45**, 820–831.

18. M. E. Himmel, M. Tucker, J. Baker, C. Rivard, K. Oh and K. Grohmann, *Biotechnology and Bioengineering Symposiom No. 15*, 1985, pp. 39–58.
19. J. Shen, X.-S. Wang, M. Garcia-Perez, D. Mourant, M. J. Rhodes and C.-Z. Li, *Fuel*, 2009, **88**, 1810–1817.
20. A. Isaac, V. Barboza, F. I. Sket, J. R. M. D'Almeida, L. A. Montoro, A. Hilger and I. Manke, *Biotechnol. Biofuels*, 2015, **8**, 1.
21. P. N. Ciesielski, M. F. Crowley, M. R. Nimlos, A. W. Sanders, G. M. Wiggins, D. Robichaud, B. S. Donohoe and T. D. Foust, *Energy Fuels*, 2014, **29**, 242–254.
22. A. Bridgwater, *J. Anal. Appl. Pyrolysis*, 1999, **51**, 3–22.
23. M. S. Mettler, A. D. Paulsen, D. G. Vlachos and P. J. Dauenhauer, *Energy Environ. Sci.*, 2012, **5**, 7864–7868.
24. R. Grout, K. Malhorta, P. Ciesielski, K. Gruchalla, B. Donohoe and M. Nimlos, *8th US National Combustion Meeting Organized by the Western States Section of the Combustion Institute*, 2013, #070CO-0341.
25. J. N. Reddy and D. K. Gartling, *The Finite Element Method in Heat Transfer and Fluid Dynamics*, CRC Press, 3rd edn, 2010.
26. O. C. Zienkiewicz, R. L. Taylor and P. Nithiarasu, *The Finite Element Method for Fluid Dynamics*, Butterworth-Heinemann, 2013.
27. P. Basu, *Biomass Gasification, Pyrolysis and Torrefaction: Practical Design and Theory*, Academic Press, San Diego, CA, 2nd edn, 2013, pp. 56,58.
28. E. Ranzi, M. Corbetta, F. Manenti and S. Pierucci, *Chem. Eng. Sci.*, 2014, **110**, 2–12.
29. A. Anca-Couce, R. Mehrabian, R. Scharler and I. Obernberger, *Energy Convers. Manage.*, 2014, **87**, 687–696.
30. A. Trendewicz, R. Evans, A. Dutta, R. Sykes, D. Carpenter and R. Braun, *Biomass Bioenergy*, 2015, **74**, 15–25.
31. C. Di Blasi and C. Branca, *Ind. Eng. Chem. Res.*, 2001, **40**, 5547–5556.
32. R. S. Miller and J. Bellan, *Combust. Sci. Technol.*, 1997, **126**, 97–137.
33. F. A. Agblevor and S. Besler, *Energy Fuels*, 1996, **10**, 293–298.
34. J. E. White, W. J. Catallo and B. L. Legendre, *J. Anal. Appl. Pyrolysis*, 2011, **91**, 1–33.
35. P. R. Patwardhan, J. A. Satrio, R. C. Brown and B. H. Shanks, *Bioresour. Technol.*, 2010, **101**, 4646–4655.
36. N. Kuzhiyil, D. Dalluge, X. Bai, K. H. Kim and R. C. Brown, *ChemSusChem*, 2012, **5**, 2228–2236.
37. R. Font, A. Marcilla, E. Verdii and J. Devesa, *Ind. Eng. Chem. Res.*, 1990, **29**, 1846–1855.
38. F. Thurner and U. Mann, *Ind. Eng. Chem. Process Des. Dev.*, 1981, **20**, 482–488.
39. C. Di Blasi, *Combust. Sci. Technol.*, 1993, **90**, 315–340.
40. A. M. C. Janse, R. W. J. Westerhout and W. Prins, *Chem. Eng. Process.*, 2000, **39**, 239–252.
41. K. Papadikis, S. Gu and A. V. Bridgwater, *Fuel Process. Technol.*, 2010, **91**, 68–79.
42. C. A. Koufopanos, N. Papayannakos, G. Maschio and A. Lucchesi, *Can. J. Chem. Eng.*, 1991, **69**, 907–915.

43. W.-C. R. Chan, M. Kelbon and B. B. Krieger, *Fuel*, 1985, **64**, 1505–1513.
44. A. G. Liden, F. Berruti and D. S. Scott, *Chem. Eng. Commun.*, 1988, **65**, 207–221.
45. A. K. Sadhukhan, P. Gupta and R. K. Saha, *Bioresour. Technol.*, 2009, **100**, 3134–3139.
46. G. M. Wiggins, P. N. Ciesielski and C. S. Daw, *Energy Fuels*, 2016, **30**, 4960–4969.
47. T. L. Bergman, A. S. Lavine, F. P. Incropera and D. P. Dewitt, *Fundamentals of Heat and Mass Transfer*, John Wiley & Sons, 7th edn, 2011, pp. 299–304.
48. S. V. Glass and S. L. Zelinka, *Wood Handbook: Wood as an Engineering Material*, Forest Products Laboratory, Madison, WI, 2010, pp. 4.1–4.19.

CHAPTER 12

Economic Comparison of Various Pathways to Pyrolysis-Based Fuels

TRISTAN R. BROWN[a]

[a]SUNY College of Environmental Science & Forestry, 320 Bray Hall, 1 Forestry Drive, Syracuse, NY, USA 13210
*E-mail: trbro100@esf.edu

12.1 Introduction

The revised Renewable Fuel Standard (RFS2) was created by the United States Congress as part of the broader Energy Independence and Security Act of 2007. A previous blending mandate, the RFS1, had been overwhelmingly successful, and the RFS2 was designed to build upon the success of its predecessor. Whereas the RFS1 had just required that a single category of biofuel—ethanol (almost entirely made from corn starch)—be blended with gasoline prior to being sold at retail fuel stations, the RFS2 includes four categories defined by both feedstock and lifecycle greenhouse gas (GHG) emissions relative to gasoline. Furthermore, the RFS2 also requires much larger volumes of biofuels to be blended for consumption than the RFS1 had required, reaching 136 billion liters (36 billion gallons) by 2022 (see Figure 12.1).

The RFS2's most ambitious requirement is for biofuels produced from lignocellulosic feedstock to become the largest blending category by 2022.

Green Chemistry Series No. 50
Fast Pyrolysis of Biomass: Advances in Science and Technology
Edited by Robert C. Brown and Kaige Wang

Published by the Royal Society of Chemistry, www.rsc.org

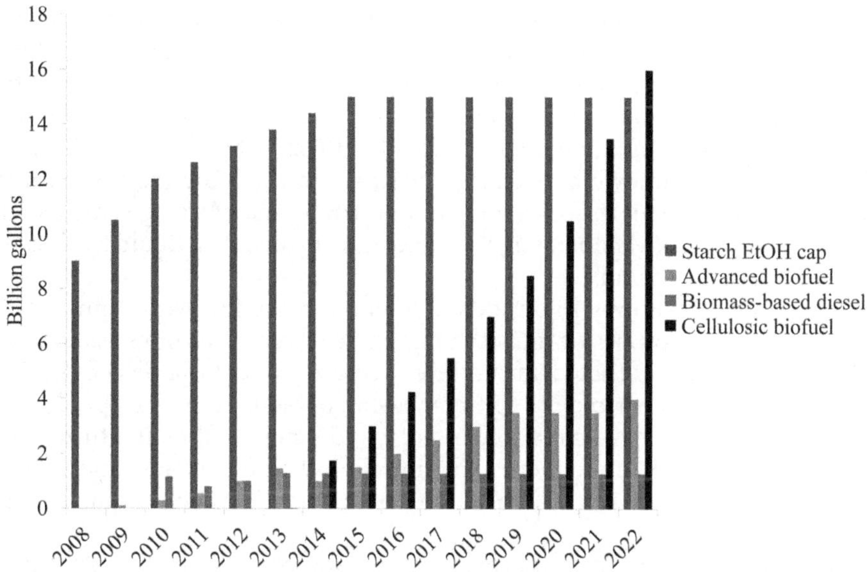

Figure 12.1 Annual RFS2 volumes.[34]

Specifically, the blending volume mandated for the category increases from nothing in 2009 to 60.6 billion liters (16 billion gallons) in 2022. Cellulosic biofuels differ from corn ethanol and soya biodiesel, which have been the largest contributors to the RFS2 by volume to date, in that they are produced from lignocellulosic biomass such as agricultural residues, short-rotation dedicated energy crops, and forest residues. Unlike feedstocks for first-generation biofuels, lignocellulosic biomass has a relatively low fertilizer demand compared to corn. The biomass also serves as a source of process heat and power when converted to biofuels. Cellulosic biofuels have much lower GHG emissions on a lifecycle basis than gasoline due to these advantages, and the RFS2's cellulosic biofuels category reflects this by requiring qualifying fuels to achieve a 60% reduction to GHG emissions *versus* gasoline.

The RFS2 also differs from the RFS1 in that it allows multiple biofuel types to participate under the mandate. This expansion reflects technological advancements in the area of advanced biofuels production and the limitations of the fuel ethanol that has been and continues to be the primary biofuel consumed in the U.S. Fuel ethanol is an imperfect substitute for petroleum since it only replaces a single refined product, gasoline. Its incompatibility with the existing midstream transportation fuel infrastructure has required the building of new production and logistics infrastructure capable of handling it. Its inability to meet other refined product demand has further required investment in other biofuel platforms such as biodiesel, which requires its own infrastructure.

Fuel ethanol's technical constraints broadly limit its use to blends of 10 vol% with gasoline (blends of up to 15 vol% are permitted by regulators but consumer acceptance of this higher blend has been very limited to date). The so-called "E10 blend wall" was first reached in 2013, and possibly earlier, at a time when corn starch was the primary ethanol feedstock. The inability of the existing transportation fuel infrastructure to utilize higher blend rates has incurred high compliance costs under the RFS2. Achievement of the biofuel blending volumes mandated by the program therefore requires either additional nationwide gasoline consumption at a time of rising vehicle fuel efficiency or the conversion of lignocellulosic feedstocks to a liquid hydrocarbon rather than ethanol.

The pyrolysis pathway represents a potential route for overcoming both of the major constraints of fuel ethanol. The conversion of biomass to bio-oil is a novel process that yields an intermediate product that is chemically very different from petroleum. The conversion of bio-oil into final products, or upgrading, employs similar technology and equipment as is utilized by conventional petroleum refineries yielding similar products. Gasoline, diesel fuel, jet fuel, petrochemicals, and even asphalt have all been produced *via* bio-oil upgrading.

Seven distinct pyrolytic approaches for converting biomass into fuels and chemicals have been described in the literature: (1) Fast pyrolysis and fluidized catalytic cracking of bio-oil; (2) fast pyrolysis and *ex situ* vapor-phase catalytic upgrading; (3) fast pyrolysis and bio-oil hydroprocessing; (4) *in situ* catalytic pyrolysis and bio-oil hydroprocessing; (5) hydropyrolysis and vapor-phase hydroprocessing; (6) slow pyrolysis and syngas upgrading; and (7) gasification of bio-oil and syngas upgrading. This chapter discusses the economic feasibility of the major routes that are available for the conversion of biomass to fuels *via* the pyrolysis pathway.

12.2 Pathway Economics

12.2.1 Fast Pyrolysis and Fluidized Catalytic Cracking of Bio-Oil

Some petroleum refiners employ fluid catalytic cracking (FCC) units to crack heavy refined products comprised of long-chained compounds into lighter, fuel-range compounds. Given the many similarities between heavy refined products and bio-oil in terms of viscosity and chain-length, the economics of bio-oil upgrading to fuels *via* FCC has been assessed as a potential biofuel production pathway. A single FCC upgrading step in the form of severe deoxygenation is employed to yield alkenes and aromatics rather than fuel-grade alkanes. While a separate hydrotreating step is not necessary as part of this pathway variant, its inclusion has been found to increase hydrocarbon yields.[1]

Brown *et al.*[2] calculates the 20 year internal rates of return (IRR) for five different scenarios in which whole bio-oil and the water-soluble fraction of bio-oil are upgraded to alkenes and aromatics *via* fast pyrolysis and FCC upgrading. Three upgrading scenarios are considered for the two types of bio-oil: a single FCC step; a low-severity hydrotreating step followed by a FCC step; and a low-severity hydrotreating step, high-severity hydrotreating step, and FCC step. The analysis finds that each additional upgrading step incurs additional capital and operating costs, the former because of the additional equipment requirement and the latter because of the need to purchase hydrogen. The additional steps also generate higher yields of alkenes and aromatics at the expense of coke production, however, resulting in a positive correlation between total project investment and IRR. Furthermore, the water-soluble bio-oil fraction is identified as a higher-value upgrading feed than the whole bio-oil.

12.2.2 Fast Pyrolysis and *ex situ* Vapor-Phase Catalytic Upgrading

Bio-oil can also be upgraded when it is still in the vapor phase prior to quenching. The choice of catalyst and timing of the upgrading step has an impact on the process reactions and products. *Ex situ* vapor phase catalytic upgrading involves a single reaction step in which the bio-oil vapors are partially deoxygenated after they have left the pyrolysis reactor so as to minimize the subsequent hydroprocessing step's hydrogen consumption and reaction severity.

The U.S. Department of Energy's (DOE) National Renewable Energy Laboratory (NREL), which employs a standardized techno-economic analysis methodology to assess the economics of a variety of cellulosic biofuel pathways, calculated the minimum fuel selling prices (MFSP) of two fast pyrolysis and catalytic upgrading processes in a 2015 report.[3,4] A fast pyrolysis and *ex situ* vapor phase upgrading scenario is compared with an *in situ* vapor phase upgrading, or catalytic fast pyrolysis, scenario. Both scenarios also include a hydroprocessing step of the catalytically-upgraded bio-oil.

Assuming a 10% IRR, 30 year plant life, 40% equity financing, \$80 MT^{-1} (where MT is metric ton) feedstock cost and 2000 metric ton per day (MTPD) feedstock capacity, the results find that an *ex situ* vapor phase catalytic upgrading step employed between the fast pyrolysis and hydroprocessing steps yields a MFSP of \$0.87 LGE^{-1} (\$3.31 GGE^{-1}, where LGE is liter gasoline equivalent and GGE is gallon gasoline equivalent). The *ex situ* scenario incurs a total project investment of \$589.6 million due to the need for additional equipment required by the additional upgrading step, which results in improved process chemistry.

An analysis conducted by researchers at Iowa State University and the DOE's Pacific Northwest National Laboratory (PNNL) of fast pyrolysis and *ex situ* vapor phase catalytic upgrading and hydroprocessing calculates an expected MFSP of \$4.20 gal^{-1}.[5] The higher MFSP compared to the NREL analysis of the same pathway is explained by the fact that it is a stochastic rather than deterministic analysis. This results in a lower average bio-oil yield even though the range of yields captured by the probability distribution includes the yield assumed by the NREL analysis.

12.2.3 Fast Pyrolysis and Bio-Oil Hydroprocessing

Hydroprocessing has long been employed by petroleum refiners as a means of cracking higher molecular weight compounds into fuel-range molecules while also removing heteroatoms such as nitrogen, sulfur and oxygen and saturating carbon bonds *via* hydrogen donation. Raw bio-oil contains both high molecular weight compounds and large amounts of oxygen, two characteristics that limit its utility as transportation fuel. Hydroprocessing has been adapted as a method of stabilizing, cracking, and deoxygenating raw bio-oil as part of upgrading to hydrocarbon products. Hydroprocessing consists of a low-severity hydrotreating step followed by a higher-severity hydrocracking step. During the hydrotreating step the bio-oil is stabilized by removing some oxygen by reaction with high-pressure hydrogen to yield water that is removed from the process stream. The process also removes other heteroatoms such as chlorine, nitrogen, and sulfur. The partially-deoxygenated bio-oil is further reacted with hydrogen under more severe conditions to crack large molecules into fuel-range molecules and remove any remaining oxygen.

The success of hydroprocessing by the petroleum refining industry resulted in it being among the earliest of the bio-oil upgrading processes to be analyzed for its economic prospects. An early analysis by Bridgewater[6] calculated a MFSP for bio-based diesel fuel produced *via* fast pyrolysis and hydroprocessing that was 158% higher than the price of diesel fuel derived from \$20 bbl^{-1} petroleum. Results were further presented by analyzing several upgrading configurations, although capital and operating cost assumptions were not clearly stated.

The RFS2's creation prompted widespread research into the economic feasibility of hydroprocessing as a bio-oil upgrading pathway. A 2009 analysis conducted by researchers at the DOE's PNNL was the first comprehensive techno-economic analysis of the pathway.[7] It calculated a total project investment of \$329 million for a fast pyrolysis and hydroprocessing biorefinery with a feedstock capacity of 2000 MTPD. Assuming a 10% IRR, 20 year lifespan, and \$60.50 metric ton^{-1} (MT) cost for hybrid poplar feedstock, the analysis further calculated a MFSP of \$0.58 LGE^{-1} (\$2.21 GGE^{-1}) for hydrocarbon fuels produced *via* the pathway.

The 2010 analysis conducted by researchers at Iowa State University and the NREL built upon the 2009 PNNL analysis by assessing the MFSP of the fast pyrolysis and hydroprocessing pathway under two different scenarios: one in which merchant hydrogen was purchased by the biorefinery and one in which a portion of the bio-oil was converted to hydrogen *via* steam-reforming.[8] While the analysis did not conduct a lifecycle assessment (LCA), the two scenarios were based on the premise that merchant hydrogen was most likely to be produced *via* the fossil fuel natural gas, in which case the resulting biofuels are not 100% biorenewable. The on-site hydrogen production scenario incurred a higher total project investment of \$311 million due to the need for additional equipment compared to the merchant hydrogen scenario's cost of \$217 million. The difference in capital costs between the two scenarios was reflected by gasoline and diesel fuel MFSPs of \$0.89 LGE^{-1} (\$3.35 GGE^{-1}) for the on-site hydrogen scenario compared to \$0.61 LGE^{-1} (\$2.29 GGE^{-1}) for the merchant hydrogen scenario, assuming a 10% IRR, 20 year lifespan, and \$90 MT^{-1} stover feedstock cost. This analysis was the first to capture the trade-off between the environmental and economic sustainability of the pathway.

The 2010 analysis was the foundation for several subsequent analyses of both the fast pyrolysis and hydroprocessing pathway and variants on it. Brown *et al.*[9] updated the earlier analysis to account for changing market conditions and an expanded model that explicitly simulated the splitting and blending of the hydrocarbons produced *via* the hydroprocessing stage to gasoline and diesel fuel. A higher total project investment of \$429 million under the same assumptions as the 2010 analysis was calculated based on the need for expanded and additional equipment, resulting in a MFSP of \$0.68 LGE^{-1} (\$2.57 GGE^{-1}) for the merchant hydrogen scenario. The on-site hydrogen scenario was not modeled by the analysis.

A 2013 analysis conducted by Zhang *et al.*[10] expanded upon the on-site hydrogen production scenario of the 2010 analysis to include other products demonstrated in laboratory-scale experiments. The expanded analysis considered a 2000 MTPD fast pyrolysis and hydroprocessing biorefinery that employed a novel red oak feedstock pretreatment process to produce levoglucosan, an anhydrosugar derived from cellulose. The levoglucosan underwent hydrolysis to yield glucose, which is a major agricultural commodity that can serve as a feedstock for a number of biofuel and biobased chemical pathways. The pyrolytic lignin that remained after the levoglucosan was separated from the bio-oil was assumed to be hydroprocessed to gasoline and diesel fuel. The analysis calculated a total project investment of \$379 million for the biorefinery configuration. Unlike the previous analyses, however, this analysis calculated a 20 year IRR of 11.4% based on an assumed fuel price of \$0.75 LGE^{-1} (\$2.82 GGE^{-1}). This analysis is unique in that it demonstrated the ability of the fast pyrolysis and hydroprocessing pathway to serve as a route for the production of products other than gasoline and diesel fuel. Glucose in particular has been demonstrated as a feedstock for the production of biobased chemicals and other important petroleum-derived hydrocarbons.[11] The Zhang *et al.* analysis represents an additional step toward demonstrating the ability of the fast pyrolysis and hydroprocessing pathway to replace an increasing share of petroleum's refined products with biobased versions.

The relatively advanced state of knowledge of economic feasibility for fast pyrolysis and hydroprocessing has allowed it to be used in quantifying the effects of different policy scenarios on cellulosic biofuel production. A 2011 analysis by Brown *et al.*[12] utilized the results of the Iowa State University and NREL analysis[8] to compare the economic feasibility of the fast pyrolysis and hydroprocessing pathway to that of the slow pyrolysis pathway under a national cap-and-trade policy scenario based on the American Clean Energy and Security Act of 2009. That analysis found that although the co-product biochar yielded by both scenarios has value as a carbon sequestration mechanism under the cap-and-trade scenario, the predicted carbon prices were much too low to make the slow pyrolysis pathway competitive with the fast pyrolysis and hydroprocessing scenario due to the high value of the liquid fuels that it produces.

Petter and Tyner[13] utilized the fast pyrolysis and hydroprocessing analysis of Brown *et al.*[9] to assess the impact of that pathway's uncertainty on private investment risk. The results of the assessment found that the pathway's exposure to commodity price volatility and uncertain biofuel market values cause high investment risk that discourages private investment in cellulosic biofuels. It also found that public policy tools such as the use of reverse auctions by governments can, by shifting some of this risk to the public sector, do more to reduce private investment risk than existing capital subsidies such as non-refundable tax credits. A separate analysis utilizing the same fast pyrolysis and hydroprocessing model supports the finding by Petter and Tyner[13] on commodity price volatility by its own finding that the economic feasibility of the pathway is especially sensitive to region-specific factors such as feedstock type and regional market conditions.[14]

12.2.4 *In situ* Catalytic Pyrolysis and Bio-Oil Hydroprocessing

In situ catalytic pyrolysis employs catalysts in the pyrolysis reactor itself to achieve partial deoxygenation prior to subsequent upgrading. This differs from the pyrolysis and *ex situ* vapor phase upgrading pathway in which the first upgrading reaction occurs outside of the pyrolysis reactor. The process economics of *in situ* catalytic pyrolysis can be improved relative to fast pyrolysis and downstream hydroprocessing by reducing the amount of hydrogen required to produce hydrocarbon fuels. This improved hydrogen efficiency comes at the expense of reduced carbon efficiency since much of the carbon in the feedstock is converted to CO, CO_2, and coke in the reactor. The pathway's economics are most advantageous compared to fast pyrolysis when hydrogen prices are high and fuel prices are low.

The economic feasibility of the catalytic pyrolysis and hydroprocessing pathway has not received as much attention as that of other pyrolytic pathways. Researchers at Iowa State University calculated the MFSP of a 2000 MTPD facility converting hybrid poplar feedstock to hydrocarbon fuels *via* what they term "mild catalytic pyrolysis" and hydroprocessing.[15] A MFSP of \$3.69 gal^{-1} is calculated for the pathway assuming a 10% IRR, 30 year facility lifespan, and feedstock cost of \$96.57 MT^{-1}. The assessment also assumed that the hydrogen employed by the hydroprocessing stage was produced on-site from off-gas rather than from natural gas, necessitating a higher total project investment. This would result in a higher MFSP compared with a scenario in which there is a lower hydrogen price and higher fuel price. It also found that the biochar and coke produced *via* the pathway only had value if an expensive co-generation unit was included, which increased total project investment.

Thilakaratne *et al.*[16] calculated the MFSP of a unique 2000 MTPD facility utilizing defatted microalgae rather than lignocellulosic biomass as feedstock for the production of hydrocarbon fuels and chemicals. A MFSP of \$1.49 LGE^{-1} (\$5.64 GGE^{-1}) is calculated assuming a 10% IRR, 30 year facility lifespan, and feedstock cost of \$66 MT^{-1}. A substantial advantage of this pathway was its use of a waste product that incurred a reduced feedstock cost. The defatted feedstock has a high moisture content that must be reduced prior to conversion, however, increasing the MFSP. The analysis calculated that employment of a thermal drying step added up to 9.4% to the MFSP, although this contribution could be reduced to 3.4% by instead utilizing a mechanical drying step. The use of defatted microalgae instead of lignocellulosic feedstock also impacted process yields by making hydrocarbon chemicals (benzene, toluene, and xylene) the primary products rather than hydrocarbon fuels.

NREL has calculated the MFSP for the catalytic fast pyrolysis pathway, which it refers to as an *in situ* scenario, based on a 10% IRR, 30 year plant life, 40% equity financing, \$80 MT^{-1} feedstock cost and 2000 MTPD feedstock capacity, of \$0.91 LGE^{-1} (\$3.46 GGE^{-1}).[3,4] While the need for less equipment compared the *ex situ* vapor phase upgrading scenario that is also analyzed results in a lower capital cost of \$546 million for the *in situ* scenario, the *in situ* scenario's MFSP is the higher of the two since the increased equipment needs of the *ex situ* scenario results in superior process chemistry.

An analysis conducted by researchers at Iowa State University and PNNL of the catalytic fast pyrolysis pathway calculates an expected minimum MFSP of \$4.28 gal^{-1}.[5] This result is higher than the MFSP calculated by the same analysis for the *ex situ* vapor phase upgrading scenario due in part to the higher fuel yields that are achieved by the catalytic fast pyrolysis scenario, the impact of which is partially offset by the *ex situ* scenario's higher total project investment. Furthermore, the Iowa State University and PNNL analysis finds that there is less uncertainty to the *ex situ* pathway's MFSP result due to lower fuel yield variability across experimental trials.

12.2.5 Hydropyrolysis and Vapor-Phase Hydroprocessing

Hydropyrolysis, or fast pyrolysis in the presence of high pressure hydrogen and a hydroprocessing catalyst, is another route for producing partially-deoxygenated bio-oil prior to upgrading. Oxygen is removed as water, resulting in a higher carbon efficiency for the pathway than for catalytic pyrolysis without hydrogen, in which the oxygen is removed as carbon dioxide and carbon monoxide. Hydropyrolysis only requires a single hydroprocessing step rather than the two steps commonly employed by fast pyrolysis, resulting in reduced equipment requirements. Reactor pressurization requires specialized equipment, however, so the reduced number of process steps does not necessarily result in lower capital costs for the pathway.

Singh *et al.*[17] calculated the MFSP for the hydropyrolysis and hydroprocessing pathway under multiple hydrogen procurement scenarios. The analysis used a previous assessment of the fast pyrolysis pathway[18] as the starting point for calculating capital and operating costs. For a 550 MTPD facility utilizing wood chips with a \$94 MT^{-1} feedstock cost and \$107 million total project investment and 20 year facility life, the analysis calculated MFSPs for hydrocarbon fuels of \$2.88 gal^{-1} or higher for the hydrogen procurement scenarios. The lowest MFSP was based on hydrogen derived from natural gas. The lowest MFSP for a fully renewable scenario based on hydrogen derived from wind power was roughly \$3.80 gal^{-1}.

Researchers at NREL have also calculated MFSPs for hydrocarbon fuels produced *via* the hydropyrolysis and hydroprocessing pathway.[19] The total project investment for a 2000 MTPD facility with on-site hydrogen production from hydroprocessing off-gases was assessed to be \$264 million. The hydrogen plant was the single-largest contributor to the total project investment, leading the authors to identify it as an important cost-reduction source. Assuming a woody biomass feedstock cost of \$79.17 MT^{-1}, 10% IRR, and 30 year plant life, the analysis calculated a MFSP for the base design case of \$1.64 gal^{-1}. The low MFSP compared to the others quoted in this chapter can be attributed to a combination of the pathway's high process yields and assumed 30 year plant life (as opposed to the 20 year plant life assumed by many other studies).

12.2.6 Slow Pyrolysis and Syngas Upgrading

Unlike the fast pyrolysis pathway, the slow pyrolysis pathway does not yield significant quantities of liquid products. A recent analysis calculated the MFSP of fuel methanol synthesized from syngas produced *via* the slow pyrolysis of biomass.[20] A 2200 MTPD facility with a total project investment of \$671 million utilizing woody biomass at a cost of \$50 MT^{-1} was calculated to have a MFSP of \$0.94 LGE^{-1} (\$3.57 GGE^{-1}) assuming a 10% IRR and 25 year plant life. The fuel methanol's MFSP was reduced if the biochar produced *via* the pathway was assumed to have value, although the sensitivity is low. The analysis only considered methanol as a liquid fuel output, but it should be noted that the economics of converting syngas to hydrocarbon fuels has been the subject of substantial research.[21]

12.2.7 Gasification of Bio-Oil and Syngas Upgrading

Bio-oil upgrading to liquid fuels is complicated by its high viscosity, acidity, and complex constituent compounds. Bio-oil gasification and subsequent upgrading of syngas to liquid fuels has been proposed as a method of overcoming these challenges. One of the first analyses to consider this integrated pathway compared three distributed scenarios in which stover feedstock was converted to bio-oil *via* fast pyrolysis at multiple locations and then gasified and upgraded to fuel *via* the Fischer–Tropsch synthesis (FTS) at a centralized facility.[22] The analysis found that the smallest (5 MTPD feedstock capacity per facility) distributed system achieved the lowest MFSP by incurring the lowest feedstock transportation costs, the effect of which offsets the system's relative lack of economies of scale. Of the three distributed systems, only the largest (500 MTPD capacity per facility) had a MFSP that was higher than that of a stand-alone 2000 MTPD gasification and FTS facility.

Manganaro and Lawal[23] calculated the MFSP of a similar pathway in which agricultural residue was converted to bio-oil *via* fast pyrolysis, gasified *via* autothermal reforming, and upgraded to fuels *via* FTS. A 2000 MTPD facility employing the pathway with a \$61.20 MT^{-1} feedstock cost, 8% IRR, and \$500 MT^{-1} co-product biochar price was calculated to have a MFSP of \$0.99 LGE^{-1} (\$3.74 GGE^{-1}). It should be noted that biochar is unlikely to have such a high value under all but the highest of carbon prices, however, and that the MFSP is very sensitive to the biochar price.

12.3 Pathway Comparisons

Efforts to compare the economic feasibility of different pyrolysis pathways and even the same pathways across different analyses face two major challenges because of the diversity of assessment methodologies and assumptions. The presentation of assessment results *via* a variety of financial metrics creates a third challenge to comparisons. Each of these challenges can potentially be overcome by replicating the individual assessments as a discounted cash flow model and changing the methodology, assumption, or financial output so as to ensure uniformity where appropriate across assessments (*e.g.* calculating an IRR or NPV rather than MFSP by assuming a predetermined fuel price).[24] Many analyses do not provide sufficient data to accurately replicate the results, however, in which case their utility is reduced in subsequent comparisons.

Various methodologies are employed to calculate both capital and operating costs, a situation that is further complicated by a lack of commercial-scale development of pyrolysis biorefineries to date.[25] Capital cost calculations frequently utilize methodologies that have been accepted by the petroleum refining industry due to their track record and equipment similarities. A typical analysis begins by employing software such as AspenPlus™ or ChemCAD™ to calculate equipment needs *via* mass and energy balances. The equipment is priced *via* literature sources, supplier quotes, and/or engineering estimates to calculate the total purchased equipment cost (TPEC). Differences in dollar basis years are accounted for by employing a basic inflation factor or, preferably, using a specific cost index such as the Chemical Engineering Plant Cost Index[26] to derive the appropriate factor.

The analyses diverge when choosing a methodology for calculating total installed equipment costs (TIEC). Some analyses employ a simple Lang factor that is multiplied by the total equipment cost to calculate the TIEC.[27] A more detailed methodology employs Hand factors, which vary according to equipment type, to calculate the TIEC. The most complex approach utilizes Module factors, which exist according to both equipment category and subcategory, to arrive at the TIEC when summed across individual equipment components. The capital cost is scaled up or down using a scaling factor, although the assumed factor also differs across analyses. Confusingly, the term Lang factor is frequently employed to illustrate the difference between the TIEC and the TPEC even when another factor is actually used to calculate the TPEC.

The TIEC only includes the costs of purchasing and installing the biorefinery's equipment. It does not include indirect costs, such as engineering, construction, and legal fees, contingency funds, working capital, and the cost of the land on which the facility is constructed. Here, too, the assumed amounts differ across analyses, although most assessments use fixed percentages of equipment or other costs that are based on industry experience. The sum of these various direct and indirect costs is the total project investment, which represents the facility's total capital cost.

Operating cost calculation methodologies also vary across analyses. Operating costs include both variable costs, which are a function of the facility's stream factor (*e.g.* raw material costs and co-product credits), and fixed costs, which are a function of the facility's capacity (*e.g.* labor, maintenance, and general overhead). While fixed costs are often calculated as a percentage of the facility's capital cost, the variable costs are derived from a wide variety of sources. Of these one of the most important is feedstock cost given the high sensitivity of the economic assessment to this cost. An accurate feedstock cost is specific to both the type of feedstock and the location of the biorefinery. Due to the lack of a commercial-scale market for many proposed biorefinery feedstocks such as agricultural residues and dedicated energy crops, however, a simple market price frequently cannot be employed. Different prices are often assumed for similar feedstock types within the same geographic location as a result. Of those pyrolysis assessments reviewed here, assumed feedstock costs range from a low of \$50 MT^{-1}[20] to a high of \$96.57 MT^{-1}[15] after accounting for differences in cost basis years.

The assumed feedstock type is also important due to the sensitivity of non-feedstock costs to geographic location. In addition to feedstock cost, other raw material costs, capital cost, labor cost, fuel price, and even tax rate are all sensitive to the assumed location. One analysis calculates a 20 year NPV range for a 2000 MTPD facility employing the fast pyrolysis and hydroprocessing pathway of \$79.5 million to \$165.5 million, depending on the location of the biorefinery within the U.S.[14] This effect likely becomes especially pronounced when differences in pathways are considered. For example, a hydropyrolysis and hydroprocessing facility could be expected to derive greater value from being located in a region with inexpensive natural gas, such as Oklahoma, than would a similarly-sized catalytic pyrolysis and hydroprocessing facility.

Finally, different facility lifespans are assumed across analyses (see Tables 12.1 and 12.2). While this assumption does not impact the capital or operating costs, it has a large impact on the economic assessment result because the discounting period is based on it. Other things being equal, a longer discounting period results in a larger IRR or NPV, or in a lower MFSP. A 20 year lifespan has been the most common assumption in assessments conducted to date, although recent analyses make use of 25 year[20] or even 30 year

Table 12.1 Results of selected pyrolysis techno-economic analyses for cellulosic biofuel production.

Source	Pathway	Capacity (MTPD)	TPI ($MM)	IRR (%)	MFSP ($/GGE)	MFSP ($/LGE)	Notes
Dutta *et al.* (2015)[4]	FPC	2000	589.6	10	3.31	0.87	*Ex situ* pathway configuration
Wright *et al.* (2010)[28]	FPH	2000	217	10	2.29	0.60	External hydrogen, *n*th plant
Zhang *et al.* (2013)[10]	FPH	2000	379	11.4[a]	3.04	0.80	Monosaccharide co-product, higher IRR
Brown *et al.* (2013)[9]	FPH	2000	429	10	2.57	0.68	External hydrogen, *n*th plant
Thilakaratne *et al.* (2014a)[15]	CPH	2000	457	10	3.69	0.97	
Thilakaratne *et al.* (2014b)[16]	CPH	2000	344	10	5.64	1.49	Mechanical dewatering
	CPH	2000	290	10	6.81	1.80	Thermal drying
Tan *et al.* (2014)[19]	HPH	2000	286	10	1.82	0.48	
Shabangu *et al.* (2014)[20]	SP	2000	671	10	3.57	0.94	No biochar revenue
Manganaro and Lawal (2012)[23]	FP & FTS	2000	231	8	3.30	0.87	$500/MT biochar credit

[a]FPC = fast pyrolysis and catalytic vapor phase upgrading; FPH = fast pyrolysis and hydroprocessing; CPH = catalytic pyrolysis and hydroprocessing; HPH = hydropyrolysis and hydroprocessing; SP = slow pyrolysis and methanol synthesis; FP & FTS = fast pyrolysis and Fischer–Tropsch synthesis.

Table 12.2 Major assumptions used in selected pyrolysis techno-economic analyses for cellulosic biofuel production.

Source	Pathway	TPI/TPEC	Stream factor (%)	Feedstock cost ($/MT)	Biorefinery life (years)	Notes
Dutta *et al.* (2015)[4]	FPC	1.86	90	88	30	*Ex situ* vapor phase upgrading configuration
Wright *et al.* (2010)[28]	FPH	5.46	90	90	20	External H_2, *n*th plant
Zhang *et al.* (2013)[10]	FPH	5.46	90	86.52	20	Monosaccharide co-product, higher IRR
Brown *et al.* (2013)[9]	FPH	5.46	90	83	20	External H_2, *n*th plant
Thilakaratne *et al.* (2014a)[15]	CPH	5.10	90	96.57	30	
Thilakaratne *et al.* (2014b)[16]	CPH	5.10	91	66	30	Microalgae remnant feedstock
Tan *et al.* (2014)[19]	HPH	3.19	96	85.89	30	
Shabangu *et al.* (2014)[20]	SP	4.00	91	48.99	25	No biochar revenue
Manganaro and Lawal (2012)[23]	FP & FTS	N/A	96	61.20	N/A	$500/MT biochar credit

lifespans.[15,19] While longer facility lifespans can be justified on the basis of actual experience in the chemicals and refining industries, the use of different assumed lifespans across analyses generates substantial differences to assessment results that are not specific to the assessed pathways.

12.4 Deterministic *versus* Stochastic Analyses

Most techno-economic analyses of pyrolysis pathways are based on deterministic process and discounted cash flow models. Deterministic models are relatively simple to build and operate because they quantify the impact of a single changed variable on the model's result at a time. For example, a deterministic model calculates the change to MFSP, IRR, or NPV that results from increasing the feedstock cost by 50%. The output of a deterministic model is therefore a single point estimate that operates as a function of a single stated value for each variable. Sensitivity analyses, which define the variables to which the economic result is most sensitive to and quantify this sensitivity, are frequently employed to illustrate the relative importance of each variable, whether assumed or based on data.

A major shortcoming of deterministic models is that they only calculate an economic result for a single point in time. Sensitivity analyses provide additional information but are still limited to quantifying the impact of a single variable change at a time. The methodology's static nature does not accurately reflect the volatility that characterizes energy prices and cellulosic biofuel pathways that are untested at scale. Stochastic, or uncertainty, analyses are a better methodology for assessing the economics of pathways under these conditions. A downside is that the methodology often employs complex software such as Oracle Crystal Ball™ in addition to common spreadsheet software such as Microsoft Excel™. This disadvantage is offset by the capabilities that stochastic analyses bring to techno-economic assessments.

Stochastic analyses consider a range of interactions between variables to quantify the impact on the economic assessment's result. The primary constraint to the number of variables that can be considered is computing power rather than methodology. The economic result is only slightly sensitive to many variables and, after being identified *via* sensitivity analysis of a deterministic techno-economic model, these variables can safely be excluded from the stochastic analysis. The next step is to develop a probability distribution function (PDF) for those variables to which the economic result is determined to be most sensitive. Unlike the point estimates employed by deterministic models, PDFs represent a range of possible values, to each of which a specific probability has been assigned. This permits the calculation of the probability that a value will be above or below a specific value within that distribution. PDFs are therefore an appropriate means of simulating uncertain values such as future commodity prices or yields that are based on small-scale experimental trials. Whereas a point estimate necessarily establishes a probability of one for the value in question (regardless of whether that value is actually known with complete certainty), a PDF reflects the value's true uncertainty.

Future energy prices are more accurately simulated *via* PDFs rather than point estimates due to their high degree of historical volatility. The change to the price of natural gas over time is one example of a variable that is more accurately quantified *via* a PDF. Its monthly average price at the Henry Hub has ranged from a low of \$1.73 $MMBTU^{-1}$ to a high of \$12.69 $MMBTU^{-1}$ over the last decade (see Figure 12.2).[29] The monthly average spot price of U.S. gasoline has ranged from a low of \$0.93 gal^{-1} to a high of \$3.28 gal^{-1} over the same period.[30] The economic result of the fast pyrolysis and hydroprocessing pathway is very sensitive to both prices because the former is an important raw material cost and the latter determines the market value of the fuels that it produces. Using the midpoint value of the range (\$7.21 $MMBTU^{-1}$ in the case of natural gas) as a point estimate is misleading, however, since it is substantially higher than the average (\$4.47 $MMBTU^{-1}$) value over the same period.

Using the mean price is also inappropriate because the value is skewed by short periods of very high prices. The median value of \$3.94 $MMBTU^{-1}$ is substantially lower than both the mean and the midpoint. Best-fit analysis with Oracle Crystal Ball™ identifies the historical monthly natural gas price for the 2007–2016 period as having an asymmetrical lognormal distribution (see Figure 12.3). While there is a roughly 50% likelihood that the natural gas price is under \$4 $MMBTU^{-1}$ when employing this PDF, there is an equal likelihood that it is above that value. More importantly, while there is a 50% likelihood that the price is equal to \$4 $MMBTU^{-1}$ or less, there is a 49% likelihood that the price is between \$4 $MMBTU^{-1}$ and \$13 $MMBTU^{-1}$, and a 1% likelihood that it is above \$13 $MMBTU^{-1}$. The asymmetrical nature

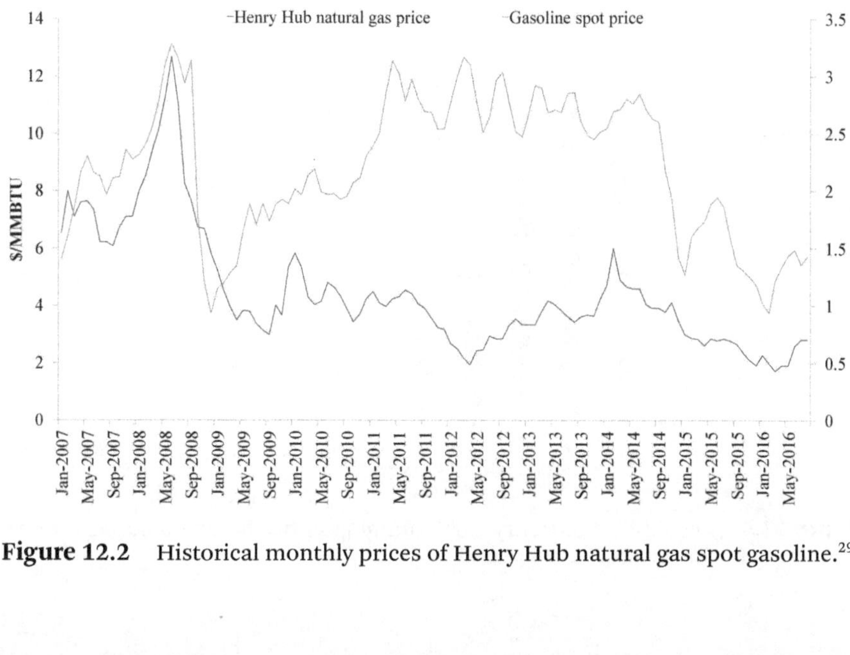

Figure 12.2 Historical monthly prices of Henry Hub natural gas spot gasoline.[29,30]

of this distribution means that the use of a point estimate in assessments of pathway economics ignores the likelihood that the actual future price will be above the assumed future price.

A second advantage of stochastic analyses is that they allow for the impact of simultaneous changes to variables on the economic assessment's result to be quantified. Monte Carlo analysis is employed to run the stochastic model through a predetermined number of simulations (10 000 simulations are common when sufficient computing power is available). This permits the interaction of each variable for which a PDF is available and the quantification of their cumulative effects on the economic assessment's result. Such a capability is not just important when simulating energy prices that, in addition to being volatile, also frequently move in tandem. It is equally important when simulating uncertain process conditions such as catalyst performance, consumption factors, and yields. In the case of the fast pyrolysis pathway, a reduction in the bio-oil yield can coincide with an increase in the syngas and/or biochar yield, in which case the reduction to the pathway value provided by the bio-oil is partially offset by an increase to the value provided by the co-products. A stochastic analysis can capture this relationship, providing a more accurate quantification of the impact of the yield change on the economic result than is possible when changing a single variable at a time.

Uncertainty in economic assessments of pyrolysis pathways is not just limited to the variable operating costs. A recent review of energy facility project capital costs identifies no fewer than six PDFs for the relationship of actual capital costs compared to projected capital costs, including two symmetrical and four asymmetrical distributions.[31] Furthermore, larger cost overruns are reported for unconventional energy projects. When applied to

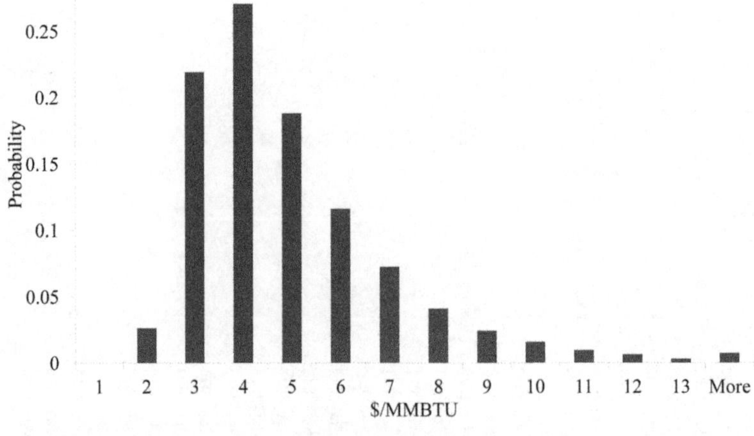

Figure 12.3 Best-fit PDF of Henry Hub natural gas price based on monthly prices for the 2007–2016 period.

stochastic techno-economic models of cellulosic biofuel pathways, including pyrolysis, the analysis calculates 20 year NPV cumulative distribution function ranges of up to \$500 million just from the capital cost uncertainty. To date this type of uncertainty has not been included in stochastic techno-economic models of pyrolysis pathways despite its large impact on the economic result.

Finally, stochastic analyses permit a more accurate representation of the operating and financial conditions that a facility employing a pyrolysis pathway can expect to encounter. A facility's cost of capital is a function of its cash flow volatility: a higher degree of cash flow volatility is associated with a higher cost of capital since the volatility increases the likelihood that the facility will be temporarily unable to cover its costs with revenues.[32] A recent analysis finds that the costs of capital assumed by many techno-economic analyses of cellulosic biofuel pathways are much lower than those that are most likely to be incurred due to the cash flow volatility of those pathways, and that the analyses overestimate economic returns as a result.[33] The analysis further finds that the costs of capital for cellulosic biofuel pathways are especially sensitive to just a few variables such as feedstock costs and fuel prices.

12.5 Conclusions

Pyrolysis pathways have been the subject of a growing number of economic assessments. Techno-economic analyses have been the primary methodology employed to conduct these assessments. A wide range of MFSPs are reported across the literature. Important differences with regard to feedstock costs, IRR, stream factors, facility lifespans, and even capital cost calculation methodologies between studies make comparisons between pathways difficult. In general, however, the economic assessments have identified feedstock cost and fuel prices to be two of the variables to which the economic result is most sensitive to. Furthermore, while facility capital costs can be quite large for pyrolysis pathways, higher capital costs can improve a facility's economic feasibility when they reduce operating costs and/or improve the yields of higher-value products.

Some recent techno-economic analyses of pyrolysis pathways employ stochastic models in addition to deterministic model. Stochastic modeling represents an important contribution to the study of pyrolysis economics due to the pathways' high level of uncertainty. Energy prices have historically exhibited a high degree of volatility while unconventional energy facility capital cost under- and over-runs are widely reported in the literature. Stochastic models allow these uncertainties to be quantified and their impacts on pathway economic assessments to be captured. They also permit higher-accuracy representations of the financial conditions that pyrolysis facilities are likely to encounter as part of the commercialization process.

References

1. T. P. Vispute, H. Zhang, A. Sanna, R. Xiao and G. W. Huber, Renewable chemical commodity feedstocks from integrated catalytic processing of pyrolysis oils, *Science*, 2010, **330**(6008), 1222–1227.
2. T. R. Brown, Y. Zhang, G. Hu and R. C. Brown, Techno-economic analysis of biobased chemicals production *via* integrated catalytic processing, *Biofuels, Bioprod. Biorefin.*, 2012, **6**(1), 73–87.
3. A. Dutta, J. A. Schaidle, D. Humbird, F. G. Baddour and A. Sahir, Conceptual Process Design and Techno-Economic Assessment of *Ex Situ* Catalytic Fast Pyrolysis of Biomass: A Fixed Bed Reactor Implementation Scenario for Future Feasibility, *Top. Catal.*, 2016, **59**(1), 2–18.
4. A. Dutta, A. Sahir, E. Tan, D. Humbird, L. J. Snowden-Swan and P. Meyer, *et al.*, *Process Design and Economics for the Conversion of Lignocellulosic Biomass to Hydrocarbon Fuels: Thermochemical Research Pathways with* In Situ *and* Ex Situ *Upgrading of Fast Pyrolysis Vapors*. Golden, CO (United States); 2015, Mar, [cited 2015 Oct 1], available from, http://www.osti.gov/scitech/biblio/1215007.
5. B. Li, L. Ou, Q. Dang, P. Meyer, S. Jones and R. Brown, *et al.*, Techno-economic and uncertainty analysis of *in situ* and *ex situ* fast pyrolysis for biofuel production, *Bioresour. Technol.*, 2015, **196**, 49–56.
6. A. V. Bridgwater, Production of high grade fuels and chemicals from catalytic pyrolysis of biomass, *Catal. Today*, 1996, **29**(1–4), 285–295.
7. S. B. Jones, C. Valkenburg, C. W. Walton, D. C. Elliot, J. E. Holladay and D. J. Stevens, *et al.*, *Production of Gasoline and Diesel from Biomass* via *Fast Pyrolysis, Hydrotreating and Hydrocracking: A Design Case [Internet]*, Richland, Pacific Northwest National Laboratory, 2009, available from, http://www.pnl.gov/main/publications/external/technical_reports/PNNL-18284.pdf.
8. M. M. Wright, J. A. Satrio, R. C. Brown, D. E. Daugaard and D. D. Hsu, *Techno-Economic Analysis of Biomass Fast Pyrolysis to Transportation Fuels*, Nov 2010.
9. T. R. Brown, R. Thilakaratne, R. C. Brown and G. Hu, Techno-economic analysis of biomass to transportation fuels and electricity *via* fast pyrolysis and hydroprocessing, *Fuel*, 2013, **106**, 463–469.
10. Y. Zhang, T. R. Brown, G. Hu and R. C. Brown, Techno-economic analysis of monosaccharide production *via* fast pyrolysis of lignocellulose, *Bioresour. Technol.*, 2013, **127**, 358–365.
11. G. W. Huber and J. A. Dumesic, An overview of aqueous-phase catalytic processes for production of hydrogen and alkanes in a biorefinery, *Catal. Today*, 2006, **111**(1–2), 119–132.
12. T. R. Brown, M. M. Wright and R. C. Brown, Estimating profitability of two biochar production scenarios: slow pyrolysis *vs.* fast pyrolysis, *Biofuels, Bioprod. Biorefin.*, 2011, **5**(1), 54–68.
13. R. Petter and W. E. Tyner, Technoeconomic and policy analysis for corn stover biofuels, *ISRN Econ.*, 2014, **2014**, 515898.

14. T. R. Brown, R. Thilakaratne, R. C. Brown and G. Hu, Regional differences in the economic feasibility of advanced biorefineries: Fast pyrolysis and hydroprocessing, *Energy Policy*, 2013, **57**, 234–243.
15. R. Thilakaratne, T. Brown, Y. Li, G. Hu and R. Brown, Mild catalytic pyrolysis of biomass for production of transportation fuels: A techno-economic analysis, *Green Chem.*, 2014, **16**(2), 627–636.
16. R. Thilakaratne, M. M. Wright and R. C. Brown, A techno-economic analysis of microalgae remnant catalytic pyrolysis and upgrading to fuels, *Fuel*, 2014, **128**, 104–112.
17. N. R. Singh, D. S. Mallapragada, R. Agrawal and W. E. Tyner, Economic analysis of novel synergistic biofuel (H2Bioil) processes, *Biomass Convers. Biorefin.*, 2012, **2**(2), 141–148.
18. M. Ringer, V. Putsche and J. Scahill, *Large-Scale Pyrolysis Oil Production: A Technology Assessment and Economic Analysis*, National Renewable Energy Laboratory, Golden, 2006, available from, http://www.nrel.gov/docs/fy07osti/37779.pdf.
19. E. C. D. Tan, T. L. Marker and M. J. Roberts, Direct production of gasoline and diesel fuels from biomass *via* integrated hydropyrolysis and hydroconversion process-A techno-economic analysis, *Environ. Prog. Sustainable Energy*, 2014, **33**(2), 609–617.
20. S. Shabangu, D. Woolf, E. M. Fisher, L. T. Angenent and J. Lehmann, Techno-economic assessment of biomass slow pyrolysis into different biochar and methanol concepts, *Fuel*, 2014, **117**, 742–748.
21. T. R. Brown, A techno-economic review of thermochemical cellulosic biofuel pathways, *Bioresour. Technol.*, 2015, **178**, 166–176.
22. M. M. Wright, R. C. Brown and A. A. Boateng, Distributed processing of biomass to bio oil for subsequent production of Fischer Tropsch liquids, *Biofuels, Bioprod. Biorefin.*, 2008, **2**(3), 229–238.
23. J. L. Manganaro and A. Lawal, Economics of Thermochemical Conversion of Crop Residue to Liquid Transportation Fuel, *Energy Fuels*, 2012, **26**(4), 2442–2453.
24. T. R. Brown and M. M. Wright, Techno-economic impacts of shale gas on cellulosic biofuel pathways, *Fuel*, 2014, **117**(PARTB), 989–995.
25. T. R. Brown and R. C. Brown, Techno-economics of advanced biofuels pathways, *RSC Adv.*, 2013, **3**(17), 5758.
26. Anon, Economic Indicators, *Chem. Eng.*, 2011, **118**(y), 63.
27. T. R. Brown, Capital cost estimating, *Hydrocarbon Process.*, 2000, **79**(10), 93–100.
28. M. M. Wright, D. E. Daugaard, J. a. Satrio and R. C. Brown, Techno-economic analysis of biomass fast pyrolysis to transportation fuels, *Fuel*, 2010, **89**, S2–S10.
29. Anon, *Henry Hub Golf Coast Natural Gas Spot Price ($/MMBtu)*, Energy Information Administration, Washington, DC, U.S. Department of Energy, 2016, [cited 2016 Jan 1], available from, http://www.eia.gov/dnav/ng/hist/rngwhhdd.htm.

30. Anon, *U.S. Gulf Coast Conventional Gasoline Regular Spot Price FOB*, Energy Information Administration, U.S. Department of Energy; 2016, [cited 2016 Jan 1], available from, http://www.eia.gov/dnav/pet/hist/LeafHandler.ashx?n=PET&s=EER_EPMRU_PF4_RGC_DPG&f=M.
31. T. R. Brown, A critical analysis of thermochemical cellulosic biorefinery capital cost estimates, *Biofuels, Bioprod. Biorefin.*, 2015, 412–421.
32. A. Damodaran, Estimating Risk Paramters and Costs of Financing, *Investment Valuation: Tools and Techniques for Determining the Value of Any Asset*, John Wiley & Sons, Hoboken, 3rd edn, 2012. p. 183.
33. T. R. Brown and M. M. Wright, A Framework for Defining the Economic Feasibility of Cellulosic Biofuel Pathways, *Biofuels*, 2014, **5**(5), 579–590.
34. R. Schnepf and B. D. Yacobucci, *Renewable Fuel Standard (RFS): Overview and Issues*, Washington, DC, Congressional Research Service, 2013, available from, http://www.fas.org/sgp/crs/misc/R40155.pdf.

Subject Index